Developments in
Analytical Methods in
Pharmaceutical, Biomedical,
and Forensic Sciences

Developments in Analytical Methods in Pharmaceutical, Biomedical, and Forensic Sciences

Edited by

G. Piemonte
F. Tagliaro and
M. Marigo
University of Verona
Verona, Italy

and

A. Frigerio
Italian Group for Mass Spectrometry
in Biochemistry and Medicine
Milan, Italy

Plenum Press • New York and London

Library of Congress Cataloging in Publication Data

International Conference on Developments in Analytical Methods in Pharmaceutical,
    Biomedical, and Forensic Sciences (1986: Verona, Italy)
    Developments in analytical methods in pharmaceutical, biomedical, and forensic
sciences.

    "Proceedings of an International Conference on Developments in Analytical Methods
in Pharmaceutical, Biomedical, and Forensic Sciences, held June 25–27, 1986, in
Verona, Italy"—T.p. verso.
    Includes bibliographies and index.
    1. Chemistry, Pharmaceutical—Technique—Congresses. 2. Chemistry, Analytic—
Technique—Congresses. 3. Chemistry, Clinical—Technique—Congresses. 4. Chemistry,
Forensic—Technique—Congresses. I. Piemonte, Giuseppe. II. Title. [DNLM: 1.
Biology—methods—congresses. 2. Chemistry, Analytical—congresses. 3. Chemistry,
Pharmaceutical—congresses. 4. Forensic Medicine—methods—congresses. 5. Pharma-
cology—methods—congresses. QV 744 I593d 1986]
RS401.I56   1986   610.1'543   87-7180
ISBN 0-306-42695-1

Proceedings of an International Conference on Developments in Analytical
Methods in Pharmaceutical, Biomedical, and Forensic Sciences,
held June 25–27, 1986, in Verona, Italy

© 1987 Plenum Press, New York
A Division of Plenum Publishing Corporation
233 Spring Street, New York, N.Y. 10013

Printed in the United States of America

PREFACE

   The papers collected in this volume were presented at an
International Conference that, with the same heading, was held at
the Verona University, Italy, in June 1986. The meeting was
organized by the Institute of Forensic Medicine and the Laboratory
of Medical Research of the University in cooperation with the
Italian Group for Mass Spectrometry in Biochemistry and Medicine.

   The aim of the symposium was bringing together people, working in different branches of the wide field of modern analytical
sciences, for promoting inter-disciplinary discussions and
exchange of experiences.

   Actually it was felt that most of the analytical problems
that very often have to be faced in quite different fields (chemistry, pharmacology, medicine, biology) have similar solutions,
that could be made much easier by closer contacts among researches
of these disciplines.

   Original papers and invited rewiews presented during the
3 days of the conference by leading experts gave an up-to-date
outline of the modern analytical methods applied in pharmaceutical, biomedical and forensic sciences and a glimpse of the future
perspectives.

   One wishes that the scientific information given at the
conference spread out over the number of participants by means of
the book of proceedings. Of course, this book could serve the
interest of those for whom is intended only if it is published in
a reasonable time. For this reason, in order to avoid a significant delay, the editors had to sort out the papers for publication
not only on the basis of the scientific level, but also on the
basis of the quickness of manuscript preparation and delivering.
We hope that readers will agree with this choice.

   We wish to thank Miss Ariella Stubelj and Miss Claire Liggins
for their invaluable technical assistance and countless efforts in
the preparation of this volume.

University of Verona   The Editors
June, 1987

CONTENTS

HIGH PERFORMANCE LIQUID CHROMATOGRAPHY IN DRUG LEVEL MONITORING -

AN OVERVIEW

K. Macek

Institute of Physiology
Czechoslovak Academy of Sciences
Prague, Czechoslovakia

In this overview I would like to summarize the present situation and trends in application of high-performance liquid chromatography (HPLC) in the area of pharmacokinetics and I would also
like to discuss exploiting of the obtained results in drug level
monitoring. Unfortunately this is a very broad theme which is
covered by several thousands of papers that have appeared during
the recent years. Therefore I can hardly do more than to keep on
the surface of these problems.

The interest of chromatographers in drug analysis can be
traced from the early fifties. In those early days chromatography
was applied to following drug synthesis, to assaying their purity,
to establishing drug stability in pharmaceutical preparations, to
the analysis of the content of active compounds in plant material
and last but not least for toxicological purposes. Of the different chromatographic techniques, paper and thin-layer chromatography were the methods that were used most frequently in the
fifties and sixties[1] while column chromatographic techniques were
used mostly for the preparative purposes or for purification of
compounds. A distinct switch occured in the seventies when the
instrumentation, particularly in column liquid chromatography,
caused a deflection from planar techniques. Moreover, the increased sensitivity of the new techniques opened the possibility
of assaying compounds in body fluid which until then was very
difficult to analyze[2].

Looking at the data compiled about the application of chromatography to different categories of compounds (Table 1) it is
obvious that the first place is occupied today by chromatography
of drugs[3]. The number of papers devoted to drug analysis increased
within the last 15 years by 771%: in 1985 there were 1,559 papers
devoted to drug analysis out of the total 10,569 papers dealing
with chromatography. Of the individual techniques the main
attention is focussed on HPLC followed by planar techniques and
gas chromatography. From the viewpoint of the material analyzed
(Table 2) the main attention in drug analysis was paid to
bioanalytical applications.

The intensity of therapeutical but also of toxicological effect of most drugs depends on the actual concentration of the drug in the site where its effect is expected. The time course of this concentration is determined by biological susceptibility of the organism to the drug. This is affected not only by the amount of drug administered and by the way of its application, but also by its absorption, by its distribution within the body, by its metabolic pathway which may be altered by concomitant administration of other drugs, and, finally, by its elimination from the organism. The effect of drugs is to a considerable extent influenced also by genetic factors and by the individual variability. Here the overall body weight, age, sex, constitution, the state of circulation, the state of hepatic and kidney functions and, of course, the pathologic situation of the organism should be considered. It is well known that for any patient there exists a particular dose of drug which should be administered to obtain the optimum concentration of the drug in blood and in the site where the effect of the drug is desirable. The dosage which is optimal for one patient may be in another patient below the desired serum concentration and the therapeutic effect may be lost. On the other hand, in another patient the same dosage could be far too high and could lead to undesirable side-effects.

In order to determine the optimum dosage of a drug in a particular patient it is necessary to know in detail the drug pharmacokinetics and beyond that the pharmacological properties of the drug and its metabolites. Today such requirements are put upon

Table 1.   Applications of Chromatographic Techniques in 1985

Class of compounds	Total Papers 1985	1970	Papers in 1985 LC	GC	PC	Increase in %
1 Drugs	1559	179	892	288	379	771
2 Proteins and Enzymes	1213	412	1184	8	21	194
3 Environmental Analysis (Incl. Pesticides)	672	119	205	348	119	464
4 Inorganic Compounds	511	185	344	94	73	176
5 Amino Acids and Peptides	501	112	378	53	70	347
6 Lipids	443	161	149	89	205	175
7 Organic Acids	405	122	225	110	70	231
8 Carbohydrates	400	116	285	48	67	244
9 Amines	359	57	233	62	64	529
10 Steroids	351	150	188	48	115	134
11 Hydrocarbons	348	70	120	184	44	397
12 Nucleic Acids and their Constituents	297	227	244	13	40	30
13 Vitamins	234	82	182	27	25	185
14 Antibiotics	212	37	143	8	61	472
Total Papers	10569	3732	6632	2282	1655	183

Table 2.   Applications of Chromatography of Drugs in 1985
           according to the Material

Applications	Total Papers	LC	GC	PC
Non-Biological	648	337	60	251
Biomedical	911	555	228	128

all newly introduced drugs and this is the reason why so much
interest is focussed on this topic. Based on these data it is possible to decide whether it is necessary to carry out the determination of drug levels as well as the estimation of metabolites in
patients at the beginning of therapy or during the whole therapeutic period. It appears that such procedures are necessary in drugs
with a low therapeutic index. The importance of such an approach
is particularly important in all cases where it is dealt with
therapy of fatal situations, like in intensive care units, in the
administration of immunosuppressives in organ transplantation and
in all situations of chronic drug administration. Particularly the
last point is nowadays considered a world problem as the number of
individuals ingesting drugs on a chronic base can be counted in
millions.

Obtaining of pharmacokinetic data and drug level monitoring
cannot be materialized without a suitable analytical background.
All methods dealing with biological material have for
pharmacokinetic studies to meet several criteria that can be
summarized as follows:

1. Sensitivity. A suitable method should allow the assay of a
   drug in the nanogram and sometime in the picogram range per
   1 ml of the body fluid. The method should be capable of
   carrying out the analysis in 10 to 1,000 microliters of
   sample.
2. Selectivity. The method must be selective not only for the
   parent drug, but for its metabolites as well as the
   therapeutic response might be the result of the action of
   both the drug and its active metabolites. The assay should
   not be interferred with endogeneous compounds present in the
   biological material, with exogeneous compounds administered
   in food and with other drugs that are administered
   simultaneously with the analyzed drug. For the purpose of
   pharmacokinetic studies specific methods come sometime in
   use: such methods put usually high demands upon
   instrumentation and therefore for routine drug monitoring
   are being abandoned.
3. Recovery. The applied methods should yield a high recovery
   during the isolation from biological material, generally in
   the 80 to 100 per cent range.
4. Accuracy and precision. Both these are a necessary condition
   for a suitable method.
5. Speed of analysis. The analytical results should be obtained
   within a short period of time. The capacity should be of the
   order 10 to 100 samples per day. Therefore methods that could
   be automatized are preferred.

6. <u>Intrumentation</u>. Current attempts are directed to methods that
    can be carried out with the common equipment of a clinical
    chemistry laboratory.
7. <u>Expenses</u>. Last, but not least, it is necessary to keep the
    expenses with a drug level assay within reasonable limits.

Currently, the most popular methods in the area of
pharmacokinetics are chromatographic procedures. In newly
introduced drugs with insufficient knowledge about their metabolic
pathways combination with mass spectrometry is necessary for
either identification or for the determination of the metabolite
structure. In drug monitoring there is a competition between
chromatography and different types of immunoassays, like
radioimmunoassay (RIA), enzyme multiple immunoassay techniques
(EMIT) or fluorescence-polarized immunoassays (FPIA). The advantage of these methods is the high speed of analysis and a high
sensitivity with low demands upon the sample size. Lower
selectivity belongs to their disadvantages: these methods
generally fail to distinguish parent drugs from their metabolites
and therefore the results are generally higher. Another
disadvantage are the relatively high costs per assay and finally
the application of these procedures is limited only to selected
categories of drugs.

HPLC TECHNIQUES

The application of HPLC to drug analysis in biological
material exhibits particular specificities in which it is different from the application to other categories of compounds and
from the application to non-biological material. Time limitations
allow me to focus only on selected problems that are being solved
in the bioanalysis of drugs. We shall talk about sample preparation, detectors, derivatization and separation of enantiomers.

<u>Sample Preparation</u>

Though the sample preparation is not a part of the very
chromatographic procedure, it should attract our attention as it
provides the success or failure of the analysis and, moreover, it
frequently requires more time than the chromatographic separation
itself[4,5].

Sample preparation involves extraction of the analyzed drug
or the metabolites from biological material and the simplification
of the obtained extract by removing the ballasts that could
interfere with the chromatographic assay and decrease the lifetime
of the analytical column. This part of the assay is known under
the name clean-up. It is generally true that the lower the
concentration of the drug in the analyzed sample and the more
general method of detection is used, the more complex should be
sample preparation.

The extraction of a drug from the biological material is
guided by its properties. The choice of the procedures is different for whole blood, for plasma or serum, for tissues, urine, or
saliva. Plasma and serum are analyzed most frequently. Here it is
necessary to respect the fact that drugs can be present either in
their free form or bound to proteins, lipoproteins, etc. In addition it is necessary to expect the presence of metabolites
inclusive conjugates. All these compounds will differ in their

solubility: it is foreseen that they will be more hydrophilic and therefore for their extraction of the whole procedure must be modified. A certain measure of effectivity of extraction is the recovery of drug added to a biological material, most conveniently by making use of radioactively labelled drug.

In the past drug extraction and extract purification was materialized by using organic solvents and pH adjustment. Such procedures usually require reextraction, centrifugation and sample evaporation and are therefore very time demanding. Additional problem arises with the extraction of hydrophilic compounds like quaternary bases and some metabolites. The losses in this category of procedures are usually high. Some solution can be seen in the application of the volatile extraction buffers[6] or in the application of ion-pair extraction with counter ions, e.g., tetrabutylammonium hydroxide[7].

Because of these reasons more recent papers deal mainly with the so-called solid-phase extraction in which small columns packed with a sorbent with chemically bonded reversed phase or, less frequently, with an ion exchanger or normal phase sorbents are used both for extraction and prepurification[8]. Operation with these cartridges can be manual - off-line - or, in large series of samples, it can be automated and carried out on line.

In the off-line process the sample of body fluids is usually deproteinized, e.g., by acetonitrile, centrifuged, the supernatant is collected in a disposable short column attached to a syringe or to a specialized device. With such a device a number of samples can be worked out simultaneously. Solid phase extraction can process 10 samples in about 15 minutes. By attachment to the AASP (Automated Advanced Sample Processor) it is possible to elute automatically the adsorbed drugs directly to the analytical column[9].

Because of the extreme demands put on the clinical biochem-istry laboratories on-line techniques using column switching have gained in their popularity. A whole number of instrument designs was suggested, however, the principle is the same as described by, e.g., Karger et al.[4] (Fig. 1). The sample, e.g., serum, is loaded via 6-port switching valve on a short (cca 5 cm long) extraction column (C) containing often reversed-phase sorbent (particle size 25-40 micrometers). The drug is enriched at the top of this ex-traction column. Buffer from wash pump (P_2) is then pumped in the same direction through the extraction column C to wash proteins and other ballasts. The switching valve V_2 is next rotated and mobile phase from pump P_1 flows through the extraction column onto the analytical column BCC. Automated backflushing is usually preferred over forward flushing.

This category of the on-line techniques is sometimes called direct application of sample. The sample can be applied to the column without a precolumn only exceptionally. This is in those situations where the concentration of the analyzed drugs in the biological material is high, in the analysis of urine or saliva. With serum or plasma at least deproteination is absolutely necessary. Ultrafiltration is more and more often used for this purpose. If the so-called micellar mobile phases are used the samples can be applied to the column directly. As long as solu-bilization of proteins occurs via a surfactant in the mobile phase, there is no column clogging and, in addition, the

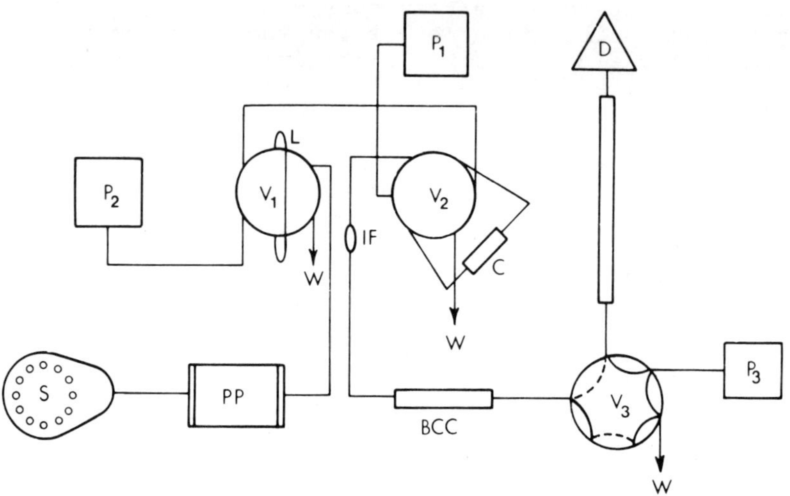

Fig. 1. Automated on-line sample cleanup HPLC system with
solid-phase extraction (Karger et al.[4]). S, autosampler;
PP, peristaltic pump; L, sample loop; V_1, V_2, V_3, 6-port
switching valves; W, waste; P_1, P_3, mobile phase pumps;
P_2, wash pump; EC, extraction column; AC, analytical
column; IF, in-line filter; D, detector.

surfactant monomers appear to displace the drug bound to the
protein[10].

Another problem to be emphasized is the analysis of conjugates. These metabolites differ from the parent drug by their
distinct hydrophilicity. The situations in which there is a simple
way of parallel estimation of the drug, its metabolites and
conjugates are very rare. In most cases free drugs and metabolites
are assayed and concomitantly the original sample is either
enzymatically or acid hydrolyzed and the amount of conjugates is
calculated from the difference of individual type of metabolites.
The application of biospecific affinity chromatography in this
sense appears very promising. A small column is packed with the
sorbent with, e.g., bound ß-glucuronidase or mixed sulphatase –
ß-glucuronidase; such columns are capable to hydrolyze on-line
glucuronidates and sulphates of analyzed drugs. This arrangement
can be used in the precolumn mode for more convenient separations
or, in the post-column mode, to be able to use electrochemical or
other type of detections that cannot be carried out with
conjugates[11].

<u>Detectors</u>

Considerable attention in HPLC is paid today to detectors.
The goal of these efforts is to increase the sensitivity of
detection and the selectivity of the assay. These tendencies are
particularly valid for pharmacokinetic studies and drug monitoring. At the beginning of the seventies UV absorption at 254 nm
was one of the most popular ways of drug detection. In those
instances where this procedure was not applicable, spectral
photometers came in use and the wavelength was decreased to

200-210 nm: in this region most organic compounds exhibit a distinct absorbancy. Unfortunately this approach resulted in a selectivity decrease because a number of endogeneous compounds and many ballasts absorb at this wavelength as well. Therefore more selective detectors were looked for. Fluorimetric detectors proved very useful in this respect. Since most drugs, however, do not display native fluorescence they had to be converted into suitable fluorescent derivatives first.

Because of the complexity of samples analyzed the main attention is paid recently to two categories of detectors: electrochemical detectors and diode array detectors. Electrochemical detectors are based on the conductivity changes or they can be used for detection of compounds which can be electrolytically oxidized or reduced at a working electrode. Amperometric, coulometric and polarographic detectors are the main types falling into this category. Their advantage is high sensitivity - frequently in the picogram range - high selectivity and simple construction which makes them relatively inexpensive. Electrochemical detectors can be used for analysis in aqueous media and are therefore applicable to reversed phase and ion-exchange chromatography. On the other hand they require strictly constant experimental conditions like flow-rate constancy, pH, ionic strength, temperature, the condition of the electrode surface, etc. Their selectivity can be further increased by using dual electrode mode[12].

The other large group of detectors is constituted by diode array detectors[13] which due to their high costs are at the moment less popular. These detectors exploit fast UV spectrum recording which is stored digitally in a micro processor where it can be recalled or handled in several ways. Diode array detectors can be used mainly in identification, peak recognition, in peak purity determination and for recording multiple chromatograms at a variety of wavelengths which can maximize the sensitivity or minimize interferences.

Pharmacokinetic studies require frequently the estimation of radiolabelled compounds. Radioactivity detection after a chromatographic column used to be carried out with separated fractions collected in a fraction collector and quantitated in a liquid scintillation counter. Today fast flow-through on-line detectors are preferred. The construction of these detectors differs in that whether heterogeneous flow cells packed with a solid scintillator or homogeneous systems where the HPLC eluate is mixed with a nongelic scintillator fluid before entering the flow cell are used[14].

In pharmacokinetic studies particularly in metabolite identification interfacing of a liquid chromatograph to a mass spectrometer is extremely important. In spite of the fact that in the recent years a lot of attention was paid to this combination still many problems persist which currently do not allow such a wide applicability as we can see with the GC-MS combination[15]. The main problems here are the extreme mass flow created by vaporizing the eluent from LC, the separation of thermolabile compounds which can not be volatilized without pyrolysis and, finally, the presence of non-volatile ballasts like inorganic compounds which can clog the interface. Though it is possible to solve some of these problems the solution itself is frequently just a compromise.

<u>Derivatization</u>

While in the earlier years of HPLC the possibility of direct chromatography of drugs without derivatization was claimed to be the main advantage over GC, today we are witnessing more and more frequently also derivatization in HPLC. The main goals of derivatization are visualized in the increase of the assay sensitivity, in the applicability of the detector which cannot be used for the underivatized compounds, in the decrease of sample complexity, in the increase of selectivity, in gaining more convenient separation conditions, and in the possibility to separate enantiomers.

Derivatization reagents currently in use offer the possibility to replace the general, however, insensitive short UV-absorption with fluorimetry. Most of the derivatization reactions increase the assay sensitivity by one or two orders with concomitant increase in selectivity. Reactions with dansyl chloride, o-phthalaldehyde, fluorescamine, trinitrobenzene sulphuric acid and with a number of other compounds are used for this purpose. The reagents just mentioned are suited for the derivatization of compounds containing nitrogen functions. In parallel, other reactions, mainly oxidation both chemical and enzymatic, are made use of.

Of the more recent procedures the method of sensitivity increase by using fluorescent ion-pair reagents (e.g., compounds of the α-phenylcinnamonitrile sulphate structure[16]) should be mentioned. Both absorption and fluorescence are frequently influenced by the pH value of the effluent. The pH optimum for the detection need not necessarily be the optimum pH for the separation. Therefore Jansen et al.[17] described a procedure offering an increase of absorbancy of barbiturates at 254 nm using post-column pH modification with ion exchangers.

In addition to the procedures offering the possibility to exploit absorption or fluorescence for detection there are also procedures in which an electroinactive drug can be converted by derivatization into an electroactive species which can be assayed by a very sensitive electrochemical detector. For drugs with a primary amino group the reaction with o-phthalaldehyde was recommended; with secondary amines the formation of metal-dithiocarbamate complexes can be successfully applied[18]. Some further possibilities regarding enantiomer separation will be mentioned in the next chapter.

All types of derivatization can be carried out either before chromatography or in those instances where it is dealt with a sensitivity increase after separation as well. In the first case derivatization can be done off-line or on-line by making use of a multicolumn system. The derivatization itself, however, must not or should not decrease the resolution of the drug and its metabolites in the system used. In post column detection it is practically exclusively dealt with influencing of the molecular properties for the subsequent detection. Everyone of the described procedures has its advantages and disadvantages. The application of the multicolumn system allows simultaneous derivatization and clean-up.

<u>Enantiomer Separation</u>

Another attractive area is the separation and assay of optically active drugs. Today it is recognized that optical isomers of some types of drugs like ß-blockers, antiinflammatory drugs and other in many instances behave differently in the body and therefore in principle they require separate determination. This can be realized with enantioselective chromatographic procedures[19]. The problem can be approached in general in three ways: separation of the enatiomers can be materialized by using chiral stationary phases, by addition of a chiral reagent to the mobile phase, or, by converting the enantiomers into diastereomers by a suitable chiral reagent before chromatography.

Currently the most popular method is the derivatization with a chiral reagent which is relatively fast, does not require special sorbents and offers generally good separations. A certain disadvantage may be seen in the fact that the reagent may cause a degree of racemization during derivatization or may react with the enantiomers at a different rate. This danger can be overcome by separating enantiomers on optically active phases which may be either chemically bound or prepared by impregnation of the sorbent with a chiral reagent. Chiral phases are commercially available[20], e.g., column based on the human plasma protein, Chiralpak, etc. Unfortunately a universal chiral stationary phase allowing the separation of all enantiomer mixtures does not exists at the moment. In working out a methodology for enantiomer separation combination with MS for a secure identification of separated peaks is frequently inevitable.

Naturally it is possible to focus on a number of other areas of HPLC which are today in the center of interest. This refers, e.g., to problems of advanced sorbents, to the problems of miniaturization, particularly the use of microbore columns[21], to the problems of complex automatization, etc. A number of these problems are still waiting for their final solution. Once this is done one can expect a flood of applications in the area of pharmacokinetics and drug level monitoring.

APPLICATIONS

At the end of this review examples of the applicability of HPLC to the categories of drugs that are most frequently chromato-graphically assayed in body fluids should be briefly outlined. From the bibliography of papers published in 1985[3] emerges that the main attention was paid to chemotherapeutic agents (213) followed by central nervous system drugs (183), cardiovascular drugs (159) and vitamins (152). Attention paid to different types of drugs is visualized in Table 3. From this table it is obvious that most papers deal with cytostatics, followed by antirheu-matics, antipyretics and analgesics, adrenergic and adrenergic blocking agents, anticonvulsants, etc.

No doubt, cytostatics are in the highlights. Current literature reports about 700,000 compounds that were tested for their possible cytostatic effects. Out of these about 50 found a practical application and in these pharmacokinetics was followed by chromatography. Monitoring of these drugs is of utmost importance because it is dealt with compounds which exhibit serious side-effects and which have to be administered at high

Table 3. Application of HPLC to Various Classes of Drugs in 1985

1	Cytostatics	89 papers
2	Antirheumatics	57
3	Antipyretics and analgesics	54
4	Adrenergic and adrenergic blocking agents	52
5	Anticonvulsants	46
6	Retinoids	45
7	Antibacterials	37
8	Antiarrhytmics	33
9	Antiasthmatics	32
10	Hypnotics and sedatives	26
11	Tranquillizers	25
12	Cephalosporins	25
13	Penicillins	25
14	Vitamin D	24
15	Hypotensives	23
16	Antidepressants	23
17	Diuretics	23

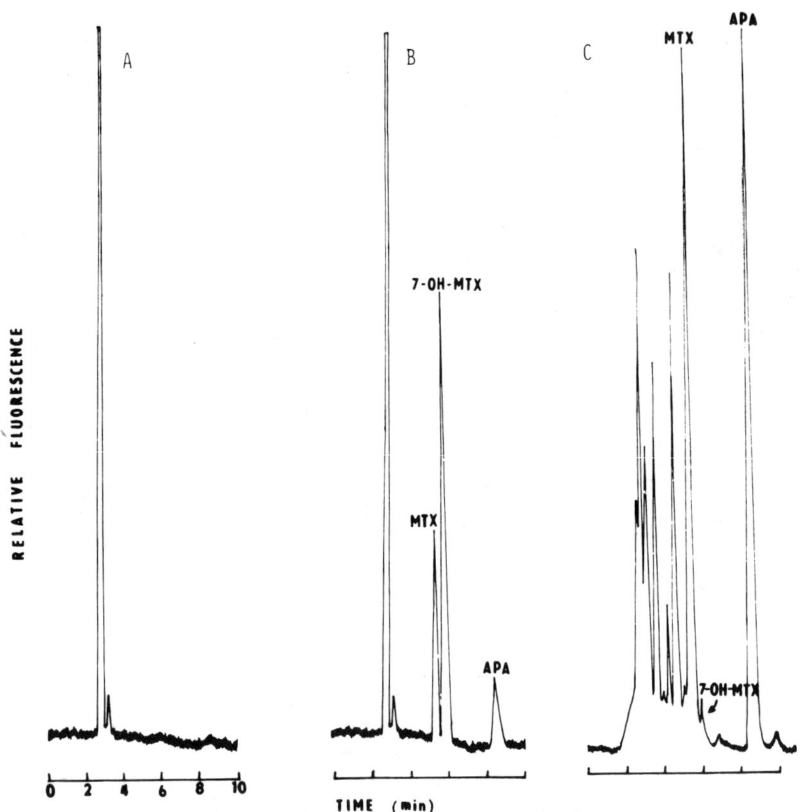

Fig. 2. Chromatogram of plasma and urine samples of a patient
with an ascitic tumour treated with methotrexate
(according to Salamoun and Frantisek[22]). A – blank plasma,
B – plasma 24 h after administration of methotrexate
(MTX), C – urine 4 h after MTX administration. 7-OH-MTX =
7-hydroxymethotrexate, APA = 2,4-diamino-N^{10}-methyl-
pteroic acid.

levels. Since the metabolites can differ both in their therapeutic
effect and toxicity, and since RIA is unable to discriminate
between the parent drug and metabolites, chromatographic methods
are clearly given preference. The estimation of methotrexate, a
compound from the category of antimetabolites with a significant
antitumor activity in acute leukemia and several other neoplastic
diseases may serve as an illustrative example. Currently high
doses of methotrexate are being used. Unfortunately such a high
dose therapy is accompanied by acute folate stress which may be
life threatening. Citrovorum factor is used to protect the patient
from such an effect. In theory citrovorum factor should be
administered immediately after circulating methotrexate has fallen
below the effective level of tumour cell kill. One of the main
metabolites, 7-hydroxymethotrexate, may cause renal damage. The
estimation of methotrexate plasma level and the level of its
metabolite is obviously of great importance for the patient. The
problem with the current methodology is the low sensitivity and
low selectivity of the assays published. Quite recently Salamoun
and Frantisek[22] described a procedure in which methotrexate and
7-hydroxymethotrexate are estimated using a post-column photo-
oxidative cleavage of both compounds to fluorescent products which
are fluorimetrically detected (Fig. 2). The time dependence of
methotrexate and its 7-hydroxy metabolite in plasma of a patient
after a single dose of 50 mg of methotrexate into the ascitic
tumor is demonstrated in Fig. 3. Further on the time curve after
oral administration of 15 mg of methotrexate is shown in this
figure.

Another drug subjected frequently to monitoring is
cyclosporin A. This compound is increasingly used for immunosuppression in patients undergoing organ transplantation.
Cyclosporin A is effective in preventing rejection of heart-lung,
liver, kidney, pancreas and bonemarrow grafts. Cyclosporin must be
administered over several months of transplantation and its
concentration must be kept at optimum level in order to provide

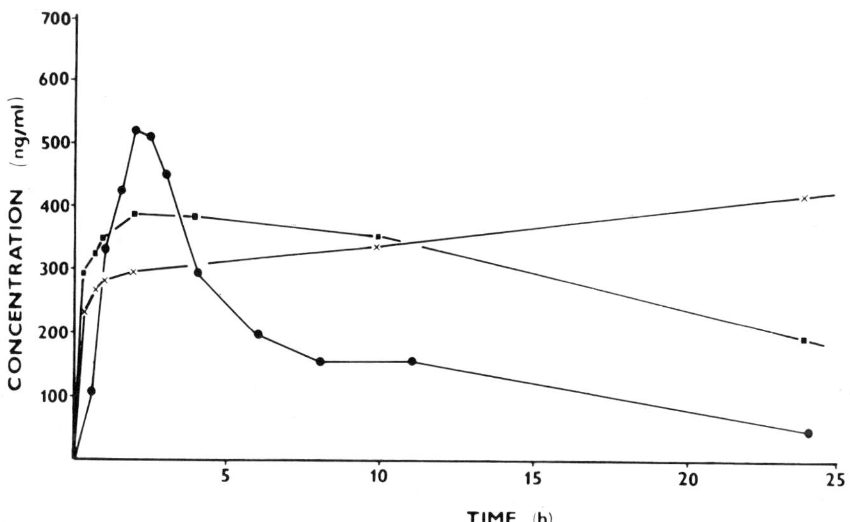

Fig. 3. Time dependence of the methotrexate (■) and 7-hydroxy-
methotrexate (x) levels in plasma after a single dose of
methotrexate into the ascitic tumour and time dependence
of the methotrexate (●) level after oral administration
(according to Salomoun and Frantisek[22]).

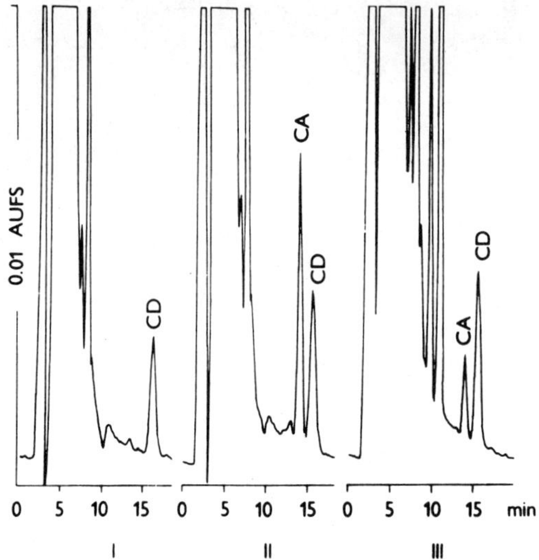

Fig. 4. HPLC of cyclosporin A in whole blood (according to Klima et al.[23]). I – blank blood containing 250 µg/l of cyclosporin D (CD) as internal standard. II – Whole blood spiked with 500 µg/ml of cyclosporin A (CA). III – Patient's blood containing 120 µl/l of CA.

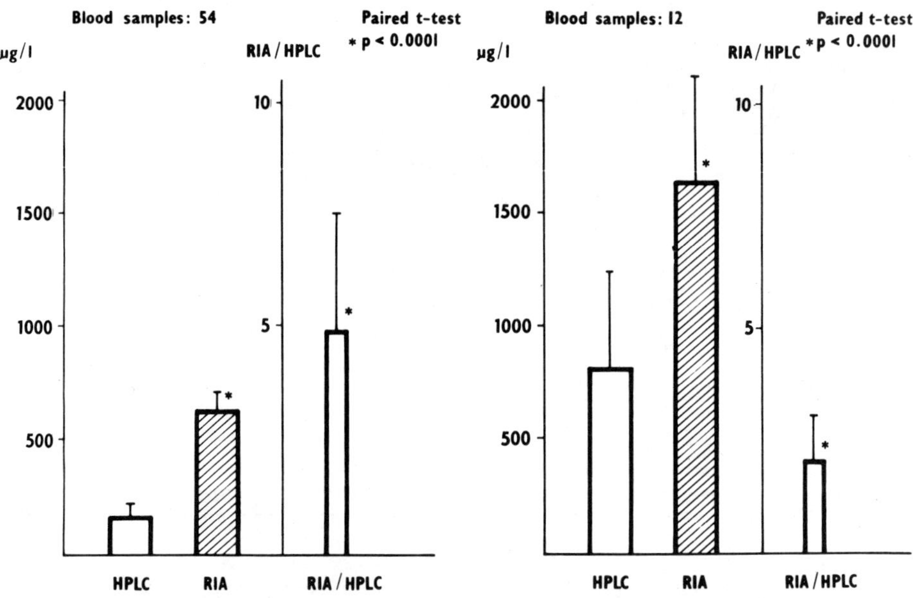

Fig. 5. Concentration of cyclosporin A in whole blood in four patients after heart transplantation (according to Klima et al.[23]). Left – patients without erythromycin and ketoconazole therapy. Right – patients with erythromycin and ketoconazole therapy.

adequate immunosuppression without significant incidence of
toxicity, like nephrotoxicity, hepatotoxicity, hirsuitism and
tremor. Regular assays of the drug at least once per week are
necessary because the drug level is influenced by the status of
liver, by the administration of other drugs (Fig. 5) and by
another possible pathological process, etc. Because the widely
used RIA methods proved insufficient as long as they determine
both cyclosporin A and its metabolites HPLC is clearly given
preference (Fig. 4). The values obtained via RIA methodology are
substantially higher compared to HPLC results (Fig. 5). Because
the cyclosporin A assay is very sensitive to the presence of
ballasts and because usually large series of analyses are required
a large number of procedures using automated on-line purification
and automated sample loading was described[10].

Another example comes from the area of enatiomeric
α-arylpropionic acid based antiinflammatory agents, like
loxoprofen[24]. Loxoprofen and its two monohydroxy metabolites,
trans- and cis-alcohols can be readily converted to respective
amides by condensation with a chiral DANE reagent. The diastereoisomers are then separated on a normal phase column (Fig. 6). In
following the time dependence of excretion of individual
enantiomers in urine (Table 4) it became obvious that there are
considerable differences in the excretion of the R and S forms.
S form is preferably excreted in urine.

At different occasions we are witnessing a discussion whether
gas or column liquid chromatography is better suited for drug
monitoring and for pharmacokinetic studies. Surprisingly literature does not offer too many papers in which both approaches would
be compared in solving a particular problem. One of the excep-

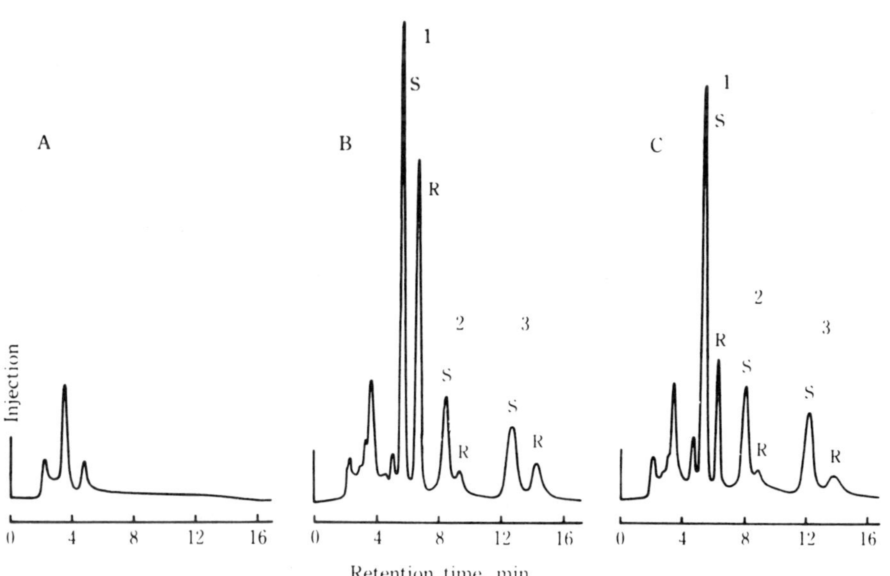

Fig. 6. HPLC of enantiomers of loxoprofen and its metabolites in
human urine (according to Nagashima et al.[24]). A - Control
urine. B - Urine 0-2 h after oral administration of 80 mg
of loxoprofen sodium. C - urine 2-4 h after loxoprofen
administration. Peaks: 1 - loxoprofen, 2 - cis-alcohol,
3 - trans-alcohol.

Table 4. Urinary Excretion of Enantiomers of the Parent Acid, trans- and cis-Alcohol following an 80 mg Oral Dose of Loxoprofen Sodium to Healthy Male Adults (according to Nagashima et al.[24])

Compound	Enantiomer[a]	Amount Excreted in Urine (mg) (n=2)				
		0-2 h	2-4 h	4-8 h	8-12 h	Total
Parent acid	S	7.99	4.82	3.36	1.09	17.26
	R	7.00	1.76	N.D.[b]	N.D.	8.76
	S %	53	73	100	100	66
trans-Alcohol	S	5.62	4.61	2.64	N.D.	12.87
	R	2.64	0.61	N.D.	N.D.	3.25
	S %	68	88	100		80
cis-Alcohol	S	3.94	2.93	1.11	N.D.	7.98
	R	0.85	0.37	N.D.	N.D.	1.22
	S %	82	89	100		87

[a]Enantiomers on the asymmetric carbon in the propionic side-chain
[b]N.D. = Not detected

tional papers in which this was done is the paper of Godbillon et al.[25] about the analysis of a ß-blocker, oxprenolol, in plasma. The results of this comparison enriched with GC-MS data are summarized in Table 5. This table reflects the real situation in making a choice of a method for drug analysis for which it is possible to apply for monitoring purposes both GC and HPLC. It emerges from the presented table that with respect to sample preparation GC is more time demanding. GC as well as its combination with mass spectrometry requires derivatization. The time required for chromatography is the shortest in GC-MS combination. However, if both sample preparation and derivatization are considered then the time parameters are better for HPLC. With regard to the capacity, that means the number of elaborated samples, the best parameters are exhibited by HPLC, on the contrary, the limit of quantitation is the lowest with GC-MS. If this comparison with the equipment costs and with the costs of chemicals are completed then it comes out that the method of choice should be HPLC.

In conclusion the fascinating expansion of column liquid chromatography in the area of pharmacokinetics and in assaying drug level in body fluids should be stressed. Many results that have been achieved in the basic research and in pharmacokinetic studies are to be transferred now to the clinical practice. This as the present situation shows is not going to be simple because many clinicians and clinical chemists prefer simple tests like RIA not respecting that such methods cannot offer as many information as separation methods. From what has been said as well as from the current development of HPLC it can be stated that it is just a question of time when column liquid chromatography will penetrate even more deeply through the laboratories of clinical biochemistry.

Table 5. Main Features of the HPLC, GC and GC-MS Methods for the Assay of Oxprenolol in Plasma (Godbillon et al.[25])

Feature	GC	GC-MS	HPLC
Tubes	Silanized glass tubes	Silanized glass tubes	Disposable poly-propylene tubes
Plasma volume (ml)	1	1	1
Internal standard	Propranolol	($^{13}C_3$)Oxprenolol	Alprenolol
Extraction	1.Extraction into dichloromethane-diethyl ether 2.Back-extraction into acidic aqueous phase 3.Re-extraction into dichloromethane-diethyl ether	1.Extraction into dichloromethane-diethyl ether 2.Evaporation	1.Extraction into ethyl acetate-diethyl ether 2.Back-extraction into acidic aqueous phase
Derivatization	1.HFBA, 1 h at room temperature 2.Extraction at pH 5 into hexane	1.HFBA, 15 min at room temperature 2.Extraction at pH 5 into hexane	None
Injection	3 μl of organic phase Manual	1-2 μl of organic phase Manual	50 μl of aqueous phase Automatic
Detection	Electron capture	Chemical ionization, negative ions	UV 222 nm
Time of analysis (min)	13	2	11
Maximum samples per day	20	50	60
Calibration curve	Valid for 1 month, calibration samples reinjected every week. Range 33-3310 nmol/l (10-1000 ng/ml)	Valid for 1 week, calibration samples reinjected every day. Range 20-1500 nmol/l (6-450 ng/ml)	Valid for 1 week Range 66-3310 nmol/l (20-1000 ng/ml)
Limit of quantitation (nmol/l; ng/ml)	33;10	20;6	66;20
Complexity of equipment	+	+++	+
Cost of instrument solvents	+	+++++	++

REFERENCES

1. K. Macek, ed, "Pharmaceutical Applications of Thin-Layer and Paper Chromatography", Elsevier, Amsterdam (1972).
2. Z. Deyl and J.A.F. De Silva, eds., "Drug Levels Monitoring", J. Chromatogr. 340:1 (1985).
3. K. Macek, J. Janak and Z. Deyl, eds., "Bibliography of Chromatography", J. Chromatogr., Bibliography Section (1970-1985).
4. B. Karger, R.W. Giese and L.R. Snyder, Automated sample cleanup in HPLC using column-switching techniques, TRAC 2:106 (1983).
5. J.C. Kraak, Automated sample handling by extraction techniques, TRAC 2:183 (1983).

6. M.A. Van Lancker, L.A. Bellemans and A.P. De Leenheer, Quantitative determination of low concentrations of adriamycin in plasma and cell cultures using volatile extraction buffer, J. Chromatogr. 374:415 (1986).

7. E. Tomlinson, Ion-pair extraction and high-performance liquid chromatography in pharmaceutical and biomedical analysis, J. Pharm. Biomed. Anal. 1:11 (1983).

8. R.D. McDowall, J.C. Pearce and G.S. Murkitt, Liquid-solid preparation in drug analysis, J. Pharm. Biomed. Anal. 4:3 (1986).

9. P. Kabra and J.H. Wall, Liquid chromatographic determination of cyclosporin in whole blood using Varian's advanced automated sample processing (AASP) unit, HPLC 86, San Francisco, Abstract No. 2733 (1986).

10. M. Arunyanart and L.J. Cline Love, Determination of drugs in untreated body fluids by micellar chromatography with fluorescence detection, J. Chromatogr. 342:293 (1985).

11. V.K. Boppana, R.K. Lynn and J.A. Ziemniak, Qualitative and quantitative determination of sulfate and glucuronide conjugates of dopamine and related compounds by HPLC-ED after in-line post-column hydrolysis with an immobilized sulfatase/ß-glucuronidase reactor, HPLC 86, San Francisco, Abstract No. 3504 (1986).

12. R. Whelpton and T. Moore, Sensitive liquid chromatographic method for physostigmine in biological fluids using dual-electrode electrochemical detector, J. Chromatogr. 341:361 (1985).

13. A.F. Fell, H.P. Scott, R. Gill and A.C. Moffat, Applications of rapid-scanning multichannel detectors in chromatography, J. Chromatogr. 273:3 (1983).

14. R.F. Roberts and M.J. Fields, Monitoring radioactive compounds in high-performance liquid chromatographic eluates: Fraction collection versus on-line detection, J. Chromatogr. 342:25 (1985).

15. R.P.W. Scott, Some future trends in gas and liquid chromatography instrumentation, TRAC 4:96 (1985).

16. K.D. Quinn, J.T. Stewart and M. Zakaria, Post-column HPLC determination of basic drugs using a new class of fluorescent ion-pair reagents, HPLC 86, San Francisco, Abstract No. 3507 (1986).

17. H. Jansen, C.J.M. Vermunt, U.A. Th. Brinkman and R. Frei, Liquid chromatographic determination of some barbiturates in body fluid using post-column pH-modification with anion exchangers, HPLC 86, San Francisco, Abstracts No. 3510 (1986).

18. P. Leroy and A. Nicolas, Determination of secondary amino drugs as their metal dithiocarbamate complexes by reversed-phase high-performance liquid chromatography with electrochemical detection, J. Chromatogr. 317:513 (1984).

19. S.G. Allenmark, Analytical applications of direct chromatographic enantioseparation, TRAC 4: 106 (1985).

20. R. Bishop, I. Hermansson, B. Jaderlund, G. Lindgren and C. Pernow, Direct HPLC resolution of racemic drugs, Intern. Lab. 16:46 (1986).

21. P. Kucera, ed., "Microcolumn High-Performance Liquid Chromatography", Elsevier, Amsterdam (1984).

22. J. Salamoun and J. Frantisek, Determination of methotrexate and its metabolites 7-hydroxymethotrexate and 2,4-diamino-N^{10}-methylpteroic acid in biological fluids by liquid chromatography with fluorimetric detection, J. Chromatogr. 378:173 (1986).

23. J. Klima, R. Petrasek and V. Kocandrle, Simple and specific isocratic liquid chromatographic procedure of cyclosporin A in whole blood compared with radioimmunoassay, HPLC 86, San Francisco, Abstract No. 2742 (1986).
24. H. Nagashima, Y. Tanaka and R. Hayashi, Column liquid chromatography for the simultaneous determination of the enantiomers of loxoprofen sodium and its metabolites in human urine, J. Chromatogr. 345:373 (1985).
25. J. Godbillon, M. Duval and G. Gosset, Determination of oxprenolol in human plasma by high-performance liquid chromatography, in comparison with gas chromatography and gas-chromatography-mass spectrometry, J. Chromatogr. 345:365 (1985).

NEW METHODS OF ANALYSIS AND CONTROL OF BIOLOGICAL FLUIDS, DRUGS AND FOODS

L. Campanella, D. Carrara, M. Cordatore, A.M. Salvi,
M.P. Sammartino, and M. Tomassetti

Department of Chemistry
"La Sapienza" University
Rome, Italy

INTRODUCTION

The activity of our research group during last year was especially devoted to analytical aspects both from the theoretical and application points of view. The aim of our efforts was to look for new analytical methods, for the optimization of other ones already known, for new applications to real matrices of well known techniques. We operated especially with electrochemical (potentiometric, amperometric)[1-7], spectrophotometric (visible, UV)[8,9], thermoanalytical (TG, DSC)[10-12] methods. In this paper the last researches, in the field of the electrochemical sensors, are reported. The developed sensors are summarized in Tables 1 and 2.

EXPERIMENTAL

Samples materials apparatus and methods

The real matrices we considered, listed in the same tables, are biological fluids of clinical interest (bile, serum, amniotic fluid), commercial drugs (antilythogenic drugs, antibiotics, drugs against anaemia and hepatic diseases containing lecithin or choline), food products (flours, cakes, starch). For the preparation of the potentiometric electrodes, both with liquid and polymeric membrane, suitable purified and characterized, home prepared ion exchangers were used. For instance, in the case of cholate[1,2,13] and benzylpenicillinate[14] electrodes, we prepared benzyldimethylcethylammonium corresponding salts. The liquid membranes were generally constituted by solutions in 1-decanol of these compounds, at concentration ranging between 1.0×10^{-2} and 2.5×10^{-2} mol/l; the polymeric membrane by PVC sebacate and 5% of the same exchangers. A scheme of the assembly of these sensors is shown in Fig. 1. The reference electrode was generally a saturated calomel electrode. The measurements were generally performed in thermostatted cells, at room temperature, in stirring conditions, by a digital electrometer (sensitivity 0.1 mV), a digital autoburette (sensitivity 0.1 ml) and a y-t recorder, with different chart speeds. For the preparation of the enzymatic

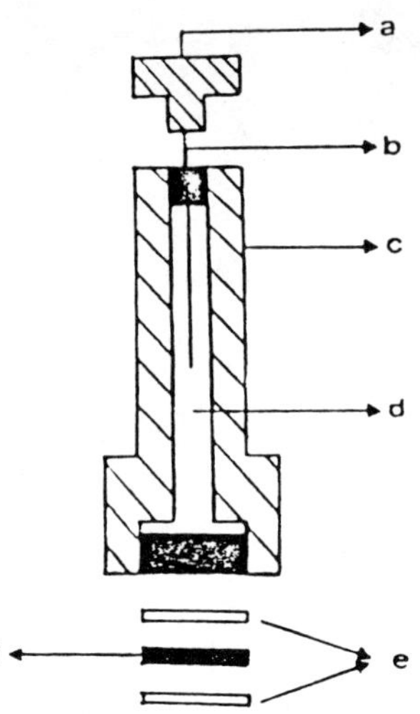

Fig. 1. Liquid or polymeric membrane ion selective electrode.
a) to potentiometer, b) reference Ag/AgCl, c) teflon
body, d) inner solution, e) teflon porous disks, f) liquid or polymeric membrane.

sensors, generally of the amperometric type[5-7,15,16], we used
commercial Clark electrodes, produced by Orion (Cambridge, Ma,
USA) or I.L. (Andover, Ma, USA) and enzymes, immobilized on
polymer nets (Fig. 2) or membranes; the apparatus for the flow
measurements, illustrated in Fig. 3, was performed by suitably
combining commercially available units; in particular the cell had
a volume of 40 μl and was thermostatted by forced circulations of
water, through a simple modification of the pH-blood gas analyzer
I.L. 213. The immobilization was carried out both by chemical
way[5,6], on nylon nets, by Hornby and Morris method, we have
recently optimized (Fig. 4) or by physical way[7,16], in cellulose
triacetate membranes by a new procedure of entrapment, developed
in our laboratory and summarized in Fig. 5. The immobilized

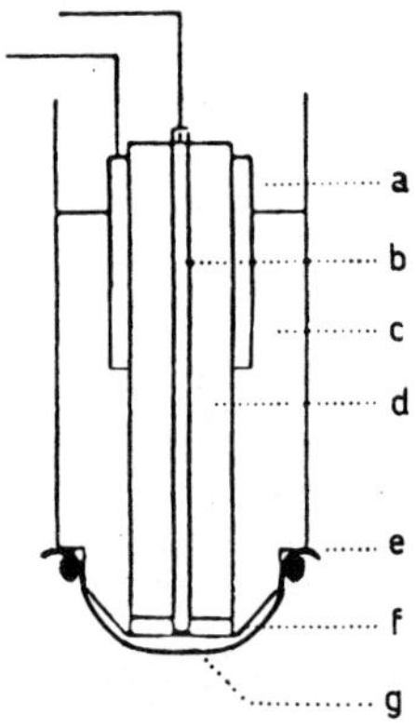

Fig. 2. Enzyme sensor.
a) Ag/AgCl anode, b) platinum cathode, c) electrolyte,
d) glass insulating support, e) rubber o-ring, f) teflon
membrane, g) nylon net, or triacetate membrane, with
immobilized enzyme.

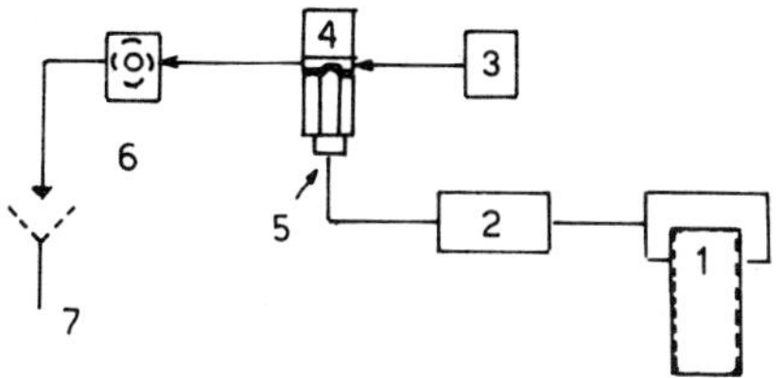

Fig. 3. Flow apparatus for enzymatic analysis.
1) recorder, 2) potentiometer, 3) sample, 4) flow-cell,
5) enzymatic sensor, 6) peristaltic pump, 7) waste.

enzymes were glucose oxidase, choline oxidase and cholesterol
oxidase. The determination could be direct, if the enzyme
substrate was present in the sample to be analyzed, otherwise a
previous hydrolysis was necessary which was obtained by enzymatic
way, in homogeneous phase: so if lecithin or acetylcholine had to
be determined in a biological matrix this had to be incubated for
20 min at 37°C, respectively with phospholipase D, or acethyl-
cholinesterase, before performing in flow conditions the

Table 1. Developed Liquid or Solid Membrane Sensors

Sensor	Method	Determined Species	Nature of the Sample
nicotinate electrode	potentiometric liquid membrane	PP vitamin	drugs
acetic acid selective electrode	potentiometric liquid membrane	acetic acid	vegetable extracts
oxalic acid selective electrode	potentiometric liquid membrane	oxalic acid	urine
cholate electrode	potentiometric liquid membrane or polymer.membr.	conjugated and unconjugated cholic acids anionic surfactants	human bile drugs fresh water
benzoate electrode	potentiometric liquid membrane	benzoic acid cholic acids	drugs drugs
benzyl-penicillinate electrode	potentiometric liquid membrane or polymeric membr.	penicillins cephalosporins	drugs drugs
lead selective electrode	potentiometric solid membrane	phosphate (pot.titr. with Pb^{++})	human bile

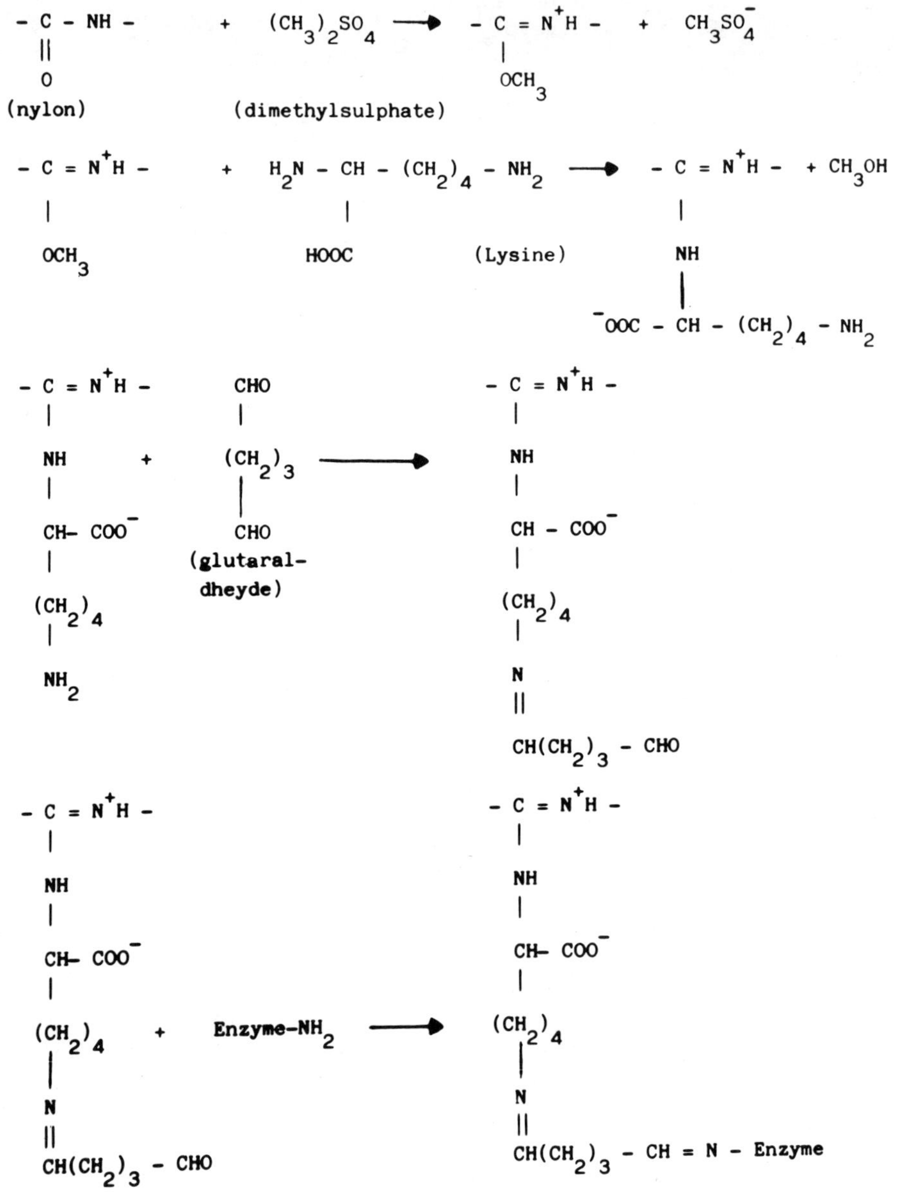

$$- \overset{\|}{\underset{O}{C}} - NH - \quad + \quad (CH_3)_2SO_4 \longrightarrow \quad - \overset{+}{C} = \overset{|}{N}H - \quad + \quad CH_3\overset{-}{SO_4}$$

(nylon)   (dimethylsulphate)

$$- \overset{+}{C} = \overset{|}{N}H - \quad + \quad H_2N - CH - (CH_2)_4 - NH_2 \longrightarrow \quad - \overset{+}{C} = \overset{|}{N}H - \quad + \quad CH_3OH$$

OCH₃   HOOC   (Lysine)

$$^-OOC - CH - (CH_2)_4 - NH_2$$

Fig. 4. Chemical enzyme immobilization on nylon net.

determination of the produced choline, according to the described
procedure. The double immobilization of two enzymes in the same
polymeric system, able to avoid the hydrolizing incubation
process, was tempted, but unsuccesfully; only in the case of the
saccharose electrode, glucose oxidase and invertase were able to
be immobilized together with satisfying results.

Table 2. Developed Enzyme Sensors

Sensor	Method	Determined Species	Nature of the Sample
cholesterol electrode	enzymatic-amperometric	cholesterol	human bile control sera
choline electrode	enzymatic-amperometric	choline	drugs biological fluids
		acethylcholine	applications in course of performing
		lecithin and choline containing phospholipids	drugs food human bile control sera amniotic fluids
glucose electrode	enzymatic-amperometric	glucose	control sera
saccarose electrode	enzymatic-amperometric	saccarose	applications in course of performing
electrode for cholic acids (in course of development)	enzymatic-amperometric	cholic acids	human bile

RESULTS

Of all the developed sensors, a complete electrochemical and analytical characterization was carried out and all the data, including precision and accuracy, on standard solutions and useful concentration range[1-7,13-16], were reported in previuos papers. In Tables 1 and 2 the species able to be determined with each electrode and the main matrices able to be analyzed are listed. On considering the methodological aspects, most meaningful results were obtained in the case of the potentiometric (liquid or

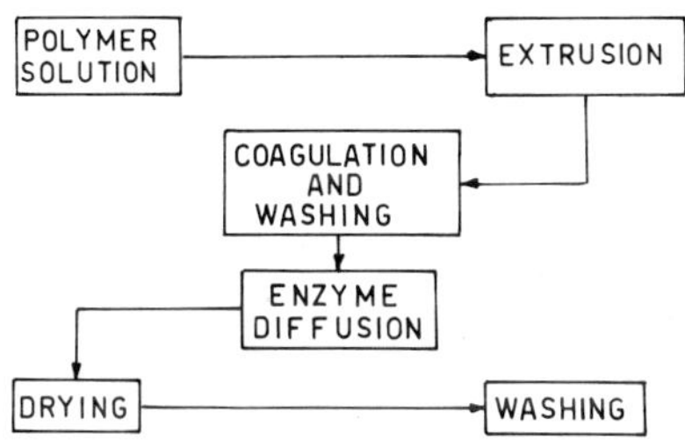

Fig. 5. Physical method, of enzyme immobilization, in cellulose triacetate membrane.

23

Table 3. Liquid Membrane Electrodes Characterization Data

	Cholate electrode	Benzylpenicillinate electrode
linearity range (mol/l)	$4 \times 10^{-5} - 10^{-2}$	$3 \times 10^{-4} - 10^{-3}$
response time (s)	< 10	≤ 15
precision on standard solution (as S.D.%)	0.7	4.3
precision on drug analysis (as S.D. %)	0.9 - 9.9	0.7 - 8.4
% inaccuracy on standard solution (by standard addition method)	+2.8 - +5.8	+0.8 - +3.3
% inaccuracy on drugs solutions (by standard addition method)	+2.0 - +8.0	-4.0 - +1.2

Table 4. Comparison between Precision Data

Species determined and nature of real matrix	Precision (as S.D.%) by enzymatic-amperometric method (physical immobilization)	Precision (as S.D.%) by enzymatic-amperometric method (chemical immobilization)	Precision (as S.D.%) by enzymatic-spectrophotometric method
choline in drugs	O.6 - 4.4	-	6.6 - 13.4
lecithin in drugs	-	0.8 - 2.3	1.8 - 12.5
lecithin in human bile	2.0	2.4	2.6
lecithin in human amniotic fluid	3.2	-	7.7

(S.D.% of lecithin or choline determinations, found by the enzymatic-amperometric method, using the physical immobilization procedure in cellulose triacetate membrane, or the chemical one on nylon net, and by the enzymatic-spectrophotometric method).

Table 5. Comparison between Inaccuracy Data

Species deter-mined and nature of the real matrix	% Inaccuracy by enzymatic-amperometric method (physical im-mobilization)	% Inaccuracy by enzymatic-amperometric method (chemical im-mobilization)	% Inaccuracy by enzymatic-spectrophotom. method
choline in drugs	-8.4 - +4.8	-	-10.1 - +5.0
lecithin in drugs	-	-4.1 - +2.3	-1.2 - +7.4
lecithin in human bile	-4.2 - +5.6	-5.0 - +5.0	-6.0 - +2.0
lecithin in human amniotic fluid	-4.6 - +2.7	-	+2.1 - -1.3

(% recovery by standard addition method of lecithin or choline
determinations, found by the enzymatic-amperometric method, using
the physical immobilization procedure in cellulose triacetate
membrane, or the chemical one on nylon net and by the enzymatic-
spectrophotometric method).

polymeric membrane) sensors. Of the several liquid membrane
electrodes prepared, those ones which found the best applications
were the cholate[2,13] and the benzylpenicillinate[14] electrodes. In
Table 3 the accuracy and precision data are shown, when these
sensors are used for analytical purpose, both on standard
solutions and on real matrices. Recently also solid (polymeric)
membranes were adopted, for these sensors. The analytical results
are not so different from those of the corresponding liquid
membrane sensors, reported in Table 3. Even greater was the
development of the enzymatic-amperometric sensors; the preparation
of a choline electrode has allowed us to determine choline or
lecithin in drugs, foods[15], sera and in two other biological
liquids: the human bile and the amniotic fluid, where these two
determinations are respectively related, to the value of the
lithogenic index[8,9] and of the foethal lung maturity[16]. In Tables 4
and 5, precision and accuracy data of the method are indicated,
both in the case of the chemical immobilization on nylon nets and
of the physical one, in cellulose triacetate membranes. In the
same tables, these data are compared with those obtained by
spectrophotometric detection. Finally, in Table 6, the comparison
between precision of spectrophotometric and potentiometric
detection is reported (the former one home optimized, in the case
of biological matrices, the latter one, developed in our laboratory) in the case of analyzing the main species contained in human
bile, important to diagnose biliar gallstones.

Table 6. Comparison of Precision Data between Electrochemical and
          Spectrophotometric Methods for Analysis of the Main
          Biliar Species

Species determined and nature of matrix	Precision (as S.D.%) by enzymatic-amperometric method (chemical immobilization)	Precision (as S.D.% by enzymatic-spectrophotometric method (Trinder color developing system)
Total cholesterol in human bile	2.3 - 6.0	4.3 - 11.1

Species determined and nature of matrix	Precision (as S.D.%) by ion selective liquid membrane electrode (pre-treatments of the sample are necessary)	Precision (as S.D.% by enzymatic spectrophotometric method (Talalay's method)
Cholic acids pool in human bile	1.5 - 5.0	1.0 - 8.1

Species determined and nature of matrix	Precision (as S.D.%) by tritration method, using a potentiometric lead selective electrode with solid membrane)	Precision (as S.D.% by chemical-spectro photometric method (Bartlett's method)
Total phosphorus in human bile	≤ 5.0	0.9 - 3.3

CONCLUSION

     The researches here briefly described (and also other ones we
propose which we have omitted to describe in detail) show that in
many cases, the analysis of determined chemical species contained
in real matrices can be carried out, by the methods we have
developed, with good precision and accuracy, fast response, no
operating difficulties and without great expenses. This is
generally true for electrochemical and for the enzymatic-spectrophotometric methods, but, for the latter case, only when no
significant chromatic interferences are present or when, if
present, they can be eliminated by masking them or by opportune
blank esperiments[6,15]. Generally the potentiometric electrodes
resulted to be particularly useful in the control of pharmaceutical matrices. On the other hand, the enzymatic sensors seem the
only ones able to be directly used, in the analysis of biological
matrices of particularly complicated nature, such as human bile
and amniotic fluid. As it concerns the immobilization methods,
both chemical and physical immobilization techniques resulted able
to yield good results. The advantage of the physical method, over
the chemical one, consists in the not use of toxic reagents. On

the contrary, the chemical technique showed the advantage of
immobilizing the enzymes in polymeric matrices, generally of
higher mechanical resistance (nylon). The choice of one or of the
other method will be adviced, time by time, by the application
which the immobilization is directed to and by the apparatus which
the laboratory can dispose of.

REFERENCES

1. L.Campanella, L. Sorrentino and M. Tomassetti, Determination
   of cholic acids by ion-selective liquid membrane electrode in
   pharmaceutical products, Anal. Lett. 15:1515 (1982).
2. L. Campanella, L. Sorrentino and M. Tomassetti, Cholate
   liquid membrane ion-selective electrode for drug analysis,
   Analist 108:1490 (1983).
3. L. Campanella, M. Tomassetti, G. D'Ascenzo, G. De Angelis,
   R. Morabito and L. Sorrentino, Potentiometric determination of
   bile phosphates using a lead selective electrode, J. Pharm.
   Biomed. Anal. 1:163 (1983).
4. L. Campanella, M. Tomassetti and M. Cordatore, Application of
   a new cholate liquid membrane electrode to the determination
   of the cholic acids pool in human bile, J. Pharm. Biomed.
   Anal. 4:155 (1986).
5. L. Campanella, M. Tomassetti, B. Rappuoli and M.R. Bruni,
   Chemical enzyme immobilization, Int. Clin. Prod. Rev.
   Sept./Oct.:44 (1985).
6. L. Campanella, M. Mascini, G. Palleschi and M. Tomassetti,
   Determination of choline-containing phospholipids in human
   bile and serum by a new enzyme sensor, Clin. Chim. Acta
   151:71 (1985).
7. L. Campanella, M.P. Sammartino and M. Tomassetti, New physical
   immobilization method for enzyme sensors, in: "Proc. of the
   2nd Int. Meeting of Chemical Sensors" Bordeaux (1986).
8. M. Tomassetti, L. Campanella, A.M. Salvi, G. D'Ascenzo and
   R. Curini, Total phosphorus determination in human bile.
   Comparison between two spectrometric methods, J. Pharm.
   Biomed. Anal. 2:417 (1984).
9. L. Campanella, G. D'Ascenzo, G. De Angelis, T. Ferri,
   M. Mascini, G. Palleschi and M. Tomassetti, Metodi elettro-
   chimici e chemiometrici nella determinazione dell'indice
   litogenico, in:"La bile: aspetti chimici, farmacologici e
   fisiopatologici", A. Roda, L. Barbara, eds., Bologna (1985).
10. M. Tomassetti, L. Campanella, L. Sorrentino and G. D'Ascenzo,
    Further applications of thermoanalytical methods to the
    detection of potassium and sodium salts of penicillins and
    cephalosporins, Thermochim. Acta 70:303 (1983).
11. M. Tomassetti, L. Campanella and G. D'Ascenzo, Thermo-
    gravimetic analysis of conjugated and unconjugated sodium
    cholates, Thermochim. Acta 78:235 (1984).
12. M. Tomassetti, L. Campanella, P. Cignini and G. D'Ascenzo,
    Thermogravimetric analysis of calcium and disodium fosfomycin.
    Analytical application for purity control, Thermochim. Acta
    84:295 (1985).
13. L. Campanella, L. Sorrentino and M.Tomassetti, Preparation,
    characterization and application to a real matrix (drugs) of a
    new liquid membrane electrode sensitive to cholate, Ann. Chim.
    74:483 (1984).

14. L. Campanella, M. Tomassetti and R. Sbrilli, Benzyl-
    penicillinate liquid membrane ion-selective electrode:
    preparation and application to a real matrix (drugs), Ann.
    Chim. (in press).
15. L. Campanella, M. Tomassetti, M.R. Bruni, M. Mascini and
    G. Palleschi, Lecithin determination in food and drugs by an
    amperometric enzymatic sensor, Food. Add. Cont. 3:277 (1986).
16. L. Campanella, M. Tomassetti, G. De Angelis, M.P. Sammartino
    and M. Cordatore, A new assay for choline containing
    phospholipids in amniotic fluids by an enzyme sensor, Clin.
    Chim. Acta (submitted).

MULTIVARIATE ANALYSIS OF CHROMATOGRAPHIC DATA: A USEFUL TOOL FOR

DRUG IDENTIFICATION

G. Musumarra, G. Scarlata, G. Romano*,
and S. Clementi**

Chemistry Department and
* Institute of Forensic Medicine
University of Catania
Catania
** Chemistry Department
University of Perugia
Perugia, Italy

INTRODUCTION

The advantages of thin layer chromatography (TLC) as a
sensitive, simple and quick method for the identification of
organic compounds are well known. However, the applications of TLC
for the identification of drugs in toxicology and related fields
have been severely limited by the problems related to: (a) the
choice of an objective criterion (i.e. an appropriate statistical
approach) which utilizes the information provided by the R_f in
different eluent systems to achieve the identification of
unknowns; (b) the selection of the minimum number of suitable
eluent systems (each providing a different piece of information).

THE METHOD

A data set suitable for a multivariate analysis consists of a
table (matrix) where a number (M) of experimental values (variables) is collected for each of the N chemical compounds (objects).
The geometrical interpretation of each object is a point in a
M-dimensional space, where each variable defines an orthogonal
axis. Accordingly, the data set has the form of N points in an M
space. Multivariate methods seek for the structure of the data,
i.e. they are aimed at recognising systematic patterns, if
present. Among these, principal components analysis (PCA) is
particularly appropriate for the analysis of chromatographic data.

The PCA using the SIMCA method[1-4] and its applications to TLC
data in different eluent systems[5-9] have been presented in detail.

In the present instance, the matrix Y with the elements y_{ik},
contains R_f or I_r values where index i is used for the chromatographic systems (variables) and index k for the compounds (objects). From this data matrix, the number of significant product

terms A and then the parameter α_i, β_{ia} and θ_{ak} in eqn.(1) are estimated by minimizing the sum of squared residuals ε_{ik}.

$$y_{ik} = \alpha_i + \sum_{a=1}^{A} \beta_{ia} \theta_{ak} + \varepsilon_{ik} \qquad\qquad (eqn.\ 1)$$

The number of significant components (A) is determined by cross validation[3].

In this model, α_i and β_{ia} are constants, dependent on the chromatographic system; θ_{ak} are the compound-dependent parameters. The deviations in the model are expressed by the residuals ε_{ik}.

In the first application of PCA to thin layer chromatographi data of drugs[5], the utility of the so-called "loadings plots" (β-β plots) for the evaluation of the information content of the eluent systems, and that of the so-called "scores plots" (θ-θ plots) for the characterization of the compounds, became immediately apparent, pointing out the potential of PCA as a suitable statistical approach both for the selection of the eluen systems and for identification of unknown drugs.

SELECTION OF THE ELUENTS

Information theory has been used for characterizing TLC separations and for comparing different solvents in the separatio of the same group of compounds[10]. Application of numerical taxonom techniques to the choice of optimal sets of solvents in TLC was also reported[11] and paper and thin-layer chromatographic separations of phenolic compounds were classified into clusters according to their selectivities[12].

Application of PCA for the evaluation of the information content of the eluent systems, presents several advantages.

In contrast with previous procedures defining the informatio content of each single eluent mixture as if it were to be used alone[10,11] or correlating two systems at a time[13-16], PCA gives a direct measure of the spanning properties of each system in combination with the others, thus directly providing information on both the minimum number of systems that are needed and the criterion for their selection. As the interdependence of TLC data is well known[17], the superior ability of PCA over regression methods in detecting multivariate patterns is expected.

PCA was applied to a matrix containing the R_f values of 55 basic and neutral drugs in 40 solvent mixtures with the purpose o selecting the minimum number of eluent systems having the maximum information content[6].

The drugs examined, belonged to various classes of compounds (tranquillizers, analgesics, natural and synthetic opiates, alkaloids, anthistamines, local anaesthetics, etc.) differing in their structural and biological properties. The eluent mixtures were chosen from those available in the literature.

PCA of this data set provided a four significant principal components (PC) model, accounting for 92% of the total variance[6]. This analysis, showing that the eluent mixtures cluster into different groups according to their information content, provides a reliable criterion for the choice of optimal eluents. Four eluents mixtures (see Table 1) could be chosen on the basis of the above criterion and of the R_f reproducibility, as a minimum set of eluents which contains practically all the information obtainable from a much larger set and can be further used for the identification of unknowns.

THE TLC MODEL

The PCA of standardized R_f values of 362 drugs in the selected four eluents (Table 1) was then carried out with the purpose of achieving a drastic restriction of the range of inquiry and hopefully identification of unknown samples. The examined compounds included substances widely used in Italy for therapeutical purposes and well known drugs of abuse which can be detected using the Dragendorff reagent and the acidified iodoplatinate solution. This PCA provided a 2 PC model, which performs a reduction of the variables, allowing a graphical representation of all compounds into a two-dimensional space.

Identification of Unknowns by the TLC Model

Identification of unknowns, provided the unknown is one of 362 compounds in the data set, can be attempted by measuring the corrected R_{fc} values in the four eluents (see EXPERIMENTAL) and fitting them into the TLC model.

The t_1 and t_2 values for each unknown can easily be calculated by equations (2) and (3):

$$t_1 = 0.0258 \ (100 \ R_{fCI} - 63.48) + 0.0186 \ (100 \ R_{fCII} - 25)$$
$$+ \ 0.0268 \ (100 \ R_{fCIII} - 9.09) + 0.0198 \ (100 \ R_{fCIV} - 41.17)$$

$$(\text{eqn. 2})$$

$$t_2 = - \ 0.0138 \ (100 \ R_{fCI} - 63.48) - 0.0294 \ (100 \ R_{fCII} - 25)$$
$$+ \ 0.0368 \ (100 \ R_{fCIII} - 9.09) + 0.0123 \ (100 \ R_{fCIV} - 41.17)$$

$$(\text{eqn. 3})$$

Then t_1 and t_2 values for the unknown substance can be fitted into the "scores" plot (Fig. 2 in ref. 8) to select the candidates for its identification.

The selection of candidates is done by defining a region of statistical relevance around the t values obtained for the unknown, in particular in the "confidence rectangle" defined by $\pm 0.20 \ t_1$ and $\pm 0.22 \ t_2$.

For example, if for an unknown the corrected R_{fc} x 100 values are 79, 9, 0 and 44 in eluents I-IV (see Table 1), their t_1 and t_2 values, calculated from eqns (2) and (3), are -0.084 and -0.034 respectively. The four candidates included in the "scores" plot of ref. 8 (which is too big to be reported here) are: Lorajmine, Etoperidone, Procaine and Metergoline. In conclusion, standardized

R_f data in four eluent systems appropriately selected to extract the maximum information available from TLC data, were found to be insufficient to achieve unambiguous identification. Measurements of a different nature, such as gas chromatographic data, are needed.

THE TLC-GC MODEL

Standardized gas chromatographic retention indices have been used for identification of drugs.

Retention indices using SE 30 as a stationary phase have been reported and inter-laboratory variation in measurement evaluated[18]

Marozzi[19] showed that the retention index (I_r) is a very reproducible measure of gas chromatographic mobilities and compared the I_r values in different stationary phases.

Computer aided search of data files containing data obtained by different analytical methods have been developed[20], and the mean list length (MLL) approach was recently applied for identification purposes to TLC and GLC data[21].

The present approach, combining PCA of the TLC model with a second analysis including I_r data, can be used as a alternative procedure aimed at unambiguous identification of unknown samples.

Addition of another variable (i.e. of the gas chromatographic retention indices) provides a better PC model which, in most cases, achieves identification.

<u>Identification of Unknowns by the TLC-GC Model</u>

For the identification of unknowns, t_1 and t_2 for the unknown substances can be calculated from eqns. (4) and (5):

$$t_1 = - 0.00286 \ (100 \ R_{fCI} - 68.56) + 0.00195 \ (100 \ R_{fCII} - 30.14)$$
$$- 0.00325 \ (100 \ R_{fCIII} - 9.87) - 0.00318 \ (100 \ R_{fCIV} - 43.23)$$
$$- 0.00202 \ (I_r - 2318.36)$$

(eqn. 4)

$$t_2 = 0.01626 \ (100 \ R_{fCI} - 68.56) + 0.00999 \ (100 \ R_{fCII} - 30.14)$$
$$+ 0.01257 \ (100 \ R_{fCIII} - 9.87) + 0.0096 \ (100 \ R_{fCIV} - 43.23)$$
$$- 0.00033 \ (I_r - 2318.36)$$

(eqn. 5)

For the compound quoted as an example in the TLC model, the I_r value is 2024 and t_1 and t_2 values, calculated by eqns. (4) and (5), are 0.554 and -0.060 respectively.

In analogy with the procedure adopted previously, the selection of candidates is done by defining a "confidence rectangle" around these t values, which in this case are given by $\pm$ 0.055 t_1 and $\pm$ 0.104 t_2.

By fitting this unknown in the "scores" plot of the TLC-GC model (Fig. 1 in ref. 9) we found two candidates in the confidence rectangle: procaine and perhexiline maleate.

By comparing these candidates with those derived from the TLC model, where all the discrimination ability of the TLC data is utilized, it is possible to choose a single compound selected by both models: procaine.

Comparison of the candidates in the TLC-GC model with those selected from the TLC model, achieve identification of unknown drugs in all the cases examined in ref. 9.

CONCLUSION

This work confirms the validity of PCA as a suitable statistical approach for the selection of the eluent systems and for the identification of unknown drugs, by means of chromatographic techniques. The present results appear to be of a great practical significance in analytical toxicology, especially when account is taken of the cost, the time the analytical instrumentation and the simplicity of the calculations required by the method.

EXPERIMENTAL

Preparation of the Sample

The examined compounds were solids or mixtures of solids. The method, however, can also be used to examine extracts from biological fluids and tissues or from post mortem samples. In this

Table 1. Standardized TLC Systems

No.	Eluent mixture (v:v)	Reference compound	100 R_{fc}
I	Ethyl acetate:methanol: 30% ammonia (85:10:5)	Morphine	25
		Strychnine	44
		Aminopyrine	70
		Cocaine	85
II	Cyclohexane:toluene: diethylamine (65:25:10)	Clobazam	15
		Aminopyrine	29
		Mebeverine	47
		Amitriptyline	60
III	Ethyl acetate:chloroform (50:50)	Caffeine	9
		Ketamine	24
		Flunitrazepam	44
		Prazepam	61
IV	Acetone*	Imipramine	20
		Pericyazine	37
		Aminopyrine	62
		Lidocaine	78

* Plates were dipped in 0.1 mol/l potassium hydroxide methanolic solution and dried.

case a preliminary TLC purification and separation using ethyl acetate:methanol:30% ammonia (85:10:5) as eluent mixture, followed by extraction with methanol from silica gel and concentration was performed. TLC or I_r analysis for each substance isolated by the above procedure was then carried out. This purification eliminated most of the interferences due to the biological matrix, giving, for each of the separated drugs, R_f values very close to those obtained for pure substances. I_r values have shown to be scarcely affected by the biological matrix[22]. The above TLC purification procedure, followed by PCA of TLC and I_r data, can be successfully applied for the identification of drugs in biological fluids. The analysis of biological extracts, however, can be complicated by the presence of metabolites not included in the examined substances.

Rf Measurements

The drug (10 mg) was dissolved as the free base or as salt in methanol (5 ml), or extracted from alkaline aqueous solution with ethyl acetate and prepared as a solution containing about 2 mg/ml of drug. No significant differences between the R_f of the free base and those of the salts were observed. Aliquots, 2-3 μl containing 4-6 μg of drug, were applied approximately 1 cm apart to 20 x 10 cm silica gel 60 F_{254} HPTLC plates (Merck, Darmstadt, FRG). The eluent compositions are reported in Table 1. For eluent IV the plates were dipped in 0.1 mol/l potassium hydroxide methanolic solution and dried before application of the drugs.

The standardization procedure suggested by Stead[16] for the correction of R_f x 100 values was adopted. The experimentally determined R_f x 100 values were converted into the corrected values (R_{fc} x 100) by a graphical method, using a six-point correction graph including the R_f x 100 values of the four reference compounds, reported in Table 1, together with the 0,0 and 100,100 points.

Ir Measurements

The methanolic solutions (2 mg/ml) of the drugs were prepared freshly according to the described procedure and aliquots of these solutions were used for the gas chromatographic determination. A glass column (2 m length, 3 mm internal diameter) filled with 1% SE 30 on Anakrom ABS (80 - 100 mesh) and a nitrogen flow rate of 50 ml/min were used throughout.

For the determination of the retention indices or Kovats indices[23], the absolute retention times were measured and compared with those for long chain hydrocarbons, determined on the same day, before and after the unknown compounds were analyzed.

The I_r values, which were calculated according to the standard procedure adopted for isothermal data[24] had, in all cases, an intralaboratory reproducibility of $\pm$ 20.

REFERENCES

1. S. Wold and M. Sjostrom, SIMCA: A method for analyzing chemical data in terms of similarity and analogy, Chemometrics: Theory and Application. A.C.S. Symp. Ser. 52:243 (1977).

2. S. Wold, in: "Evaluation and Optimization of Laboratory
   Methods and Analytical Procedures", D.L. Massart,
   A. Dijkstra and L. Kaufman, eds, Elsevier, Amsterdam (1978).
3. S. Wold, Cross-validatory estimation of the number of
   components in factor and principal component models,
   Technometrics 20:397 (1978).
4. S. Wold, C. Albano, G. Blomquist, et al., Pattern recognition
   by means of disjoint principal components models (SIMCA).
   Philosophy and methods, in: "Proceedings symposium i anvent
   statistik", Copenhagen (1981).
5. G. Musumarra, G. Scarlata, G. Romano and S. Clementi,
   Identification of drugs by principal components analysis of R_f
   data obtained by TLC in different eluent systems, J. Anal.
   Toxicol.7:286 (1983).
6. G. Musumarra, G. Scarlata, G. Cirma, G. Romano, S. Palazzo,
   S. Clementi and G. Giulietti, Application of principal
   component analysis to the evaluation and selection of TLC
   eluent systems for the thin-layer chromatography of basic
   and neutral drugs, J. Chromatogr. 295:31 (1984).
7. G. Musumarra, G. Scarlata, G. Romano, S. Clementi and
   S. Wold, Application of principal component analysis to TLC
   data for 596 basic and neutral drugs in four eluent systems,
   J. Chromatogr. Sci. 22:538 (1984).
8. G. Musumarra, G. Scarlata, G. Cirma, G. Romano, S. Palazzo,
   S. Clementi and G. Giulietti, Qualitative organic analysis.
   Part 1. Identification of drugs by principal components
   analysis of standardized thin-layer chromatographic data in
   four eluent systems, J. Chromatogr. 350:151 (1985).
9. G. Musumarra, G. Scarlata, G. Romano, G. Cappello, S. Clementi
   and G. Giulietti, Qualitative organic analysis. Part 2.
   Identification of drugs by principal components analysis of
   standardized TLC data in four eluent systems and of retention
   indices on SE 30, J. Anal. Toxicol., in press (1987).
10. D.L. Massart, The use of information theory for evaluating the
    quality of thin-layer chromatographic separations,
    J. Chromatogr. 79:157 (1973).
11. H.D. Clercq and D.L.Massart, Evaluation and selection of
    optimal solvents and solvent combinations in thin-layer
    chromatography. Application of the method to basic drugs,
    J. Chromatogr. 115:1 (1975).
12. L. Nagels, Study of paper and thin-layer chromatography of
    phenolic substances by statistical methods, J. Chromatogr.
    209:377 (1981).
13. A.C. Moffat and K.W. Smalldon, Optimum use of paper, thin-
    layer and gas liquid chromatography for the identification of
    basic drugs, J. Chromatogr. 90:9 (1974).
14. P. Owen, A. Pendlebury and A.C. Moffat, Choice of thin-layer
    chromatographic systems for the routine screening for neutral
    drugs during toxicological analyses, J. Chromatogr. 161:187
    (1978).
15. P. Owen, A. Pendlebury and A.C. Moffat, Choice of thin-layer
    chromatographic systems for the routine screening for acid
    drugs during toxicological analyses, J. Chromatogr. 161:195
    (1978).
16. A.H. Stead, R. Gill, T. Wright, et al., Standardised thin-
    layer chromatographic systems for the identification of drugs
    and poisons, Analyst 107:1106 (1982).
17. K.A. Connors, Use of multiple R_f values for identification by
    paper and thin-layer chromatography, Anal. Chem. 46:53 (1974).
18. A.C. Moffat, Use of SE-30 as stationary phase for the gas
    liquid chromatography of drugs, J. Chromatogr. 113:69 (1975).

19. E. Marozzi, V. Gambaro, E. Saligari, et al., Use of retention index in gas chromatographic studies of drugs, J. Anal. Toxicol. 6:185 (1982).
20. R.D. Maier and J. Derksen, Computereinsatz in der praxisbezogenen forensich-toxicologischen Analytik, Z. Rechtsmed. 92:159 (1984).
21. P.G.A.M. Schepers, J.P. Franke and R.A. de Zeeuw, System evaluation and substance identification in systematic toxicological analysis by the Mean List Length approach, J. Anal. Toxicol. 7:272 (1983).
22. M. Bogusz, J. Wijsbeek, J.P. Franke, R.A. de Zeeuw and J. Gierz, Impact of biological matrix, drug concentration, and method of isolation on the detectability and variability of retention index values in gas chromatography, J. Anal. Toxicol. 9:49 (1985).
23. E. Kovats, Gas chromatographische Characterisierung organischer Verbindungen. Teil 1: Retentionsindices aliphatischer Halogenide, Alkohole, Aldehyde und Ketone, Helv. Chim. Acta 41:1915 (1958).
24. E. Marozzi, V. Gambaro, F. Lodi and A. Pariali, La ricerca chimico tossicologica sistematica generica in tossicologia forense. Nota III. Raffronto tra gli I_r e i Δ I_r ottenuti in programmata e in isoterma, Il Farmaco 32:330 (1982).

QUANTITATIVE EVALUATION WITH IMAGE PROCESSING SCANNER

M. Prosek, M. Medja, J. Korsic, and R.E. Kaiser*

Analytical Laboratory, Research Department
LEK Pharmaceutical and Chemical Works
Ljubljana, Yugoslavia
* Institute for Chromatography
Bad Duerkheim, FRG

INTRODUCTION

TLC is a very suitable microanalytical method because it is quick, simple, and inexpensive.

Today with a normal scanner on-line, connected to a personal computer such as Apple//e we can make more than 5000 samples, (20,000 lanes) per month. This number of determinations can't be enlarged with normal scanners. As it is possible to prepare practically an unlimited number of TLC plates, densitometry remains the slowest part in QTLC. For this reason we think that future development of QTLC lies in the multisensor detectors and in-line computer evaluation.

In our laboratory we constructed an Image Processing (IP) scanner. The results obtained during last year are very promising nevertheless the IP scanner has been constructed from a relatively simple in expansive sensor and a medium priced personal computer.

METHODS

Quantitative evaluation of QTLC is done with a IfC-LEK TLC plate program pack (Fig. 1) using Micron-Eye digital camera connected to an Apple //e microcomputer (Fig. 2). Micron-Eye camera uses optic RAM with 256 x 256 pixels as a sensor. It can be used only in VIS because it has no quartz optic.

During one scan 128 images with 128 x 256 pixels are collected in order to obtain enough information for 15 lanes with 4 to 5 spots. Evaluation of a plate is concluded in 5 minutes, 90 seconds is necessary for scanning and about 2 to 3 minutes for calculation.

The quantitative evaluation was tested with a mixture of indicators with different colors. They were spotted on a HPTLC plate (No.: 5628, Merck, Darmstadt, FRG) in different concentrations.

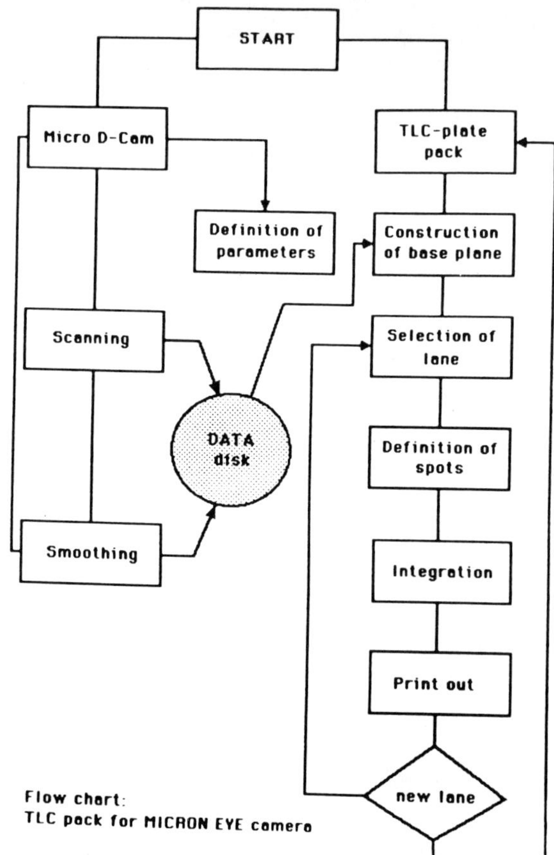

Fig. 1. Flow chart of TLC-plate program pack.

A special scanning device was constructed. A HPTLC plate was installed on the table over the lamp (fluorescent lamp Silvania G8T5) and scanning conditions were determined; from short exposition time when nearly all pixels were black, to the longest time when all pixels were white. In this range 128 images with different exposition (soaking) time were taken. These images were cumulated in a memory. Fig. 3 shows the image of a TLC plate just as the scan had been taken. New sets of data were constructed with the special smoothing procedure from these measurements which were full of noise. Meanwhile the geometric distortion of the camera was repaired and the data were stored on the data disk.

Integration of corrected data set (32k) was done with special programs which offer automatic or manual integration. The program for automatic integration finds all cardinal points of base plane, and constructs and substracts it from the image (Fig. 4). Then, it finds the positions of lanes and integrates them (Fig. 5). The program for manual data processing, allows the operator to enter adequate parameters manually, using keyboard. Integrated values are printed out.

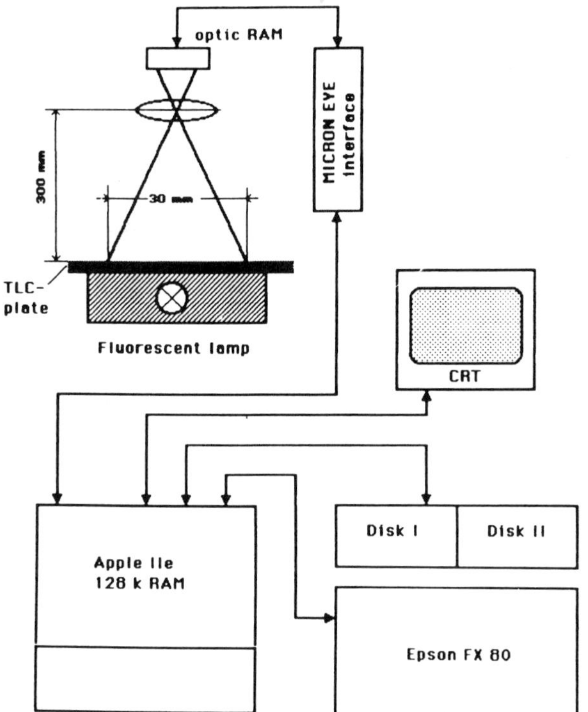

Fig. 2. Hardware used in quantitative evaluation of TLC with
        Micron-Eye digital camera.

Fig. 3. The image obtained as a sum of 128 images, taken at dif-
        ferent exposition time. Distortion in Y direction is
        seen, because each image sensing element of 1532 Optic
        RAM is twice as wide as long.

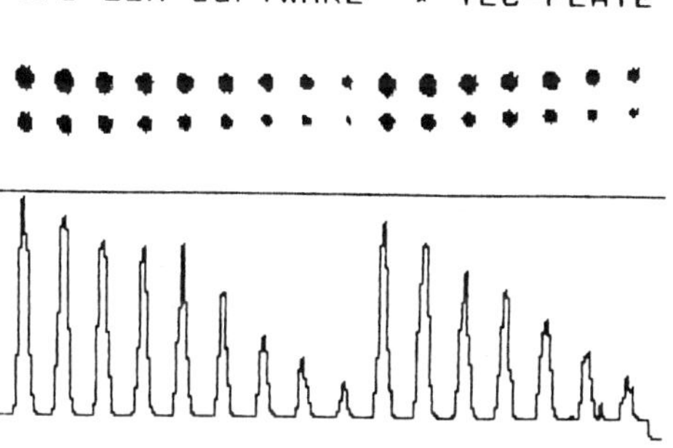

Fig. 4. In the upper part of a screen corrected data of a TLC
        plate are shown, in the lower part sum of all measure-
        ments in Y direction is given.

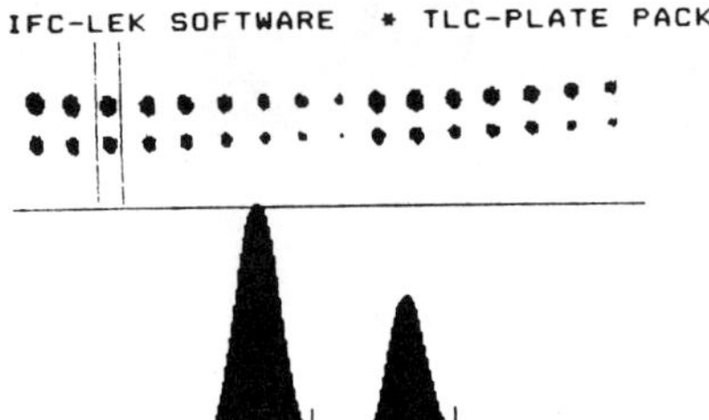

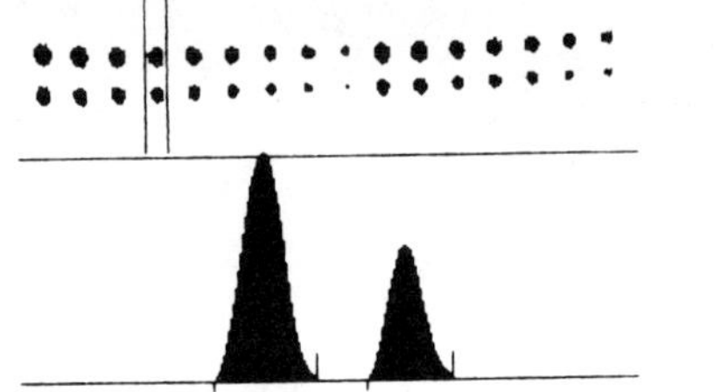

Fig. 5. Print out of integrated lane, spot positions, lane position, peak start, peak end and base line are shown.

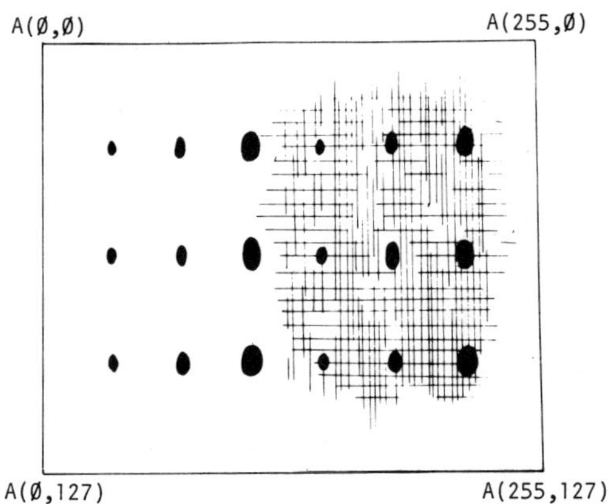

Fig. 6. The image of TLC-plate.

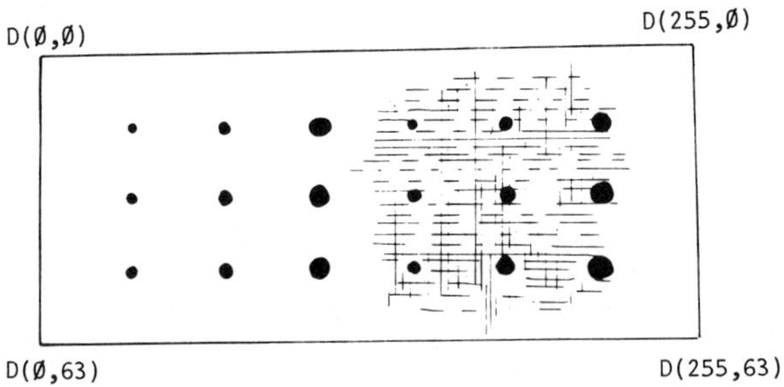

Fig. 7. Corrected image from Fig. 6.

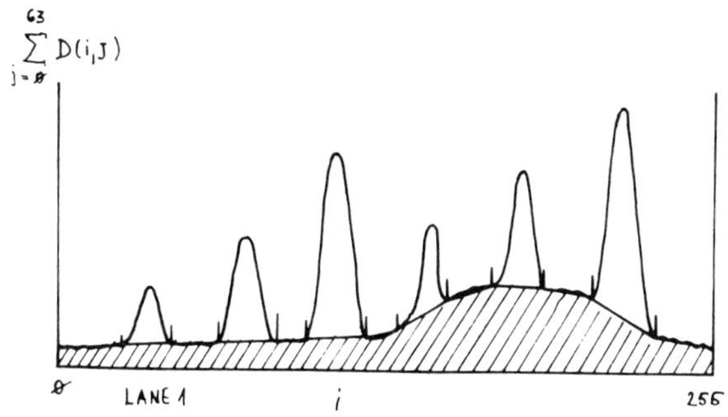

Fig. 8. Integral of all points in Y direction.

INTEGRATION ALGORITHMS

Sensor in camera are not uniform in both directions, therefore the image after the scan (Fig. 6), must be corrected in order to get corrected spacing in X and Y directions (Fig. 7).

Quantitative evaluation of an image from TLC plate can be done in many ways. We developed a simple, quick method, which yields very good results in linear and circular chromatography.

Linear Chromatography

The first determination of lane positions on the plate takes place. They are detected with integration of all data points in Y direction, equation 1.

$$x(i) = \sum_{j=\emptyset}^{63} \qquad \text{(equation 1)}$$

We got a special set of data, which is shown in Fig. 8. From this set of data, positions of lanes on a plate are detected. Between detected lanes base plane is constructed and substracted from the image. It is calculated so that points, which determine lane position in integral set of data X(i), are also taken as a base plane points in each row in Y direction, D(i,j). In each row, from 1 to 64, base line is calculated and subtracted from image

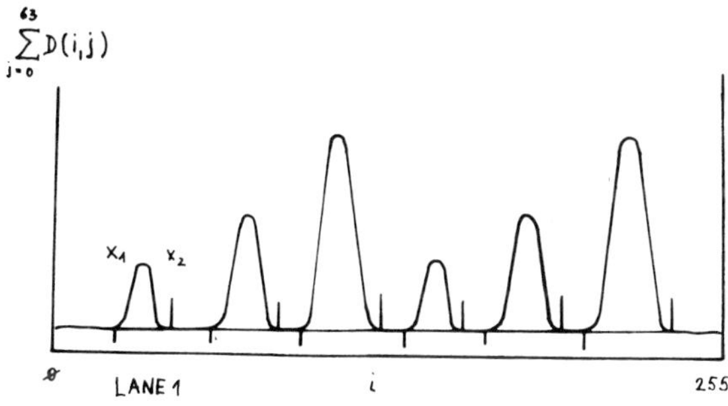

Fig. 9. The same plate as in Fig. 7, now without the influence of base line.

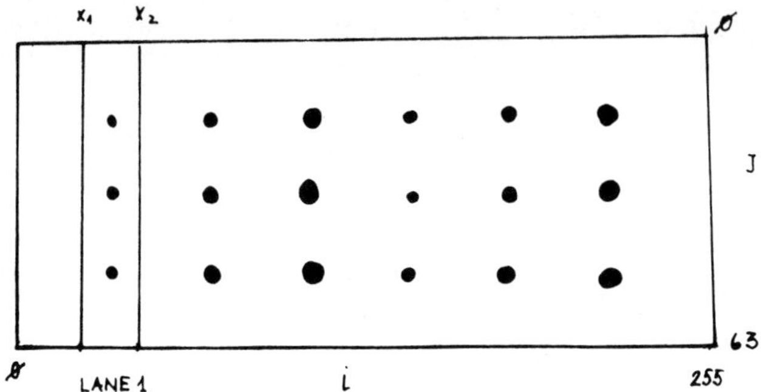

Fig. 10. Detection of lanes.

(Fig. 9), equation 2.

$$D(i,j) = D(i,j) - BL(i,j) \qquad \text{(equation 2)}$$

$$BL(i,j) = D(x1,j) + [D(x2,j)-D(x1,j)] / (x2-x1) * (i-x1)$$

With this procedure information on the plate, without influence of base-line is obtained. The next step is integration of lanes. Once again positions of lanes are determined (Fig. 10). All data in X direction in one lane are integrated and from them, base plane is subtracted, equation 3.

$$Y(j) = \sum_{i=x1}^{x2} D(i,j) - [D(x1,j) + D(x2,J)]/2 \qquad \text{(equation 3)}$$

With this procedure 64 points which represent two dimensional integral of a lane on a plate are obtained. As this number of points is not enough, interpolation is done in order to get 256 points. This chromatogram is integrated according to the normal integration procedure used in other integration packs, done in our laboratory. Cardinal points are detected, base line is constructed and peak areas are calculated (Fig. 11).

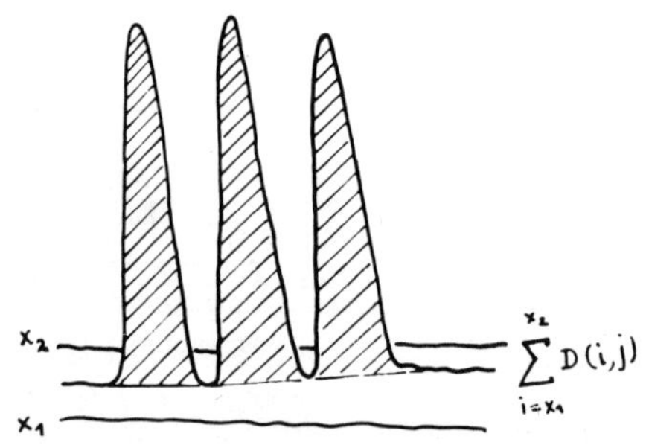

Fig. 11. Integration of one lane.

<u>Circular Chromatography</u>

Quantitative evaluation of circular and anticircular chromatogram with image process scanner is very simple and it is really a great improvement compared to classical mode of scanning. Calculation is performed similarly as with linear technique, instead of X and Y axis radius R and angle O are used.

CONCLUSION

These results even with restriction of low priced hardware shows that the future of quantitative planar chromatography is in image processing systems, nevertheless producers of TLC equipment are strongly against it.

COMBINATION OF COMPUTING INTEGRATOR AND PERSONAL COMPUTER

M. Prosek, A. Medja, and J. Korsic

Analytical Laboratory, Research Department
LEK Pharmaceutical and Chemical Works
Ljubljana, Yugoslavia

INTRODUCTION

Modern analytical laboratory needs a data system capable of collecting, handling, and reporting results from analytical instruments, because it has to produce reliable results in the shortest possible time and cost effective.

Some people are still sure that a centralised multi-channel control system, which does not only control the instruments, but also handles and reports data from the whole analytical laboratory, is the best solution for analytical information system.

Working in the analytical field, we see that distributed computer support can give better results in the shortest time and a much lower starting price. The best solution is obtained by using personal computers on-line connected to a different analytical instrument. Each system is independent in its work, and as such it can be developed without interfering with other systems in a lab. When each individual system is concluded it can be connected in a net using special net program.

In our laboratory we are using Apple II+, Apple //e, IBM PC, HP 85A and HP 85B personal computers as controllers of different analytical instruments. Communication among the instruments and computer is done via IEEE 488 interface, and among the computers via serial interface. Besides this, some special controllers such as MP3000 and CCM from Milton Roy (Rochester, NY, USA) are used in HPLC. Some old instruments are connected to a computer using ADC, DAC and binary I/O.

COMPUTING INTEGRATORS

In this paper we present new levels of performance using inexspensive, easy to use IC-10B computing integrator. The combination of a personal computer with one or more integrators can be very succesfully used as simple hardware, capable to support distributed Laboratory Information and Management System (LIMS). Data acquisition, integration and post-run calculation are done with CI-10, all other data processing are managed with personal computer (Fig. 1).

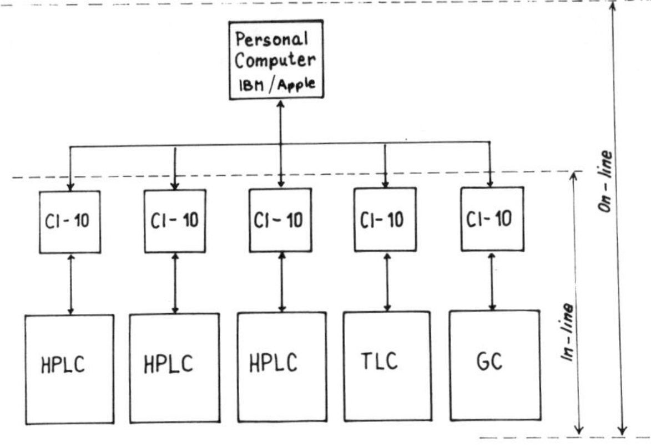

Fig. 1. Combination of personal computer with more CI-10 integrators through IEEE 488 link is shown. Each integrator
is connected to one chromatographic system GC, TLC or
HPLC, and it can be used independently of other systems
in a net.

Computing integrator CI-10B has Motorola 6809 microprocessor,
32K RAM, about 12K ROM, keyboard with 40 dedicated functional
keys, including numeric keypad, 16 character LED alphanumeric
display and Voltage-Frequence type of AD Converter with 100 Hz
sampling rate and -5 mV to +1 V range. Furthermore, it is equipped
with different interfaces, Centronics 8 bit parallel interface for
printer-plotter, IEEE-488 interface for communication with other
devices, autosampler interface with all signals for synchro-
nization, six independant ON/OFF triggers and remote start/stop
connector (Fig. 2).

Built-in software performs integration according to the
selected time programmable functions. These functions are
autozero, noise suppression, treshold, skim ratio, peak width
change, baseline set, valley to valley, horizontal baseline,
integrate set, inhibit integration, negative peak, digitize,

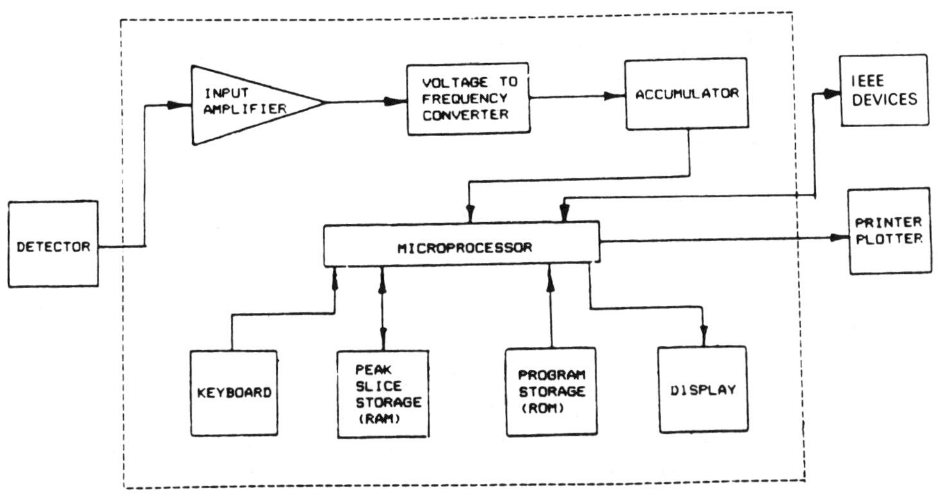

Fig. 2. Details of CI-10 integrator.

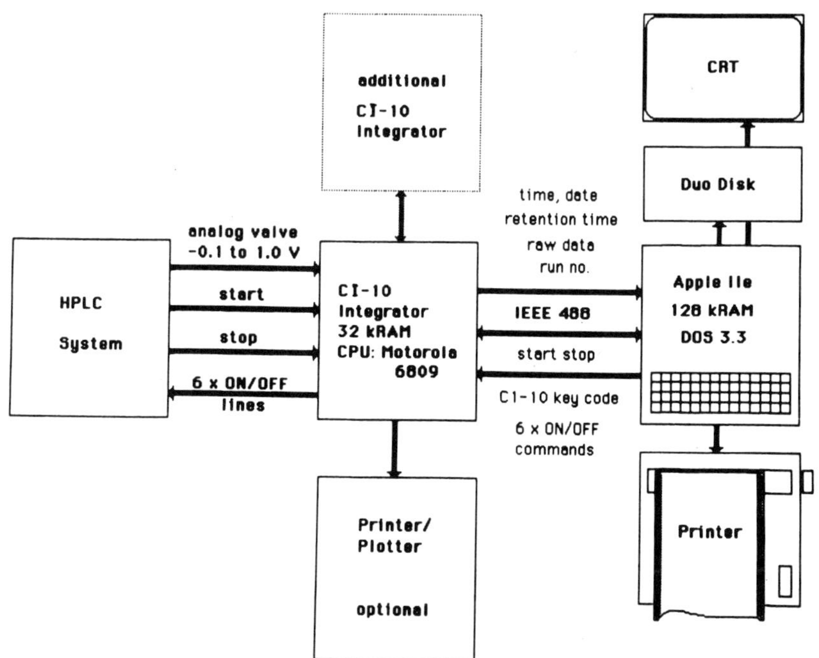

Fig. 3. Connection of CI-10 to HPLC system, which gives ON-LINE
and IN-LINE possibilities.

external triggers 1 to 6. It has nine built in analysis methods
(Method 1-9) one sequence file (Method 0), clock and calendar.

Post-run calculation offers: Area/Height %, Normalization,
Internal Standard, External Standard, Auto-Calibration, Response
Factor and Retention Time up-dating.

Combination of CI-10 integrator and computer (Fig. 3)
provides in-line and on-line handling and reporting data from a
run. In-line mode gives results according to the integration
parameters selected on CI-10 and the integration algorithm
prepared by the producer.

On-line mode used CI-10 as the universal controller and AD
converter. Integration and data manipulation are performed
according to the power of the selected personal computer and the
knowledge of the user. Data acquisition can be done in real time
mode, when each measuring point is transferred to a computer, or
in data logging mode, when the whole batch of data is sent to a
computer after the run. The first method is used when data rate is
slow and only one IC-10 is connected to a computer. The second
method can be used in the case of very fast data acquisition rate,
and in the case of several CI-10 connected to one computer.

Besides the programs prepared by Milton Roy which offers
simple transmission of files from CI-10 to computer and vice-
versa, we prepared a group of programs named LDC-pack for data
acquisition, integration and post run calculation on IBM PC and
Apple //e. These programs use CI-10 as a universal controller and
AD converter. Integration is performed by the use of algorithms

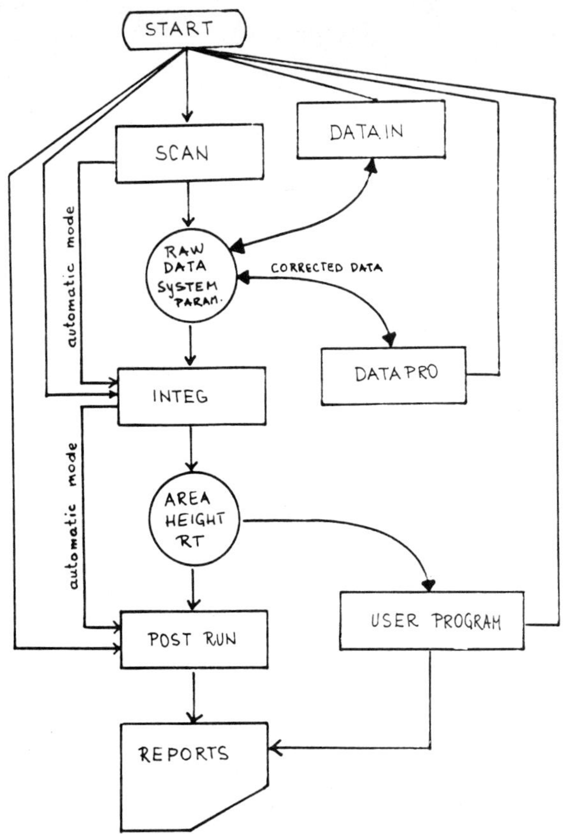

Fig. 4. Flow chart of LDC-pack prepared for post-run data manipulation.

prepared at the Institute for Chromatography Bad Duerkheim (FRG) and Lek Ljubljana (YU). The flow chart of this HPLC-pack is shown in Fig. 4.

The aim of this pack is to separate the data acquisition which can be done automatically, once the correct separation parameters have been determined, from the construction of base line and peak area allocation, which is strictly under the control of an analytical chemist.

Parameters used for measurement, integration, post-run calculation and user made subprograms are determined with DATAIN program.

Data acquisition is done with the program SCAN. Analog signal from HPLC detector (from -0.01 V to 1.0 V) is converted into 16 bit (0-FFFF) number. This number is sent into the computer through IEEE-488 interface, and chromatogram is displayed on the screen. After the run, raw data are stored on a disk for post-run integration.

Integration can be done by two different methods, automatically and manually. Two programs AUTOINT or MANINT can be selected.

Our group made the first program for manual selection of cardinal points in chromatogram in the year 1980. Today we are glad that this useful procedure became an every day routine also in other chromatographic packs because when this procedure was presented in 1980 it was considered as a useless one.

Before the integration, raw data can be tailored to suit integration requirements with the special program DATAPRO, which makes smoothing, according to Savitsky-Golay routines, and mathematical operation such as multiplication, division, substraction and addition of constat to raw data set.

The report may be produced on any type of printer, hard copy of chromatogram can be done only on a graphic printer. All information on peaks, is stored on the disk in text-file, in order to be easily accessed by the program user.

POSTRUN is the program which takes data from the disk and makes post-run calculation according to selected parameter (DATAIN program).

When we need special calculation, we can make our own program. This program must be stored on a program disk (drive 1) under the name which is stored in DATAIN program.

CONCLUSION

The combination of IBM or Apple // with one or more CI-10 integrators is one of the most suitable, inexpensive and easy programmable approaches to distributed laboratory data system.

Its modular system design, intelligent communication proto-col, high level integration algorithms, and versatile post-run software enable the operator to combine the advantage of in-line use of an integrator with the strength and sophistication of the on-line computer system, with all possibilities for construction of a LIMS or a subpart of a LIMS.

FORENSIC TOXICOLOGY: GENERAL UNKNOWN

N.C. Jain

Department of Pharmacology and Nutrition
U.S.C. School of Medicine
Los Angeles, Ca., U.S.A.

The approach to a general unknown varies with the type of
sample(s), the history and objective for that specific case, and
resources available. With that in mind, select from the following
to tailor the procedure to your requirements.

1. Check history of the subject: prior drug use; availability of
    toxins (cyanide to a silverplater, drug to a nurse, etc);
    post mortem results (needle marks may indicate opiates,
    characteristic lung necrosis indicates paraquat,etc).

2. Perform visual examination: identifiable tablets in gastric
    content; cherry red color of blood/tissue may indicate
    cyanide or carbon monoxide.

3. Smell the sample for characteristic odors such as cyanide,
    ethchlorvynol, paraldehyde, petroleum products, etc.

4. Simple color tests: FPN for phenothiazines; Forrests' reagent
    for imipramine and desipramine; Fujiwara for trichlorinated
    hydrocarbons such as chloroform and chloral hydrate; ferric
    chloride test for salicylates; sodium dithionite for
    paraquat; diphenylamine for ethchlorvynol.

5. Reinsch test for metallic poisons such as mercury or arsenic.

6. Gas chromatographic test for volatiles:
  a. direct liquid injection onto a 6' x 1/4" glass column packed
    with 0.5 % Carbowax 600 + 5 % Hallcomid M on Chrom T 40/60 at
    80°C. Will detect acetone, ethanol, methanol, propanol and
    others.
  b. headspace injection onto same column will detect toluene,
    chloroform and others. Place 1 ml fluid or 1 gm homogenized
    solid in serum bottle with septum seal, warm, inject 1 ml of
    headspace.

7. Radioimmunoassay for cocaine/benzoylecgonine, barbiturates,
    LSD, opiates, phencyclidine, methaqualone, amphetamine, THC.
    Blood and urine analysed as directed in package insert;
    tissue homogenized and diluted 1:5; bile diluted 1:5, powder
    dissolved and serially diluted.

8.   Enzyme immunoassay for any of the following: benzodiazepines,
      amphetamine, propoxyphene, methadone, cannabinoids, digoxin,
      lidocaine, disopyramide, theophylline, methotrexate,
      gentamicin, tobramycin, amikacin, netilmicin, tricyclic
      antidepressants, acetaminophen, phenytoin, primidone,
      ethosuximide, carbamazepine, valproic acid, quinidine,
      methaqualone, procainamide/metabolite. Clear samples such as
      urine and serum analysed as directed in package insert. Other
      samples can be diluted or extracted and dissolved in a more
      appropriate matrix.

9.   Systematic extraction and isolation of the various chemical
      fractions, followed by whatever techniques are available for
      identification.
      Proteinaceous tissues such as liver or blood may be subjected
      to any of several protein precipitation techniques. Care must
      be exercised in this choice because not one method is optimal
      for all toxins. Refer to "Clarke" for descriptions of Stas-
      Otto's tungstate, ammonium sulfate, and hydrochloride acid
      methods. Several methods are discussed[1]. Also available is
      enzymatic precipitation using the protease subtilisin
      Carlsberg. The homogenized tissue is incubated with
      subtilisin Carlsberg, 1 mg/g original tissue, at pH 7.4 for
      one hour at 55°C[2,3].
      The resulting solution can be analysed as described in
      Table 1.
      The analyst may opt to omit protein precipitation and go
      directly to extraction.

10.   Choice of TLC solvent systems and sprays: many systems are
      published for both special and general purpose. One of the
      better systems is Davidow's (ethyl acetate: methanol:
      ammonium hydroxide 85:15:10). Another good place to start is
      the system listed in "Clarke". In this lab, the following
      would probably have the greatest utilization.
  a. Chloroform: acetone (9:1) for barbiturates. Spray with
      potassium permanganate to detect unsaturated barbs such as
      secobarb and to differentiate between the
      amobarb/talbutal and butabarbital/butalbital pairs which
      are almost indistinguishable otherwise. Follow with mercu-
      ric chloride and diphenylcarbazone to detect the rest such
      as pentobarb and phenobarbital.
  b. Chloroform: acetone (8:2) for the neutrals. Split the
      residue so more than one spray can be used. Mercuric
      chloride will detect glutethimide and its metabolites.
      Furfural followed by hydrochloric acid detects the carbamates
      such as meprobamate, carisoprodal, etc. Spraying serially
      with bleach, phenol, and then starch also will detects
      glutethimide.
  c. Methanol: ammonia (either 99:1 or methanol alone with a
      beaker of conc. ammonium hydroxide in the tank) for the
      bases. Useful sprays are iodine, iodoplatinate,
      Dragendorff's, followed by sodium nitrate. Among the numerous
      drugs which can be detected in this system are morphine,
      codeine, propoxyphene, cocaine. Special purpose sprays may
      also be desirable: FPN for phenothiazines and
      imipramine/desipramine; ninhydrin for primary amines such as
      amphetamine. Prior to attempting use of published systems,
      in-house data should be developed so that intelligent inter-
      pretation of results is possible.

Table 1. Procedure for Systematic Extraction and Isolation of
General Unknown

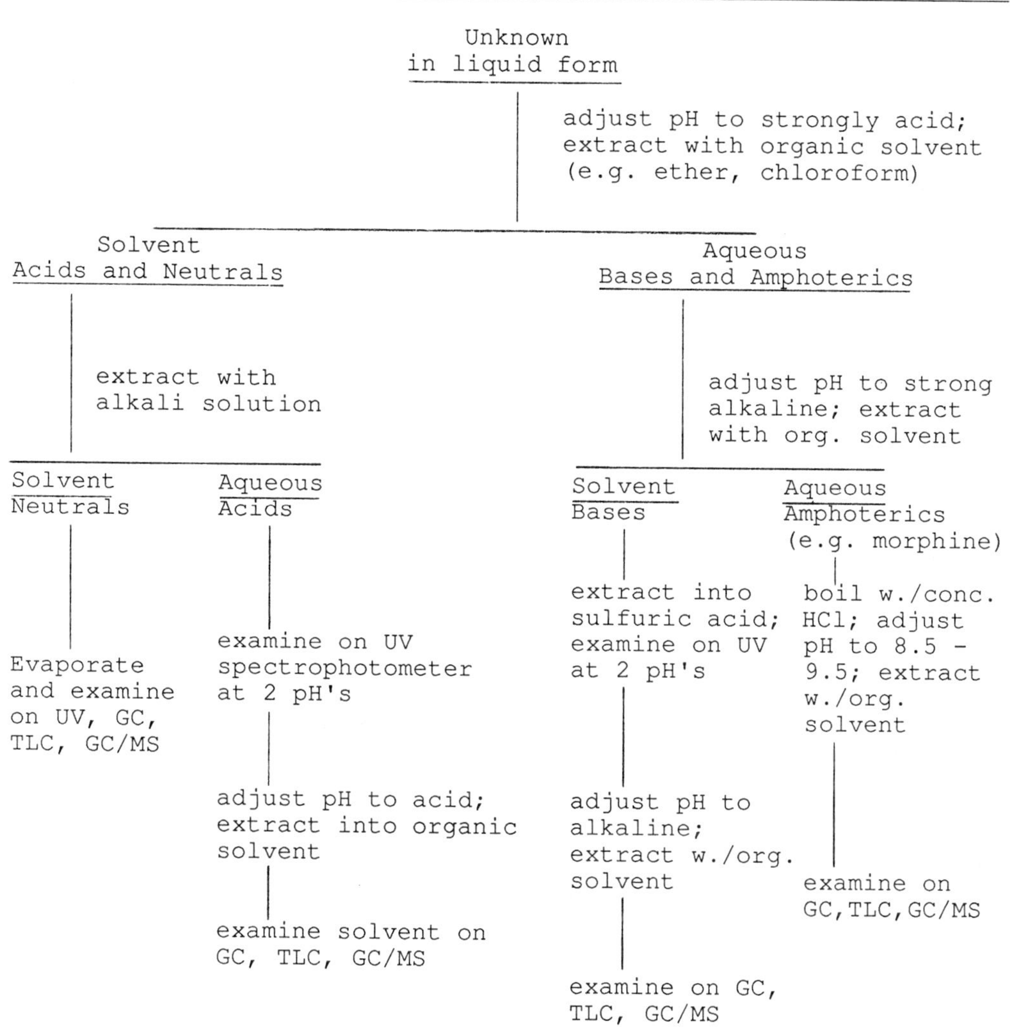

11. Choice of GC systems depends on whether packed or capillary
columns are in use and on whether the detector is FID, NPD,
ECD, MS, etc. Much published data is available to assist in
this choice. In this lab, our largest data bank is for 2' and
6' columns packed with 2.5% SE-30 operated over a wide range
of temperatures. The same flexibility is available with
capillary columns which are temperature programmed. Many
compilations of retention data for drugs have been published
recently. Much of this data has been systematized so that it
is not essential to develop a complete data base in-house[4-7].

12. Derivatization used to enhance sensitivity may make the
difference between detecting or missing some drugs. A variety
of reagents are available commercially for this purpose. In
addition, noting the reactivity of a compound with several

derivatizing agents may provide useful structural
information.

13. Ultraviolet spectroscopy is frequently overlooked as a good
source of information. The usefulness of this tool depends on
having access to a good set of systematically obtained data.
Substances of interest should be analysed under standard
conditions. It is advantageous to run the spectra at two or
three pH's so that shifts in absorbance may be noted. Many
drugs have spectra which are very characteristic, such as
phenothiazines. The utility of this method has been enhanced
by the relatively recent availability of software which plots
derivative curves as well.

14. Another important piece of information which should not be
neglected is presence of metabolites. Many drugs give very
distinctive metabolite patterns, which can be a dead
giveaway. Examples are the multiple spots visualized on TLC
for phenothiazines or the three peak pattern on GC for
propoxyphene and its metabolites. Attention should be paid to
the metabolic behavior of various drugs in deciding if it is
the parent drug or the metabolites or both which will be
detectable.

REFERENCES

1. H.M. Stevens et al., Release of alkaloids from body tissues
by protein precipitating reagents, J. Forensic Sci. Soc.
(1977).
2. M.D. Hammond and A.C. Moffat, Report N° 395, Home Office
Central Research Establishment (1981).
3. M.D. Osselton, Release of basic drugs by the enzymatic
digestion of tissues in cases of poisoning, J. Forensic Sci.
Soc. (1977).
4. D.T. Stafford, Comparison of retention indices on bonded and
coated polysiloxane capillary columns, Crime Laboratory
Digest, July (1985).
5. D.J. Ehresman et al., Screening of biological samples for
underivatized drugs using a splitless injection technique in
fused silica capillary column GC, J. Anal. Toxicol. March
(1985).
6. Gas-chromatographic retention indices of toxicologically
relevant substances on SE-30 or OV-1, Special Issue of the
T.I.A.F.T. Bulletin, VCH (1985).
7. E.M. Koves and J. Wells, Evaluation of fused silica capillary
columns for the screening of basic drugs in postmortem blood,
J. Forensic Sci. July (1985).

EPIDEMIOLOGICAL INVESTIGATION AND ROLE OF THE TOXICOLOGY

LABORATORY

S.D. Ferrara and L. Tedeschi

Clinical and Forensic Toxicology Laboratory
Institute of Forensic Medicine
University of Padua
Padua, Italy

INTRODUCTION

Multi-drug abuse has become common in developed countries and is spreading to developing countries.

To develop a national policy, the nature and extent of the problem must be assessed, but this kind of assessment is, almost without exception, still one of the weakest links in the chain of action that governments take to control drug abuse.

Before examining any measurement techniques, it seems quite necessary to consider the possible range of behaviours and events commonly subsumed under the general categories "Drug Abuse Problem".

When the term "Problem" is used, behaviour or human condition needs to be changed for someone's benefit (the individual, some group of people, or a society at large). Therefore, we can ask: What is the Problem? Is the "Drug Abuse Problem" breaking the law? Is the Problem property crime? Is the social Problem undesirable lifestyle? Is there a Problem of relationship between the individual and his family? Are medical emergencies and deaths the problem? Or is the "Drug Abuse Problem" concerned with how many people fall into each of the above categories or how many rapidly new persons enter the various problem groupings?

The answer is, indeed, that we must be interested in all parts of the "Drug Abuse Problem" but, since the direct measurements of the drug abuse problem do not exist, measures are used that are generally accepted as indicative of whether the drug abuse problem is getting bigger or smaller even through the measures are not generally used as determinations of absolute problem size. Therefore, in order to arrive at the most feasible way which could lead to estimating prevalence and incidence of drug abuse, it is usually suggested to use:

HYPNOTICS	TRANQUILIZERS	ANALGESICS & NARCOTICS	ANTIDEPRESSANTS
ALLOBARBITAL	CLONAZEPAM	ACETYL SALICYLIC ACID	AMITRIPTYLINE
AMOBARBITAL	CHLORDIAZEPOXIDE	SALICYLIC ACID	DESIPRAMINE
APROBARBITAL	K-CLORAZEPATE		IMIPRAMINE
BARBITAL	DIAZEPAM	CODEINE	TRIMIPRAMINE
BUTABARBITAL	DEMOXEPAM	MORPHINE	
BUTALBITAL	DESMETILDIAZEPAM	METHADONE	
BUTOBARBITAL	FLUNITRAZEPAM	PENTAZOCINE	**DEPRESSANTS & SOLVENTS**
CYCLOBARBITAL	FLURAZEPAM	PROPOXYPHENE	
HEPTABARBITAL	LORAZEPAM		ETHANOL
HEXETHAL	NITRAZEPAM		ACETONE
IBOMAL	OXAZEPAM	**STIMULANTS-**	ACETALDEHYDE
IDOBUTAL	TEMAZEPAM	**PSYCHOMIMETICS**	METHANOL
NEALBARBITAL			HYDROCARBONS
PENTOBARBITAL	CHLORPROMAZINE	AMPHETAMINE	
PHENOBARBITAL	PROMAZINE	METHAMPHETAMINE	
SECOBARBITAL	THIORIDAZINE	COCAINE	
TALBUTAL	TRIFLUOPERAZINE	PHENCYCLIDINE	
VINALBARBITAL		CANNABINOIDS	
	MEPROBAMATE		
METHAQUALONE			

Fig. 1. Drugs included in the analytical screening.

- data of different services dealing with drug abuse in a given community;
- general and special population surveys by means of simple random sampling, clustered and stratification sampling, cross-sectional studies[1-3].

During the late seventies and early eighties both general population surveys and studies of special population have provided new understanding of drug use and experience.

Unfortunately, it is still difficult to compare the findings because each team of investigators tends to gather a different type of data and to employ instruments and methods different from those used by other teams.

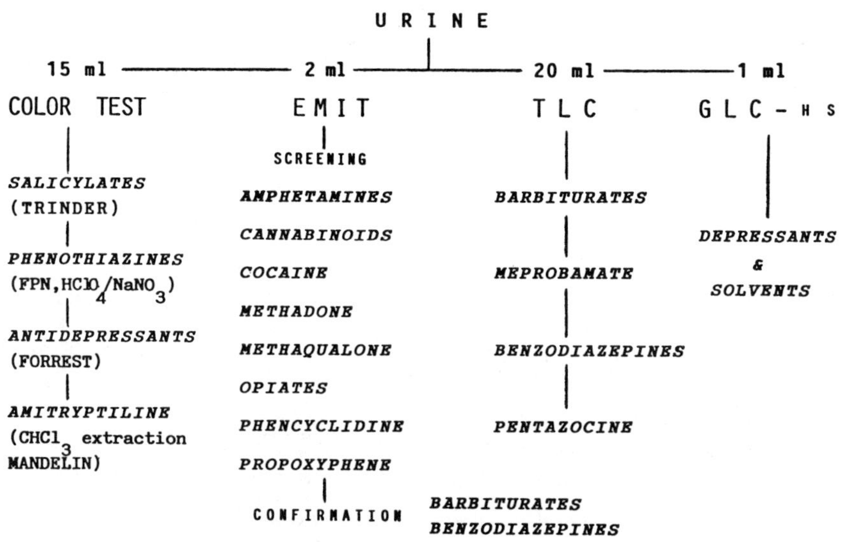

Fig. 2. Analytical approach to the toxicological assay of urine.

In international collaboration, it is necessary to reconcile the criteria and definitions to be used in different sociocultural settings in order to create a common basis for measuring the particular conditions under study, otherwise there will be no assurance that investigators in different parts of the world will study and report the same phenomena.

Fundamental requirements of the epidemiological study are[3]:

- Representativeness;
- Respondent anonymity and cooperativeness;
- Validity in estimating change;
- Sampling accuracy estimation;
- Constancy and accuracy of measurement procedures;
- Low staff-skill;
- Cost quite low;
- Economies with repetition.

TOXICOLOGY LABORATORY APPROACH

To validate the results of every study, primary role is played by the Toxicology Laboratory through analyses on biological fluids[4].

Considering the clinical acute poisoning, the drug-related accidents and deaths as main forensic toxicological interests, in such view it is reasonable to: deserve attention to limited classes of drugs and specific biological fluids; set quite a high concentration (0.5-1 µg/ml barring alcohol and cannabinoids), as a cut-off in order to differentiate between negative and positive

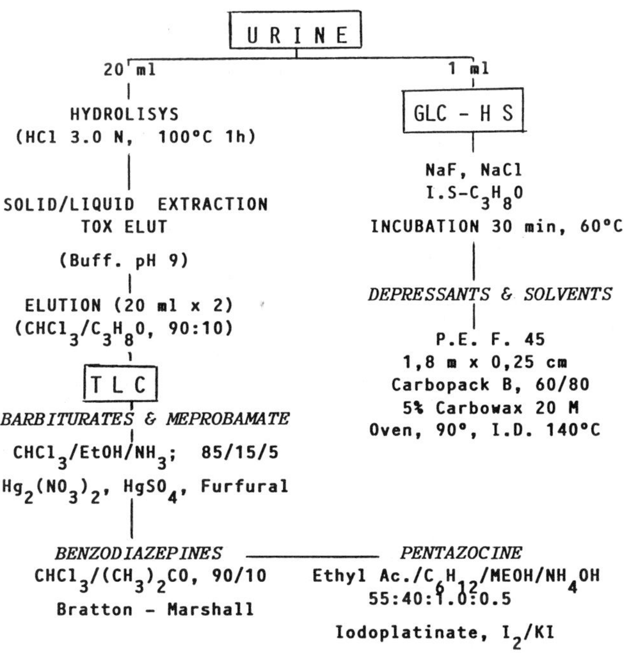

Fig. 3. Analytical scheme for the determination of barbiturates, meprobamate, benzodiazepines, pentazocine, depressants and solvents in urine using TLC or GLC.

results; conceive and adopt a methodological scheme suitable for different levels of staff-skill and equipment.

Fig. 1 shows the classes of drugs that might be included in the analytical screening, most of them investigated during the polyannual epidemiological study on "Alcohol, drugs and traffic accidents" initiated in 1978 at the University of Padua[5].

Drawing, sampling and analytical simplicity, pharmaco and toxicokinetic equivalence make saliva and urine, vitreous fluid and urine suitable for the detection of drugs listed in Fig. 1, respectively in ante and post-mortem epidemiological studies.

The small quantity of saliva and vitreous fluid which may be drawed however, restricts their use to the investigation of few drugs selected according to the epidemiological objectives.

The peculiarity of the general and special population survey, based on a great number of samples and representativeness, does not entail use of sophisticated techniques or compulsory confirmation of results, which solely depend on the scientific and technical level of the laboratory.

<u>Urine Screening</u>

Such a conceptual approach allows one to conceive two independent analytical stages, the first of which consists of the technical procedures illustrated in Fig. 2 for minimal quantities of urine.

Current approaches and experiences in drug determination in fluids rely on chromatography, enzyme immunoassay and color-test methods.

For simplicity, rapidity and minimal instrumentation, TLC, EMIT and Color-Test are techniques of choice in systematic

```
INSTRUMENT          :   HP 5790 A
DETECTOR            :   HP 5970 A (MSD)
COLUMN              :   CROSSLINKED 5% PH/ME SILICONE
                        12 MT - ID 0.2 - FILM THICKNESS 0.33 um
CARRIER             :   He, 80 CM/SEC

1. MODE INJ.        :   SPLIT 1:20
CAPILLARY CONDITIONS

TEMP. 1     TIME 1      RATE        TEMP. 2     PORT        E.M. VOLTS
   150         0.5      10.0          260         240         A + 400

2. MODE INJ.        :   SPLITLESS
CAPILLARY CONDITIONS

TEMP. 1     TIME 1      RATE        TEMP. 2     PORT        E.M. VOLTS
    50         1.0        18          270         240         A + 400

SPLIT VENT FLOW     :   100 ml/min
VALVE PROGRAM NUMBER:   2
VALVE TIME          :   0.5 min.
```

Fig. 4. Analytical conditions for GLC-MSD (SIM) confirmation tests.

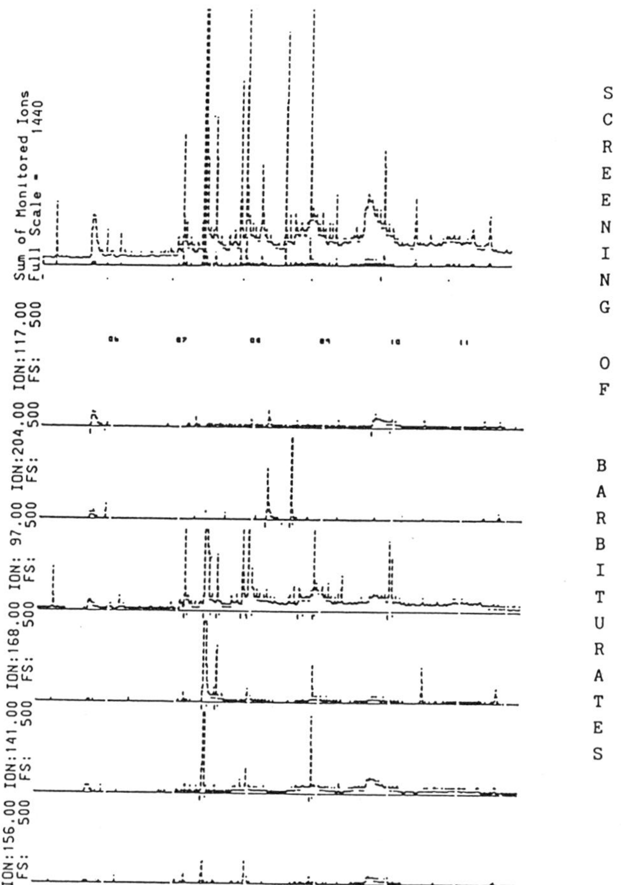

Fig. 5. GLC-MSD assay of barbiturates in a sample previously purified by TLC (see text).

toxicological analysis of acidic, neutral and basic drugs.

The validity of the reported TLC (Fig. 3) can be attributed to sensitivity and specificity of the visualization methods, based on the hydrolysis and pre-chromatographic extraction, which comes in handy for eventual confirmation by the alternative chromatographic technique.

The wide validity of EMIT® for urine screening is restricted, in case of confirmation, to amphetamines (by EMIT itself using the confirmation Kit) and a limited number of barbiturates and benzodiazepines which, if positive by TLC, do not require any other confirmation.

Similar validity may be attributed to Color Test for phenothiazines and salicilates, whereas a positive result for antidepressants should be confirmed.

GLC becomes necessary for investigating ethanol and related volatile substances adopting the conditions illustrated in Fig. 3.

<u>Urine Confirmation</u>

According to the requirements of the epidemiological study, levels of staff-skill and equipment, the confirmation of the presence of a drug or a single class of drugs may be achieved by monitoring series of various ionic masses by means of GLC-MSD (SIM).

In such a way a substance is presumed to be identified by the occurence of consistent abundance of two peaks at correct retention time in the appropriate multiple ion trace. This preliminary identification is then confirmed by a second injection of the sample extract and monitoring of additional ions specific to the drug detected. The presence of the compound in the sample is established by the appearance of all the specific ion peaks at the correct retention time and by the comparison of relative

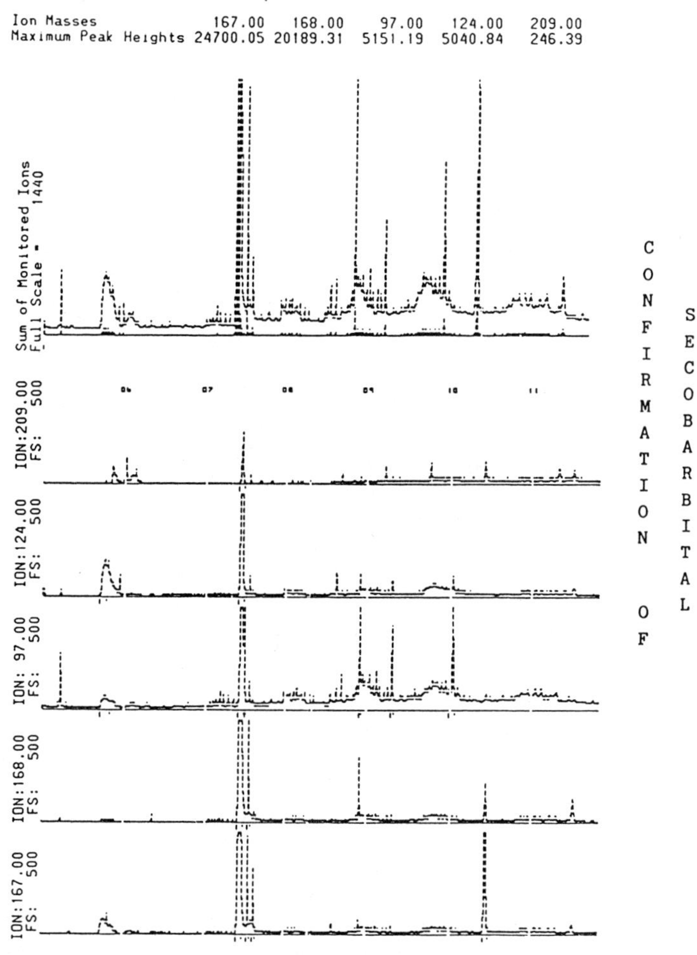

Fig. 6. GLC-MSD assay of secobarbital in a sample previously purified by TLC (see text).

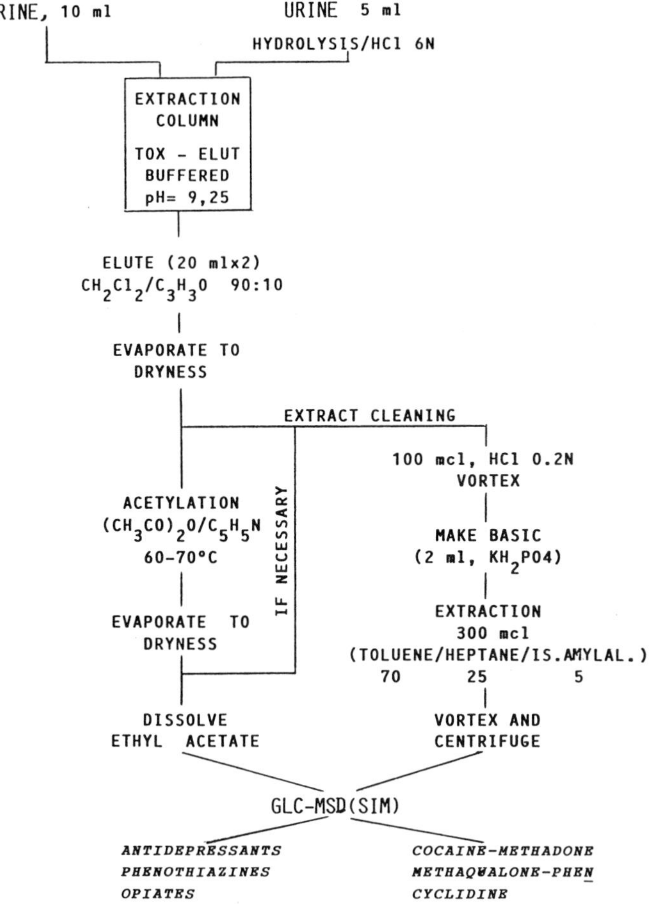

Fig. 7. Sample preparation for GLC-MSD (SIM) of antidepressants, phenothiazines, opiates, cocaine, methadone, methaqualone and phencyclidine.

Table 1. GLC - MSD (SIM) Confirmation

	Masses					
Barbiturates	156	141	168	97	204	177
	141	124	207	167	221	195
Benzodiazepines	77	230	246	249	262	
Benzophenones	109	211	227	244	248	
	86	241	249	274	260	
	63	95	109	123	285	
Meprobamate	55	83	96	114	144	
Meprobamate Artifact	56	84	101			
Pentazocine	70	110	217	285		
Pentazocine Hydrate	230	303	288	58	100	

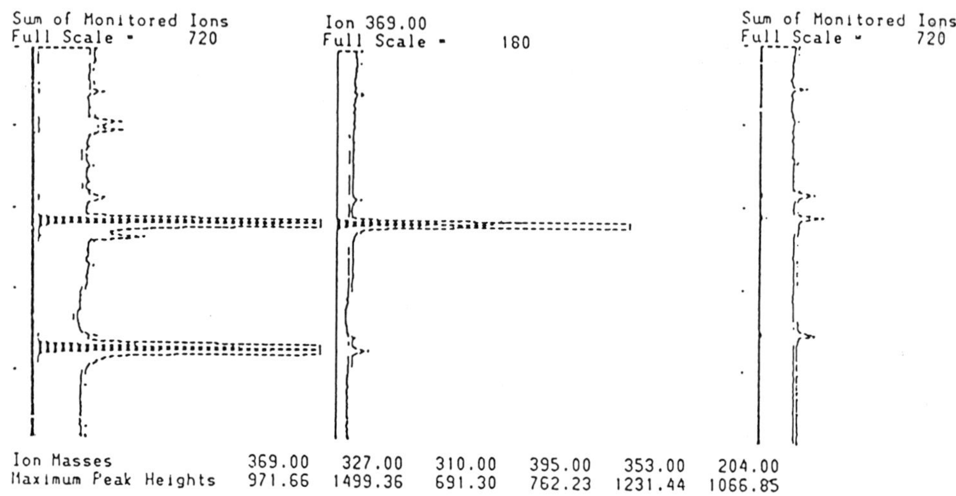

GC RUN COMPLETE, Data stored in RUN 1

*** INTEGRATION REPORT ***

Run # 1
Date Acquired: 4/23/1986

Area Threshold: 300
Slope Sensitivity: 20.0%

Peak		Scan Numbers		Area Counts			Flags			
RT	Ion	Start	Stop	Raw	Horiz	Tangt	A	B	C	D
7.13	369.00	142	161	3922	3722	3708				
7.13	327.00	142	159	5941	5647	5636				
7.13	310.00	140	158	2730	2663	2647				
8.73	310.00	250	264	470	417	414				
8.73	395.00	248	272	3923	3883	3875				
8.73	353.00	249	267	6230	6078	6075				
7.13	204.00	141	157	4158	3985	3978				

Fig. 8. Typical examples of GLC-MSD (SIM) carried out on blank and morphine containing samples.

Table 2. GLC - MSD (SIM) Confirmation

	Masses					
Antidepressants	58	71	195	202	235	
Cocaine	82	83	182	94	96	303
Ecgonine Methyl Ester	82		96	199	168	
Codeine	341	282	229	204	162	
Methadone	72	165	294	198	309	
Methadone-Metabolite	227		276	262	165	
Methaqualone	235	250	132	91		
Morphine	369	327	310	268	162	
Phencyclidine	200	91	84	243	242	
Phenothiazines	58	71	195	202	235	
Propoxyphene	58	115	208	265		
Propoxyph. Artifact	91	115	130	208		

intensities of abundance with those of standard samples analysed
in the same conditions.

The instrument and conditions employed to carry out the
confirmation analyses are almost the same for all the classes of
drugs considered (Fig. 4).

Table 1 lists the ionic masses for GLC-MSD screening and confirmation, carried out on extracts of material scratched off the
TLC plate at Rf of spots positive for barbiturates (Fig. 5, 6),
benzophenones, meprobamate and pentazocine.

GLC-MSD confirmation of the remaining classes of drugs
involves the preliminary general hydrolysis and/or on column
extraction shown in Fig. 7.

Table 2 lists the ionic masses for screening and specific
identification of antidepressants, cocaine, methadone,
methaqualone, morphine (Fig. 8), phenothiazines, propoxyphene as
parent drugs and acetyl-derivates.

To confirm the presence of cannabinoids (Fig. 9) it is necessary, on the contrary, to carry out the specific pretreatment,
extraction and derivatization shown in Fig. 10, where the ionic
masses are listed.

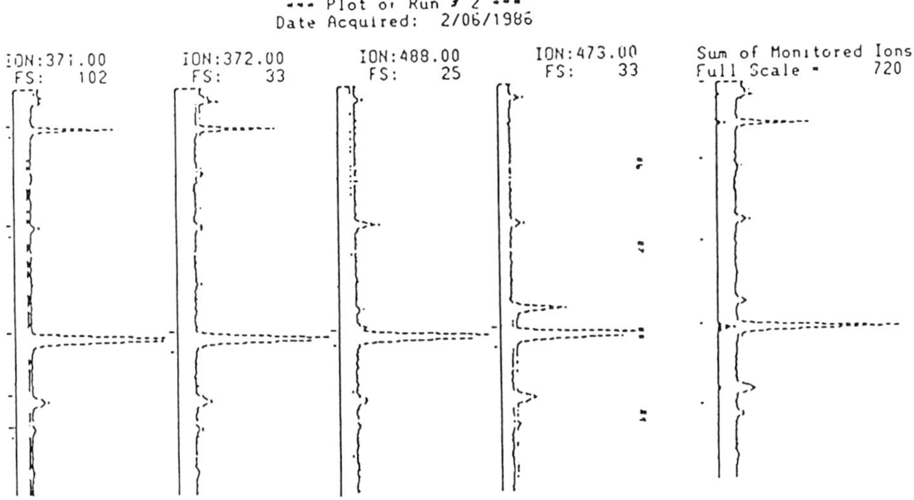


```
--- INTEGRATION REPORT ---

Run # 2                              Area Threshold:      100
Date Acquired:  2/06/1986            Slope Sensitivity:   20.0%
```

Peak		Scan Numbers		Area Counts			Flags				
RI	Ion	Start	Stop	Raw	Horiz	Tangt	A	B	C	D	
5.55	371.00	32	41	300	194	181					
6.71	371.00	115	124	134	37	28					
8.03	371.00	204	221	647	459	449					
8.75	371.00	255	266	200	76	73					
9.12	371.00	280	289	120	30	23					
8.03	372.00	204	219	213	48	141					
8.03	488.00	204	215	144	106	103					
8.03	473.00	203	218	200	148	148					

Fig. 9. Typical example of GLC-MSD (SIM) carried out on a sample
         containing THC-Metabolite.

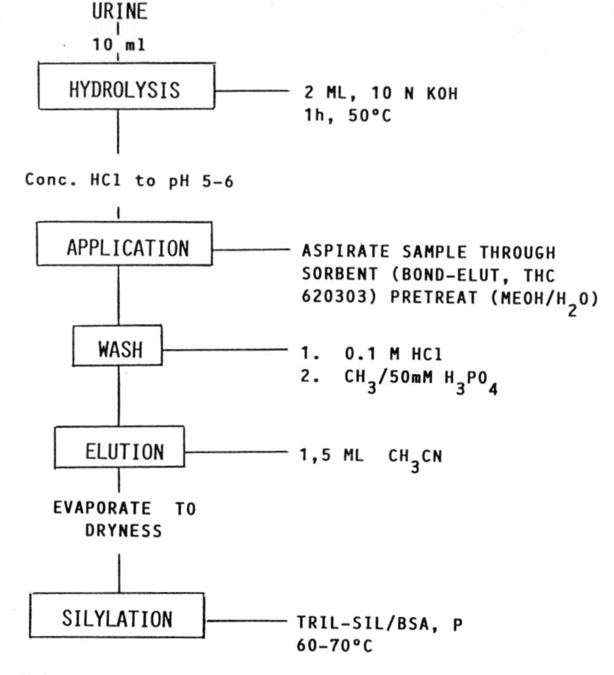

Fig. 10. Sample preparation for GLC-MSD (SIM) confirmation of THC-Metabolite.

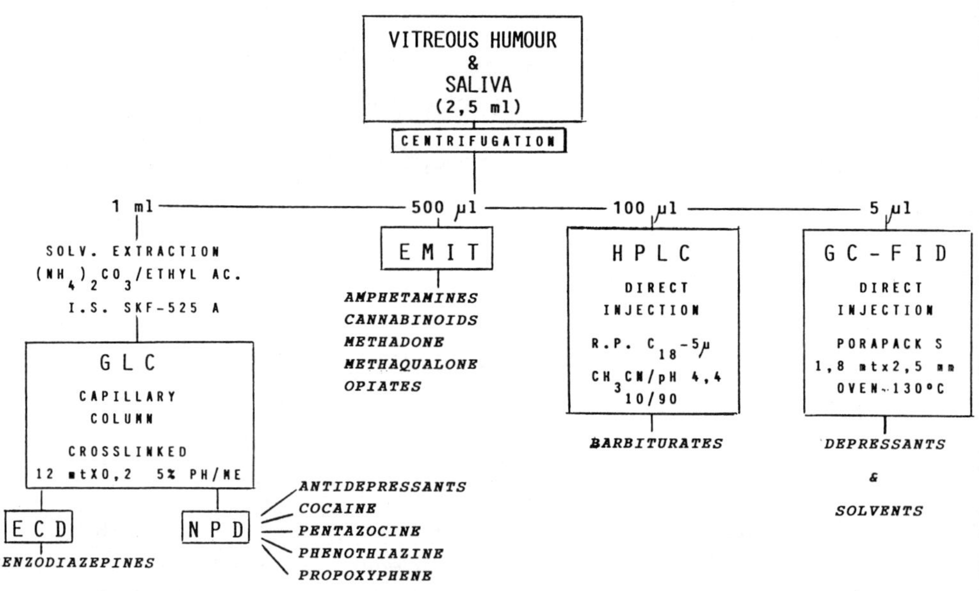

Fig. 11. Analytical scheme of toxicological assay of vitreous humour and saliva.

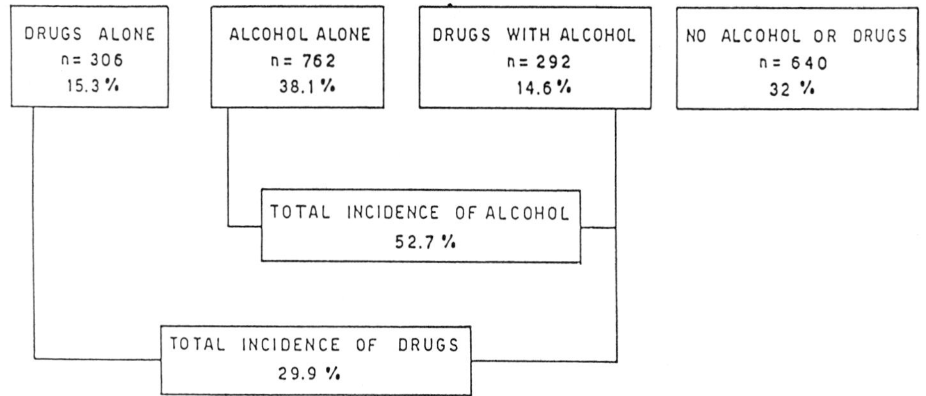

*Source: FERRARA S.D.[7], 1985

Fig. 12. Frequency of drugs and/or alcohol detected in urines of drivers involved in street accidents.

Saliva and Vitreous Fluid Screening

The small quantities and peculiarities of saliva and vitreous fluid require to adopt a different conceptual and methodological approach, like that shown in Fig. 11.

Thus, after preliminar centrifugation, very small quantities of fluids may be directly analyzed by EMIT, GC,HPLC for most of the drugs considered.

A liquid/liquid extraction is however necessary for GLC/EC

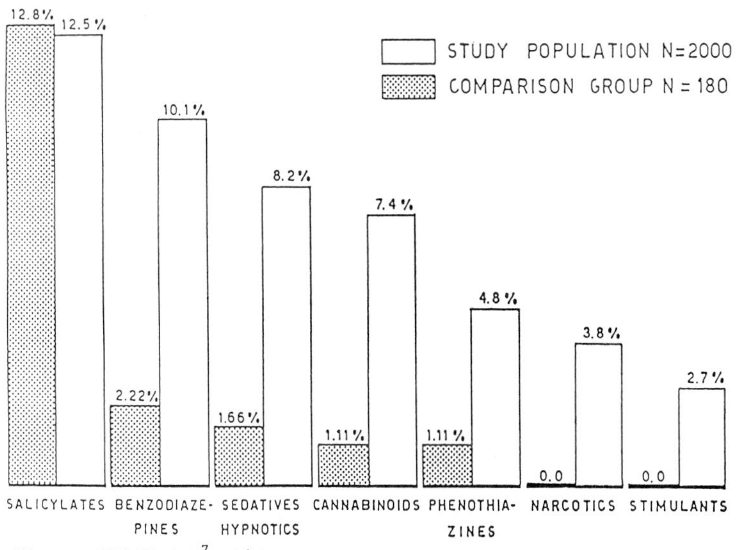

*Source: FERRARA S.D.[7], 1985

Fig. 13. Frequencies of drugs detected in urines from drivers involved in street accidents, compared with a control group.

and NP detection of benzodiazepines, phenothiazines and antidepressants.

POPULATION SURVEY

The application of the methodological approach on urine
furnished effective results in a special population survey carried
out at the University of Padua upon the relationships between the
use of alcohol, drugs and road accidents[6-8].

Fig. 12, 13 show the frequency of detections and of specific
drugs present in urine samples of drivers involved in accidents,
in comparison with a control group.

CONCLUSION

The role played by the Laboratory in epidemiological studies
demonstrates Analytical Toxicology means an interesting occasion
to carry out scientific research and it is even more valid what
Claude Bernard wrote in 1875:"… For him (the physiologist) the
poison becomes an instrument which dissociates and analyses the
most delicate phenomena of living structures and by attending
carefully to their mechanism in causing death he can learn indirectly much about the physiological processes of life".

REFERENCES

1. P.H. Person, R.L. Retka and J.A. Woodward, Toward heroin in
    problem index, NIDA, DHEW Publication No. 76 (ADM), Rockville
    (1978).
2. L.D. Johnston, Review of general population surveys of drug
    abuse,. WHO Offset Publication, No. 55, Geneva (1980).
3. B. Rexed, K. Edmondson, I. Khan and R.J. Samson, Guidelines
    for the control of narcotic and psychotropic substances, WHO,
    Geneva (1984).
4. S.D. Ferrara, J. Pikkarainen and A. Pentilla, Psychotropic
    drugs: organization and role of the laboratory, in: "Public
    Health Problems and Psychotropic Substances", J. Idanpaan-
    Keikkila, I. Khan, eds., The Government of Finland, Helsinki
    (1982).
5. S.D. Ferrara, F. Castagna and L. Tedeschi, Alcohol, drugs and
    road accidents in North-East Italy, Preliminary report in:
    "Proceedings 8th International Conference on Alcohol, Drugs
    and Traffic Safety", L. Goldberg, ed., Almquist-Wiksell
    International, Stockholm (1981).
6. S.D. Ferrara, J. Pikkarainen and M. Mattila, Psychotropics in
    traffic accidents, in: "Public Health Problems and
    Psychotropic Substances", J. Idanpaan-Heikkila, I. Khan, eds.,
    The Government of Finland, Helsinki (1982).
7. S.D. Ferrara, Alcohol, drugs and road accidents.
    Epidemiological study in North-East Italy, in "Proceedings 9th
    International Conference on Alcohol, Drugs and Traffic
    Safety", S. Kaye, G.W. Meier, eds., US Department of
    Transportation, Washington DC (1985).
8. S.D. Ferrara, Alcohol, Drugs and Taffic Safety, British
    Journal of Addiction, in press (1987).

TRACE ELEMENT ANALYSES IN FORENSICS

R. Borriello and R. Gagliano Candela*

Institute of Forensic Medicine
University of Naples
Naples
* Institute of Forensic Medicine
University of Bari
Bari, Italy

In the biological field the trace element definition concerns
a limited number of essential elements that are normally present
in the organism in small quantities and regulate enzymatic systems, hormone production or specific protein activities. In the
forensic field, on the contrary, such a definition fits all the
elements that are present in very small quantities in an heterogeneous matrix: it includes a very large number of matrices and
several areas of application.

Trace element analysis has been applied for acute poisonings,
professional expositions, iatrogenic toxicology both in forensic
(see Table 1) and clinical problems. In these areas, except for
the acute poisoning caseworks, the elemental analysis has to face
very low concentrations of analyte. This firstly proposes problems
concerning the choice of instrumentation and methodology, and the
reliability of the results[1], being often the specific signal not
much greater that the instrumental response arising from the
matrix.

The requirements for the whole analytical procedures and the
instruments employed could be briefly summarized in:

- high sensitivity (elements in the μg - ng region
  should be detectable);
- high precision and accuracy;
- non destructive technique.

In Table 2 the detection limits of the currently used techniques for trace element analysis are compared. It is clear that
the best performances of each instrument are not the same for
every element.

Therefore a laboratory that has only one of these instruments
will achieve highly reliable results for a certain number of elements, whereas in measuring other elements, or more complex matrices, it will be less reliable and, in some cases, completely
unreliable. Yet, in our experience, the choice of the instruments
devoted to forensic purposes depends mainly on the costs,

Table 1. Application of Trace Element Analyses in Forensics

                - Acute poisonings
                - Occupational exposures
                - Iatrogenic Toxicology
                - Food pollution
                - Drug characterization
                - Gunshot residues analysis
                - Fiber, soil, paint, glass characterization

---

handiness, staff training necessity, time per analysis and possibility of simultaneous multielement analyses.

Table 3 shows the advantages and the drawbacks of the analytical techniques most commonly applied in forensics.

For investigations concerning acute poisoning, occupational exposures, iatrogenic intoxications, Atomic Absorption Spectrometry (AAS) together with Inductively Coupled Plasma spectrometry (ICP) are the favourite methods, thanks to the highest sensitivity for all the elements of interest. Furthermore another important reason has been the widely reported ability to carry out rapid, quantitative, multielement analyses on a wide variety of materials[2-8].

A great number of materials can be studied by means of elemental analysis; for example, AAS and ICP have been proposed for characterization and identification of fiber, paint, glass and ink[9-11].

Otherwise, to detect and identify powder and gunshot residues Neutron Activation Analysis (NAA) and Anodic Stripping Voltammetry (ASV) have been firstly applied[12-17]. The former shows high specificity and sensitivity, and the possibility to carry out multielement analyses. Yet, high costs and instrumental complexity hamper its diffusion out of highly specialized centers. On the other hand ASV allows the determination of a number of elements with high sensitivity and much lower analytical difficulties and costs, but some problems of specificity still remain, mainly at the highest sensitivity levels. Recently the inherent simplicity and high sensitivity of Flameless Atomic Absorption Spectrometry (FAAS) have prompted several investigators to utilize this technique for the determination of trace quantities of Pb, Ba and Sb present in gunshot residues[18].

At present, AAS is mainly utilized in the controls on food pollution and in environmental and occupational toxicology.

In the field of drug characterization NAA have been applied for elements such as rare earth[19-22]. A joint research program on FAAS application to determine other elements is in progress at the forensic laboratories of Naples and Rome University.

ICP and AAS are, of course, destructive analytical techniques, nevertheless, particularly for AAS, the low detection limit allows to use very small samples for each analysis.

Table 2. Detection Limits of the Main Techniques employed for
          Trace Element Analyses

	Neutron Activation	I.C.P. Spectrometry	Atomic Absorption	Fluorescence Spectrometry		Inverse Polarography	Differential Polarography
Ag	1 ng	4 ng/ml	5 ng/ml	1200	ng	0.25 ng	
Al	1 ng	2 ng/ml	30 ng/ml	500	ng		
As	1 ng	40 ng/ml	100 ng/ml	110	ng		
Au	10 pg	40 ng/ml	20 ng/ml	1	ng	1 ng	
Ba	1 ng	0.1 ng/ml	50 ng/ml	120	ng		
Be		0.5 ng/ml	2 ng/ml				
Bi	100 ng	50 ng/ml	50 ng/ml	610	ng	10 pg/ml	200 pg/ml
Ca	100 ng	70 pg/ml	4 pg/ml	100	ng		
Cd	10 ng	2 ng/ml	30 pg/ml	400	ng	5 pg/ml	980 pg/ml
Co	1 ng	3 ng/ml	2 pg/ml	50	ng		10 pg/ml
Cr	100 ng	1 ng/ml	1.2 pg/ml	60	pg		
Cs	10 ng			150	ng		
Cu	1 ng	1 ng/ml	0.6 pg/ml	20	pg	5 pg/ml	13 ng/ml
Fe	10 µg	5 ng/ml	0.2 pg/ml	8.5	ng		1 µg/ml
Ga	1 ng	14 ng/ml	1 pg/ml	10	ng	400 pg/ml	
Hg	1 ng	200 ng/ml	80 pg/ml	240	ng		
K				520	ng		
Ir	0.1 ng						
Mg	0.1 µg	0.7 ng/ml	40 fg/ml				
Mn	10 pg	0.7 ng/ml	0.2 pg/ml	150	pg		10 ng/ml
Mo	0.1 ng	5 ng/ml	3 pg/ml	72	ng		
Na	1 ng	0.2 ng/ml	1 pg/ml				
Ni	0.1 µg	6 ng/ml	4 pg/ml	60	ng	100 pg/ml	780 ng/ml
Pb	0.1 µg	8 ng/ml	2 pg/ml	300	pg	10 pg/ml	1.5 µg/ml
Pd		7 ng/ml	4 pg/ml				
Pt	10 ng	80 ng/ml	10 pg/ml				
Rh		3 ng/ml	8 pg/ml			100 pg/ml	
Sc	10 µg			20	ng		
Sn		300 ng/ml	2 pg/ml	3.9	µg	200 pg/ml	
Sr	10 ng	20 pg/ml	1 pg/ml	70	pg		
Ti	0.1 µg	3 ng/ml	1 pg/ml	1	ng	10 ng/ml	
Tl	0.1 µg	200 ng/ml	1 pg/ml				
U	0.1 ng	6 ng/ml	50 pg/ml				80 ng/ml
W		2 ng/ml	1 µg/ml				
Zn	0.1 µg	2 ng/ml	20 fg/ml	40	pg	40 pg	

Anyway the choice of the instrumental technique will not
significantly change the final result, if sample collection and
treatment methods are adequate. This is the most binding rule of
trace element analyses in the forensic field, because they are
often performed on heterogeneous matrices for which certified
reference materials are not available. Therefore to control and
assure the quality of the analytical results, a rigorous program
concerning all the aspects of the analytical process is needed.

The first step of investigation (the collection of samples)
may introduce many factors that invalidate the precision and the
accuracy of the method. Particularly for solide matrices such as
food, biological tissue and powder, with non homogeneous
concentration of traces, the sample must be representative.

A variety of causes contribute to the sample contamination or
to the loss of analyte during collection and storage procedures:

- collection modality;
- containers quality;
- storage temperature;
- storage time.

69

Table 3. Advantages and Drawbacks of the Main Techniques for
         Trace Element Analyses

	Advantages	Drawbacks
ICP	- High source temperature	- Liquid samples are required
	- Availability of spectral lines of ions	- The emitted spectra are very rich in lines
	- Simultaneous multielement analysis	- High cost
NAA	- Good detection limits	- Scarce availability of neutron sources
	- Simultaneous multielement analysis	- Long analysis time
	- Non destructive technique	- High cost
AAS	- High sensitivity for elements of toxicological interest	- Matrix modification
	- Possibility of micro-sampling in FAAS	
ASV	- Simultaneous multielement analysis	- Limited sensitivity for many elements of toxicological interest
	- Non destructive technique	
	- Highly skilled personnel not required	
	- Low cost	

During autopsies, four general sources of sample contamination can be identified[23]:

1) Dust, pollutants, cosmetics on the surface of the body.
2) Disinfectants, talc and dust on the gloves of the pathologist.
3) Metallic corrosion products and residues on the instruments from previous dissections.
4) Dust and disinfectants on the dissecting table.

To avoid these problems, the use of instruments fabricated from high purity quartz, titanium and inert plastics has been recommended. Therefore, an autopsy collection kit comprising of

plastic forceps, precleaned plastic vials, especially prepared
titanium knives must be supplied to the collaborating pathologist
in all cases concerning trace element analyses[24].

Preservatives as reported by Sansoni and Iyengar[25], contain
significant amounts of contaminants. Hence, the use of heparin and
formalin are discouraged. Instead of chemical preservation,
samples should be immediately frozen and stored at a temperature
below $-15°C$ until analyzed.

Furthermore, according to several Authors, changes in trace
element concentration can be due to desorption from the container
or to adsorption on the container walls[26,27]. Iyengar reported[28]
postmortal changes of the elemental composition of autopsy
specimens; really, when samples are frozen, changes in the copper,
iron, manganese and zinc concentrations of liver can take place,
owing to the movement of fluid in and out of the organ.
Lyophilization prior to freezing may reduce or eliminate such
changes.

Such prevention procedures can be easily applied during the
collection of biological samples, but they are hardly carried out
when non biological specimens are brought to the laboratory by un-
skilled personnel. Janhari[29] reported the levels of some elements
in a variety of materials commonly used either in the laboratory
practices or for swabbing the hands of a suspected shooter.

Sansoni and Iyengar[25] reported data on the composition of dust
in unfiltered air from the laboratory atmosphere demonstrating
that contamination by Al, Zn, Mn, Cu is severe, if compared with
the very low concentrations that normally occur in forensic
samples.

The second step of analysis (the treatment of the samples)
must be directed to achieve the best resolution of chemical, physical and spectral interferences. The analytical technique employed
very often requires specific procedures for sample preparation.

Many conditions have to be fulfilled in order to validate an
analytical method for trace element analysis. The method should
be:

- extremely selective and fully reproducible;
- suitable for both qualitative and quantitative analyses;
- the same pre-treatment should be suitable for many elements;
- the number of pre-treatment steps should be reduced at
  minimum.

In the case of NAA and ASV techniques, little sample preparation beyond drying and homogenization is required.

The use of AAS and ICP for analyzing solid samples, needs the
destruction of the organic matrix and/or the solubilization of
residual trace elements.

Wet ashing is usually achieved by digestion of the sample
with acid that dissolves the organic residue. Wet ashing in high
pressure decomposition vessels has the advantages of reducing
losses by volatilization; it was advantageously applied in the
assay of inorganic poisons[30].

Dry ashing involves destroying the organic matrix by combustion and subsequent dissolving the trace elements from the ash. Unfortunately, samples prepared through dry ashing are susceptible to loss of trace elements by volatilization. Samples prepared by low temperature plasma ashing are less vulnerable to losses by volatilization; yet, this procedure is too slow.

Impurities in reagents have to be considered among the many sources of contamination, to which samples are exposed: for this reason only the purest chemicals should be used in trace element analysis.

There are, therefore, many potential sources of errors in preparing samples for the measurement of their trace element contents. However, it is possible to obtain reliable results provided these sources of error are known and precautions are adopted. A rigorous program of quality control is needed to confirm the reliability of the results, especially when matrices are analyzed, for which certified reference material is not available.

The last step is data evaluation; it must be stressed that the quality of the results of the examinations performed in forensics has always been a concern of individual forensic scientists[1,31,32]. In trace element analysis these problems became much bigger, because the higly sophisticated instrumentation and the too vast array of methods can be at the origin of uncontrolled intra- and inter-laboratory variability. This in turn hampers the possibility of data comparation and places a far greater burden of responsibility on the forensic scientist.

It is therefore essential both in the interest of the Justice and of the forensic scientists that the results of trace element analysis are always integrated with other analytical evidences, giving a more extensive outline of the studied material. For example, about gunshot residues, a judicial answer cannot be supported only by the results of trace metal analysis, but also by investigating other organic components of the powder.

We could have presented the analytical protocols applied in our laboratory for caseworks, but we believe that it would be much more useful the creation of a working group with other forensic scientists concerned in the problem. Really, we believe that trace element analysis, particularly as a support of casework investigations, will take a remarkable place in forensic science only if homogenous standards are applied.

Of course, for a correct management of the problem, we must point out that there is an abyss between a standard protocol application and the reliability of the results. This abyss can be bridged only by the experience of the scientist: he is still an irreplaceable "essential element" in trace element analysis.

REFERENCES

1. R. Borriello, La spettrometria di assorbimento atomico nella diagnostica tossicologica e medico-legale, in: "Dentro e Attraverso" N° VI, Rass. Med. For., Naples (1983).

2. M. Montagna, F. Avato and O. Crippa, Livelli di Hg nel tessuto umano normale. Contributo sperimentale e rassegna critica dei metodi di attacco dei campioni, Arch. Med. Leg. Ass. 10:295 (1974).

3. M. Montagna, F. Avato and O. Crippa, Verifica sperimentale del contenuto normale di Cromo nei tessuti umani, Arch. Med. Leg. Ass. 10:343 (1974).

4. I. Barni Comparini and G.A. Norelli, La diagnosi medico legale del mercurialismo: rilievi tossicologici ed epicritici sul significato del criterio chimico, Zacchia 51:416 (1976).

5. R. Borriello, E. Caccavale, D. Cittadini and G. Sciaudone, Valutazione degli indici biologici dell'intossicazione da Piombo in una popolazione di lavoratori esposti, Rass. Med. For. 20 (1982).

6. M. May, S. Zizolfi, R. Borriello and G. Sciaudone, Studio in vitro dell'effetto di alcuni psicofarmaci sul trasporto del litio negli eritrociti, in: "Atti del XXXIV Congr. Naz. Soc. Ital. Psich." Catania (1980).

7. K. Okamato, Preparation and certification of human hair powder reference material, Clin. Chem. 31:1592 (1985).

8. E.I. Hamilton, The concentration and distribution of some stable elements in healty human tissue from United Kingdom, Sci. Total. Environ. 1:341 (1972-73).

9. T. Catterick and D.A. Hinckman, The quantitative analysis of glass by ICP-OES: a five element survey, Forens. Sci. Int. 17:253 (1981).

10. R.C. Carpenter, The analysis of some evidential materials by ICP-OES, Forens. Sci. Int. 27:157 (1985).

11. R. Borriello and G. Sciaudone, Diagnosi generica di sperma su tracce mediante spettroscopia di assorbimento atomico, Rass. Med. For. 19:2 (1981).

12. J.H. Liu, The application of ASV to forensic science. The construction of a low cost polarograph, Forens. Sci. Int. 16:43 (1980).

13. R. Cornelis and J. Timperman, Evaluation of credibility of test gunfiring detection method NAA, Med. Sci. Law 14:98 (1974).

14. M. Tavani and A. Brandone, Le possibilità di impiego della analisi per attivazione neutronica nella pratica medico forense, Arch. Med. Leg. Ass. 10:28 (1974).

15. R. Gagliano Candela, Rassegna critica sulla ricerca dei residui di sparo ed attuali orientamenti, Riv. Polizia 8-10:513 (1976).

16. R. Gagliano Candela, M. Colonna and L. Strada, Identificazione dei residui dell'esplosione di colpi d'arma da fuoco corta mediante ASV, Zacchia 12:44 (1976).

17. P. Benedetti, A. Brandone, F. De Ferrari and D. Salzo, Ulteriori esperienze sulla interpretazione del reperto di residui di polvere da sparo sulla cute della mano, Riv. It. Med. Leg. 8:112 (1980).

18. S. Napoli, R. Gagliano Candela, L. Strada and T. Gagliardi, Valutazione della distanza di sparo (II). Analisi del Pb sul bersaglio mediante AAS, Riv. It. Med. Leg. 31:469 (1979).

19. G. Henke, Activation analysis of rare-earth elements in Opium and Cannabis samples, J. Radioanal. Chem. 39:69 (1977).

20. M. Shinogi and I. Mori, Multielement determination in Cannabis leaves by NAA. A comparison of Cannabis of various geographical origins in Japan, Yakug. Zashi. 98:1466 (1978).

21. A. Brandone and E. Marozzi, Caratterizzazione di stupefacenti di origine vegetale attraverso i loro contenuti in elementi in traccia determinati mediante NAA, in: "Simposio sulle droghe d'abuso: tecniche analitiche", Bologna (1981).
22. A. Brandone, E. Marozzi and M. Montagna, La caratterizzazione di sostanze stupefacenti di origine vegetale mediante analisi per attivazione neutronica, Riv. di Merceologia 23:307 (1984).
23. H. Smith, Handling of biological materials, in: "Activation analysis", J.M.A. Leniham and S.J. Thompson, eds., Academic Press, London (1965).
24. S.R. Koirtyoham and A.C. Hopps, Sample selection, collection, preservation and storage for a data bank on trace element in human tissue, Federation Proc. 40:2143 (1981).
25. B. Sansoni and G.V. Iyengar, Elemental analysis of biological materials: current problems and techniques with special reference to trace elements, in Tech. Rep. Ser. N° 197 Int. Atom. Energy Agency, Vienna (1980).
26. R. Masironi and R.M. Parr, Collection and trace element analysis of post-mortem human samples, in: Int. Workshop on Biological Specimen Collection, WHO/IAEA Research Program, Luxemburg (1977).
27. J. Versieck, Contamination induced by collection of liver biopsies and human blood, in: "Nuclear Activation Techniques in the Life Sciences", IAEA/SM 157/24:39, Vienna (1972).
28. G.V. Iyengar, Autopsy sampling and elemental analysis: error arising from post-mortem changes, J. Pathol. 34:173 (1981).
29. M. Janhari, M.S. Rao, N. Chattopadhyay, S.M. Chatterjee and A. Sen, Shooter identification: elemental analysis of swabbing materials by NAA and ASV. Forens. Sci. Int. 28:175 (1985).
30. M. Marigo, S.D. Ferrara, L. Tedeschi and P. Gori, Su una nuova metodica di distruzione del materiale biologico per la ricerca dei veleni inorganici, Zacchia 52:435 (1977).
31. F. Lodi and E. Marozzi, "Tossicologia Forense e Chimica Tossicologica", Cortina, Milano (1982).
32. G. Sciaudone, Tossicologia Forense in: "Manuale di Medicina Legale e delle Assicurazioni", P. Zangani, V.M. Palmieri and G. Sciaudone, eds., Morano, Napoli (1985).

THE STABILITY OF 7-NITROBENZODIAZEPINES IN POST MORTEM BLOOD:

ANALYTICAL PROBLEMS AND FORENSIC IMPLICATIONS

M. Montagna, M.L. Baldi*, and D. Fantoccoli

Institute of Forensic Medicine
University of Pavia
* "S. Matteo" Hospital
Pavia, Italy

INTRODUCTION

The use, misuse and abuse of benzodiazepines are common phenomena in Italy as well as elsewhere. In particular flunitrazepam
seems to be specifically abused by narcotic addicts[1].

Using gas chromatography with an electron capture detector[2]
as a method of screening the blood for benzodiazepines, we have
however observed a considerable discrepancy in the data for flunitrazepam between the addicts admitted to hospital and subjects who
had died because of intravenously administered opiates. In the
first group (period 1981-1983) some 68% of all the positive
indications for benzodiazepines was attributed to flunitrazepam;
on the other hand, no positive indications were found in the 136
cases of acute heroin deaths examined from 1973 to 1985. The data
collected in Milan are similar[3]. In the period 1980-1984, flunitrazepam was revealed in one case only, while the benzodiazepines
were found in the victims of acute narcotism in an overall
percentage of 31%. Moreover it is worthwhile remembering that
flunitrazepam and nitrazepam were not detected in blood even in
some forensic cases in which their taking would appear certain
because they had been revealed in the stomach and/or in drugs or
in syringes found near the victims.

Publications concerning the stability of benzodiazepines in
biological material are comparatively rare. Diazepam was found by
Howard[4] to be very stable in plasma when stored under varying
conditions: at -20°C for 1 year; at 4°C for 8 weeks; at room
temperature for 20 days. According to Levine[5], who performed a
study on the post mortem stability of benzodiazepines in blood and
tissues, diazepam stored in blood is stable for up to five months.
Nordiazepam appeared to degrade with time when stored at room
temperature (about 50% of loss within the first two weeks).
Flurazepam and N-1-desalkylflurazepam demonstrated a slight decrease in concentration (< 20%) in time both at room temperature
and when refrigerated. Chlordiazepoxide, because of the lability
of the N-oxide group, was found to be unstable and
norchlordiazepoxide was even more unstable. Therefore an initial

toxic chlordiazepoxide concentration of 5 mg/l can rapidly decrease within days to a concentration considered therapeutic if the body remains at room temperature. Fluoride and oxalate partially protect chlordiazepoxide from degradation, but not its demethylated metabolite.

Concerning the 7-nitrobenzodiazepines, Knop[6] reported that a plasma drug concentration of 25 µg/l disappeared within six days when stored at 20°C. Approximately 85% of clonazepam originally present remained after 20 days of storage at 1°C; about 95% remained after the same length of time when stored at -20°C. More recently Stevens[7], studying the behaviour of 56 drugs and drug-related compounds in putrefying human liver macerated over different periods of time, confirmed the stability of diazepam, flurazepam and lorazepam. On the other hand he included the benzodiazepines with N-Oxide and the nitro group, such as chlordiazepoxide, clonazepam and nitrazepam, among the molecular structures prone to putrefactive decomposition. The experimental findings of this work indicated that the destruction of labile drugs is affected by putrefaction caused by the introduction of fly-borne faecal anaerobic bacteria.

To our knowledge flunitrazepam has been investigated in re- spect to its stability in stored samples only by Ekund and Ceder[8]. The in vitro studies performed by these Authors showed that nitrazepam and flunitrazepam are reduced to their 7-amino derivatives in post mortem blood.

A reduction of the nitro group to the amino group also occurs in vivo and the 7-amino derivatives are active metabolites. For example 7-aminoflunitrazepam was the major metabolite found in human plasma together with desmethylflunitrazepam, which is further metabolized to 7-aminodesmethylflunitrazepam[9]. The concen- trations that occur in blood following therapeutic doses are of the order of some ng/ml, therefore highly selective and sensitive methods are needed[10-12].

The lack of studies concerning the stability of the 7-nitro- benzodiazepines in cadaveric blood, with particular reference to flunitrazepam, was the incentive for performing a series of experiments on the behaviour of these drugs during storing and for developing an analytical procedure for blood screening of both 7-nitro and 7-amino compounds.

EXPERIMENTAL DESIGN, PROCEDURES AND RESULTS

Preparation Of 7-Aminobenzodiazepines

Solutions of clonazepam, nitrazepam and flunitrazepam in ethanol were hydrogenated at 25°C at 1 atm. in the presence of palladium on carbon catalyst. The solutions were filtered from the catalyst and concentrated under vacuum.

7-Aminoclonazepam was crystallized from acetonitrile, then chromatographated on silica gel column (chloroform: methanol 95:5) and finally recrystallized from ethyl acetate. 7-Aminofluni- trazepam was crystallized from ethyl acetate after the washing of crud product with light petrol ether. 7-aminonitrazepam was crystallized from light petrol ether, then chromatographated and recrystallized like the 7-aminoclonazepam.

The melting points were 224°-225°, 202-203° and 234-235°C for clonazepam, flunitrazepam and nitrazepam 7-amino derivatives respectively. Analytical data (TLC, IR, UV and MS) corresponded to those described in the literature[13,14].

Stability of 7-Nitrobenzodiazepines

Faecal Contamination. To blood samples inoculated with a faecal slurry 5 µg/ml of clonazepam, flunitrazepam and nitrazepam were added. Samples were alkalinized with solid sodium bicarbonate and extracted with an excess of benzene:methylene chloride (9:1) immediately and after a storing period of 4 and 9 days at room temperature. The solvents were concentrated near to 20 µl and chromatographated on silica gel with standards of the drugs and its amino derivatives (benzene:acetone:light petrol ether:ammonium hydroxide 35:35:35:1; detection spray: Dragendorff and sodium nitrite).

Thin layer chromatography showed that the 7-amino derivatives were already detectable after 4 days. In particular, about 50% of clonazepam and nitrazepam appeared to be transformed, while only traces of flunitrazepam were still present. After 9 days the reduction of flunitrazepam and clonazepam was virtually complete and only a very small amount of nitrazepam was detectable.

Ante Mortem and Post Mortem Blood. To 1 ml of blood samples taken from living persons and from people who had died 24 hours before for a traffic accident, 100 ng of flunitrazepam were added. The specimens were stored in stoppered glass conical tubes at -20°C, at 4°C and at room temperature with and without sodium fluoride addition.

At various times, aliquots of blood were quantitated for flunitrazepam according to the following method.

Extraction - To 0.5 ml of blood, were added internal standard (flurazepam) at appropriate concentrations, 2 ml of borate buffer pH 9 and 6 ml of benzene:methylene chloride (9:1). The mixture was vortexed for 1.5 min and the organic layer was separated and evaporated. To the residue was added 1 ml of 0.05 mol/l sulphuric acid and the acid solution was washed with 3 ml of hexane then adjusted to pH 9 with solid sodium bicarbonate. Finally the mixture was extracted with 4 ml of benzene:methylene chloride on a vortex mixer for 1.5 min. The organic phase was transferred into another tube and evaporated to dryness under a stream of nitrogen. The residue was reconstituted with about 20 µl of methanol and an appropriate volume was injected into the chromatograph (system a).

The remaining extract was evaporated to dryness in a micro reaction vessel and derivatized by adding 10 µl of pyridine and 10 µl of acetic anhydride to the residue. The mixture was heated at 70°C on a water bath for 15 min. Then the reagents were evaporated under a stream of pure nitrogen and the residue was dissolved in 100 µl of benzene. Analyses were carried out according to the system b.

System a - The analyses were carried out on a Hewlett Packard 5890 gas chromatograph (Palo Alto, Ca, USA) equipped with a nitrogen phosphorus detector. The column was a flexible fused silica wide bore (10 m x 0.53 mm I.D.) coated with OV-17 (2 µm

film thickness). The flow-rates of the carrier gas (helium) and of
the auxiliary gas (helium) were 10 and 20 ml/min respectively. The
oven temperature was programmed from 230 to 270°C (3°/min);
initial time 0; final time 10 min. The detector and injector
temperatures were 270 and 200°C respectively.

System b - A C. Erba Mega Series 5160 gas chromatograph
(Milan, Italy) equipped with a cold on-column injector, an
electron capture ^{63}Ni detector and a linear temperature programmer
was used for the analyses. The column was a flexible fused silica
with SE-52 (0.15 μm film thickness). The carrier gas was hydrogen
at 0.3 kg/cm^2; nitrogen was used as the scavenger gas. The oven
temperature was programmed from 85 to 260°C (5°/min). The detector
temperature was 320°C.

Relative retention times to flurazepam were:

	System a	System b
flunitrazepam	0.94	0.89
7-aminoflunitrazepam	1.24	-
7-acetylaminoflunitrazepam	-	1.16

The flurazepam absolute retention times were 11 min in
system a and 7.48 min in system b.

Blanks of blood and of blood spiked with flunitrazepam in
different amounts were assayed along with each set of samples to
calculate the levels of the analyte in the unknowns. Using 0.5 ml
of blood sample, the detection limit was 10-20 μg/l for both
flunitrazepam and 7-amino derivative.

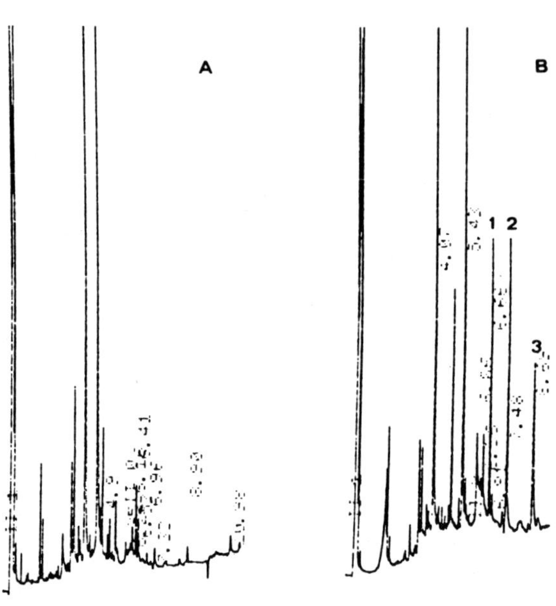

Fig. 1. Capillary gas chromatograms of A) blank blood; B) blood
spiked with flunitrazepam (25 μg/l), flurazepam (50 μg/l)
and 7-aminoflunitrazepam (50 μg/l). Peaks: 1) flunitrazepam; 2) flurazepam; 3) acetyl-7-aminoflunitrazepam.

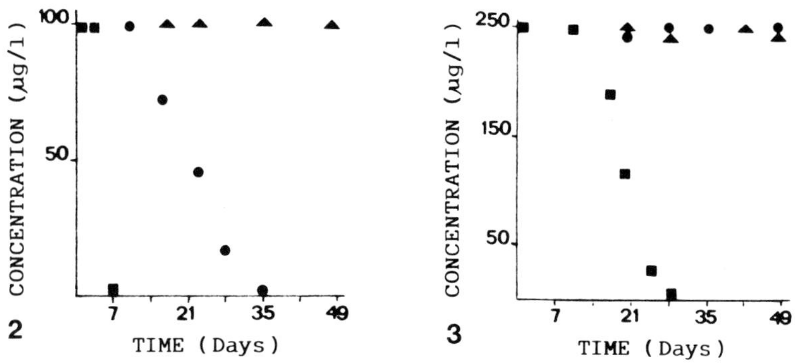

Fig. 2 and 3. Decay patterns of flunitrazepam in cadaveric
blood (2) and in plasma from an intoxicated
patient (3), and stored at -20°C (▲), +4°C (●),
and room temperature (■).

Typical capillary chromatograms obtained after extraction and
derivatisation from cadaveric blood and cadaveric blood with added
flunitrazepam, 7-aminoflunitrazepam and internal standard are
shown in Fig. 1.

The results of the experiments performed in vitro showed that
flunitrazepam is stable at -20°C. In fact the initial concentrations remained unchanged for almost 7 weeks and even for 4 months
both in ante mortem and post mortem blood samples. At 4°C, halving
of the initial levels was observed in cadaveric blood after
3 weeks. After 5 weeks flunitrazepam was not detectable in post
mortem samples while it was stable in ante mortem ones (Fig. 2).
At room temperature a quick and massive decay was observed, so in
the course of 7-11 days the levels fell under the detection limit
also in those specimens added with sodium fluoride.

Blood Taken from Intoxicated Patients. Blood specimens taken
from patients intoxicated with nitrobenzodiazepines were checked
after storing at the same conditions as in the previous section.

Concerning flunitrazepam, plasma concentrations remained
constant throughtout the time of the controls (at least 7 weeks)
when the samples were stored at -20°C and 4°C (Fig. 3). These
plasma samples proved to be relatively more stable than the blood
samples spiked in vitro with the drug.

Experiments carried out at room temperature on plasma spiked
with 2 mg/l of nitrazepam showed that in samples contaminated with
faecal material, there was a rapid decrease of drug concentration
(0,6 mg/l after 24 hours and 0.02 mg/l after 3 days), while in
uncontaminated samples nitrazepam concentrations were reduced by
half after 7 days and were undetectable after about 2 weeks
(Fig. 4). Therefore in this case too, the reduction of nitrazepam
was less rapid in the samples taken from the patients than in
samples spiked with the drug.

In the plasma of a patient intoxicated by nitrazepam, inoculated with faecal material and stored at room temperature, the
initial concentrations (2 mg/l) decreased to 50% and 2,5% after
2 and 4 days respectively. In the same uncontaminated specimen,
0.7 mg/l were still detectable after 18 days (Fig. 5).

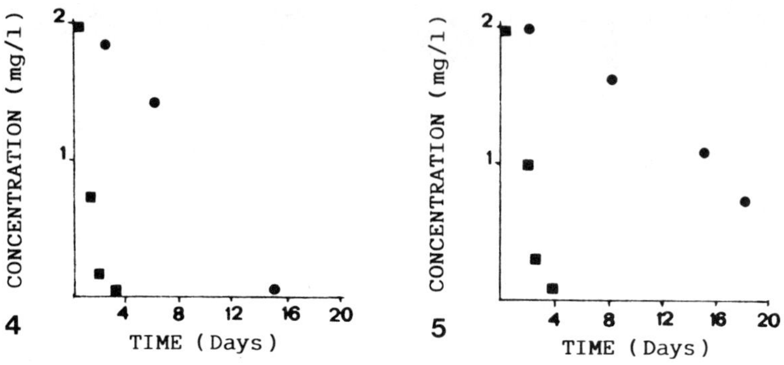

Fig. 4 and 5. Decay patterns of nitrazepam in spiked plasma (4),
and in plasma from an intoxicated patient (5)
inoculated (■) and not inoculated (●) with faecal
material, and stored at room temperature.

The decomposition product of flunitrazepam in cadaveric or in
vitro contaminated blood samples was identified by GC-MS as the
7-aminoflunitrazepam (Fig. 6). The apparatus used was an Hewlett
Packard system constituted by a 5890 capillary gas chromatograph
equipped with a fused silica column (12 m x 0.20 mm I.D.) coated
with methyl silicone (thickness 0.35 μm) and a 5970 mass spectrometer detector.

Complementary Investigations. The presence of 7-aminofluni-
trazepam in blood was investigated in some cases of narcotic
deaths in which flunitrazepam was not previously detected using
GLC-ECD screening for benzodiazepines, despite a significant
history of Rohypnol abuse. GLC-NPD and GLC-ECD analyses of the
acetylated extracts from blood samples stored for several months
in a refrigerator revealed the peak of the 7-amino and 7-acetyl-
aminoflunitrazepam respectively.

DISCUSSION AND CONCLUSIONS

The experimental findings of this work indicated that
7-nitrobenzodiazepines are labile drugs which are affected by
putrefaction and therefore by relatively high temperatures which
facilitate post mortem bacterial development and diffusion.

A virtually complete conversion of the nitro group to an
amino group was observed to occur in blood samples inoculated with
a faecal slurry over a few days when they are stored at room
temperature. The degradation was not impeded by the addition of
sodium fluoride.

The more detailed studies performed on flunitrazepam showed
that toxic concentrations may decrease rapidly with time to
therapeutic ranges or become undetectable in post mortem blood
samples when stored at room temperature. The refrigeration at 4°C
slackened the decomposition but it does not ensure good stability
of the nitrobenzodiazepine.

The 7-aminobenzodiazepines have a very low response to the
electron capture detector. This may explain why these drugs, in
particular flunitrazepam, might not be detected, unlike other

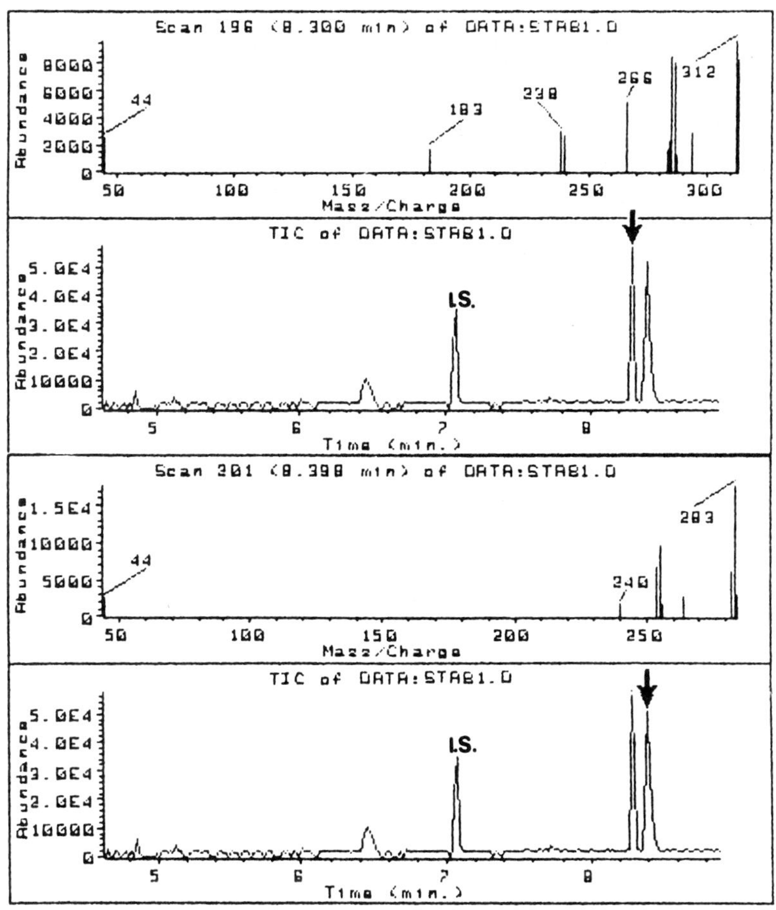

Fig. 6. Total ion chromatograms, and mass spectra obtained from cadaveric blood spiked with flunitrazepam, and stored for 24 hours at room temperature. Peak at 8.300 mins = flunitrazepam; peak at 8.398 mins = 7-aminoflunitrazepam; I.S. = internal standard.

benzodiazepines, in toxicological investigations applied to the examination of individual foresinc cases or to retrospective epidemiological studies.

A suitable method for screening blood for 7-nitrobenzodiazepines is gas chromatography, employing a nitrogen sensitive detector or an electron capture detector, provided that the amino compounds are previously derivatized, or, better, mass detected.

Because of the problems of stability mentioned above, all specimens which are suspected to contain nitrobenzodiazepines must be keept frozen and analyzed as soon as possible.

However, in forensic situations the almost 24 hours which elapse before post mortem blood samples are collected could represent a critical time and the analytical search of the amino derivatives becomes imperative for a correct definition of the cases from the toxicological point of view.

REFERENCES

1. F. Bruno and F. Ferracuti, Descrizione di alcuni casi clinici di abuso e di possibile dipendenza da flunitrazepam, Riv. Tossicol. Sper. Clin. 12:125 (1982).
2. J.A.F. De Silva, I. Bekersky, C.V. Puglisi, M.A. Brooks and R.E. Weinfeld, Determination of 1,4-benzodiazepines and -diazepin-2-ones in blood by electron-capture gas-liquid chromatography, Anal. Chem. 48:10 (1976).
3. M. Caligara, R. Mariani and E. Marozzi, Studio dell'incidenza dell'uso e dell'abuso delle benzodiazepine sulla mortalità da composti morfinici, Arch. Med. Leg. Ass. 7:183 (1985).
4. P.J. Howard, The stability of diazepam in plasma samples when stored under varying conditions, J. Pharm. Pharmac. 30:136 (1978).
5. B. Levine, R.V. Blanke and J.C. Valentour, Postmortem stability of benzodiazepines in blood and tissues, J. Forensic. Sci. 28:102 (1985).
6. H.J. Knop, E. van der Kleign and L.C. Edmunds, The determination of clonazepam in plasma by gas-liquid chromatography, Pharmaceutisch Weekblad 110:297 (1975).
7. H.M. Stevens, The stability of some drugs and poisons in putrefying human liver tissues, J. Forensic Sci. Soc. 24:577 (1984).
8. A. Eklund and G. Ceder, Determination of 7-nitro-benzodiazepines in post mortem blood, in:"International Tiaft Congress-Abstracs", Rigi Kaltbad, August 25-30, (1985).
9. G. Wendt in W. Hugin, G. Hossli and M. Gemperle, eds., "Bisherige Erfahrungen mit Rohypnol (Flunitrazepam) in der Anaesthesiologie und Intensivtherapie", Editiones Roche, Basel 1976).
10. P. Haefelfinger, Determination of nanogram amounts of primary aromatic amines and nitro compounds in blood and plasma, J. Chromatogr. 111:323 (1975).
11. Y.C. Sumirtapura, C. Aubert, P. Coassolo and J.P. Cano, Determination of 7-aminoflunitrazepam (Ro 5-4650) in plasma by high-performance liquid chromatography and fluorescence detection, J. Chromatogr. 232:111 (1982).
12. C. Crevoisier, M. Eckert, W.H. Ziegler, J.P. Cano and Y. Sumirtapura, in:" IIIrd World Congress on Biological Psychiatry-Abstracts", Stockholm, 28 June - 3 July (1981).
13. H. Schutz, Benzodiazepines a Handbook, Springer-Verlag Berlin, Heidelberg, New York (1982).
14. S. Ebel and H. Schutz, Zur Analytik von Flunitrazepam (Rohypnol), einem neuen Benzodiazepinderivat, under besonderer Beruecksichtigung der Metaboliten, Z. Rechsmed. 81:107 (1978).

HPLC/FLUORESCENCE DETECTION OF 9-ACRIDANONES AS ANALYTICAL METHOD

FOR BENZODIAZEPINES IN BIOLOGICAL FLUIDS

M. Chiarotti and N. De Giovanni

Institute of Forensic Medicine
Catholic University of Rome
Rome, Italy

INTRODUCTION

The analysis of the widely prescribed benzodiazepines in biological samples often requires sensitive and highly specific techniques, because of their low level amounts in biological specimens[1]. In some instances however, a sensitive qualitative screening technique may also provide suitable information. Indeed when the relative number of specimens containing benzodiazepines is low, and the total number requiring analysis is high, such a screening technique may be preferred[2].

EMIT® technique is now generally used for this purpose, but its low sensitivity and poor specificity render it unsuitable in some particular cases.

Other techniques, such as gas chromatography with electron capture detection, give very high sensitivity and good specificity, but the analytical procedure may be difficult for screening analyses[3,4].

Fluorimetric determination of drugs has been referred as a high sensitivity technique and we tested this system for benzodiazepine analysis.

These compounds however show a little native fluorescence and derivatization requires drastic reaction conditions because of the low reactivity of the functional groups present in the molecule: these strong conditions generally destroy the derivatizing agent. However an intense fluorescent product may be obtained from benzodiazepines after structural rearrangement of the related benzophenones, which are benzodiazepine metabolites too. The production of 9-acridanone derivatives from benzophenones is depicted in Fig. 1.

This reaction is essentially an internal nucleophilic aromatic substitution and the aminic attack is encouraged by the carbonyl group activation (electron-withdrawing effect). The R" group displacement is enhanced by its nucleophilic characteristics and by the kind of solvent. The yield of the reaction is also

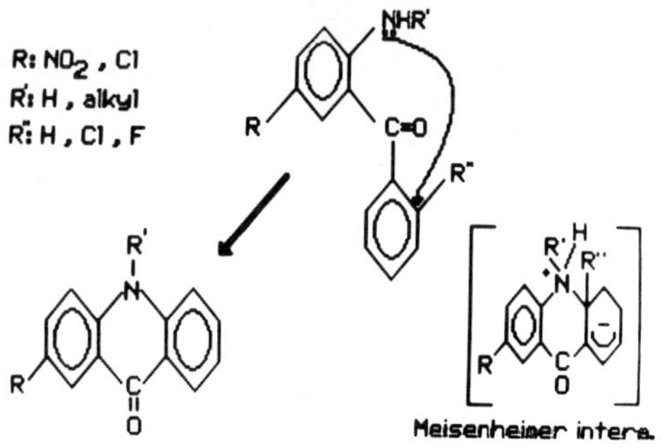

Fig. 1. Production of 9-acridanone derivatives from benzophenones.

influenced by the presence of nitro or chlorine groups in para position in respect to the amino group, because of the activating mesomeric effect of chlorine. This reaction was firstly applied to benzophenones by Fryer[5] and further applications to benzodiazepines have been described by De Silva[6] and Valentour[2]. These Authors, however, referred about direct fluorimetric determination of few benzodiazepines and no other extensive data are reported in literature.

We studied different solvents and catalysts to apply this reaction to many benzodiazepines, and the subsequent high performance liquid chromatographic separation to enhance sensitivity and specificity.

EXPERIMENTAL AND DISCUSSION

Benzodiazepines were firstly hydrolyzed by hydrochloric acid 6 mol/l to benzophenones (60 min at 100°C). The products were extracted by diethylether and the synthesis of acridanones was studied in sodium hydroxide (60 min at 120°C), dimethylsulfoxide (90 min at 120°C) or dimethylformammide (90 min at 120°C), using lead dioxide or potassium carbonate as catalysts. The reaction products were extracted with ethyl acetate after alkalinization and HPLC analysis with fluorescence detection was performed. The choice of the two aprotic solvents was suggested because these kinds of nucleophilic substitutions are promoted in non-aqueous solvents when the leaving group is an halogen.

In our experimental conditions, potassium carbonate and lead dioxide resulted as better catalysts.

When the leaving group is an hydrogen, the use of sodium hydroxide with lead dioxide becomes necessary to render the yield suitable for analytical purposes. Moreover these cyclisation conditions are acceptable for all the benzodiazepines examined, hence it is suggested as a screening tool. On the contrary the use of dimethylsulfoxide and/or dimethylformammide permits obtaining higher acridanone amounts when some benzophenones are processed (2-halobenzophenones, 2-nitrobenzophenones). These solvents must be used when specific analysis is requested. Consequently by the use of appropriate cyclisation conditions, different 9-acridone derivatives may be obtained starting from different benzophenones

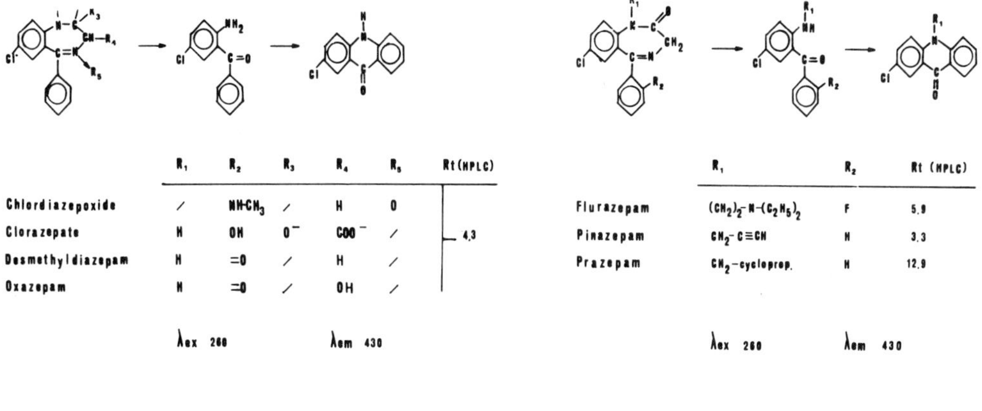

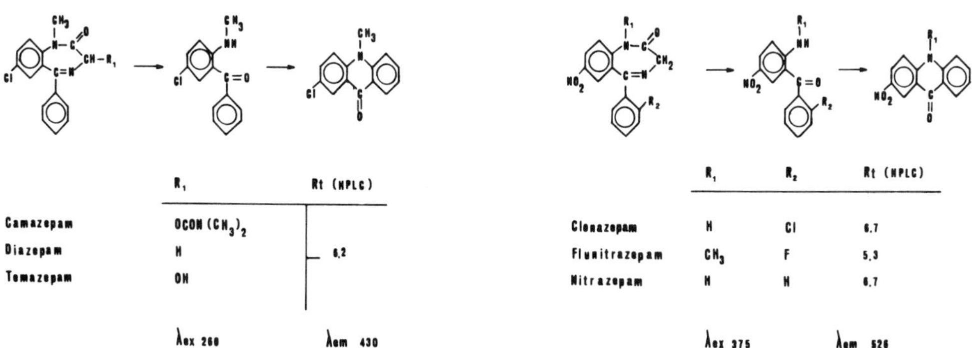

Fig. 2. Benzodiazepines (giving different acridones) classified according to their principal functional groups.

(Fig. 2). In this picture thirteen benzodiazepines (that give seven different acridones) are classified according to their principal functional groups.

The proposed structure for the acridanones was confirmed by literature data when referred[2], by chromatographic mobilities, fluorescence spectroscopy and by infrared and mass spectrometry.

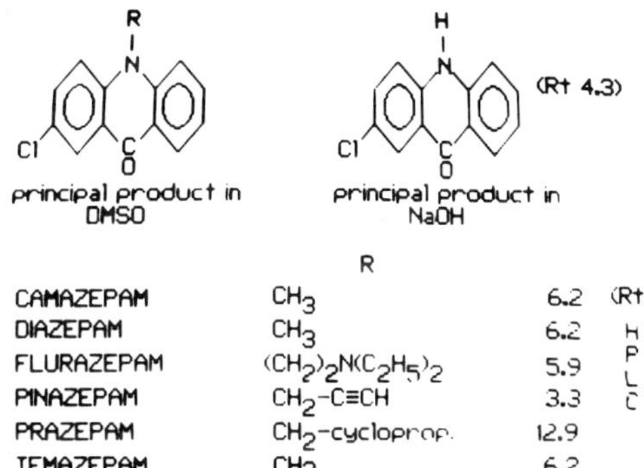

	R		
CAMAZEPAM	CH₃	6.2	(Rt)
DIAZEPAM	CH₃	6.2	H
FLURAZEPAM	(CH₂)₂N(C₂H₅)₂	5.9	P
PINAZEPAM	CH₂–C≡CH	3.3	L
PRAZEPAM	CH₂–cycloprop.	12.9	C
TEMAZEPAM	CH₃	6.2	

Fig. 3. Benzodiazepine derivatization products obtained in DMSO
    or NaOH/PbO₂.

We observed that starting from the thirteen benzodiazepines, the number of acridones obtained may be decreased from seven to three using sodium hydroxide/lead dioxide (Fig. 3). This is probably due to an N-dealkilation occurring during the nucleophilic attack, when the Meisenheimer intermediate is formed. In the strong alkaline medium, probably an Hofmann degradation takes place to give N-dealkilated acridones, while in aprotic solvent a substituted acridone is obtained[7]. This fact confirms the opportunity of sodium hydroxide use when screening is requested.

This hypothesis was confirmed by the methylation of an N-dealkilated acridanone carried out with methyl iodide and

```
INST  1  METH    1  FILE   10

RUN       1  BLANK URINE  EXTACT N 3  10 : 31.8    5 / 16 / 86

SENSITIUITIES  1023 255
```

TIME	AREA	BC	RRT	RF	C	NAME
1.33	1.6707	T	0.133	1.000	1.2265	'
1.98	2.0421	T	0.198	1.000	1.4992	'
2.17	2.1776	T	0.217	1.000	1.5986	'
2.73	1.6913	T	0.273	1.000	1.2416	'
3.03	4.9459	T	0.303	1.000	3.6309	'
3.51	56.8345	T	0.351	1.000	41.7228	'
4.50	1.7332	T	0.450	1.000	1.2724	'
5.15	5.4038	T	0.515	1.000	3.9670	'
5.43	4.1280	T	0.543	1.000	3.0304	'
5.61	2.6465	T	0.561	1.000	1.9429	'
5.73	5.1000	T	0.573	1.000	3.7440	'

```
INST  1  METH    2  FILE   15

RUN       3  URINE SPIKED WITH OXHZEPAM  EX N 4

SENSITIUITIES  1023 255
```

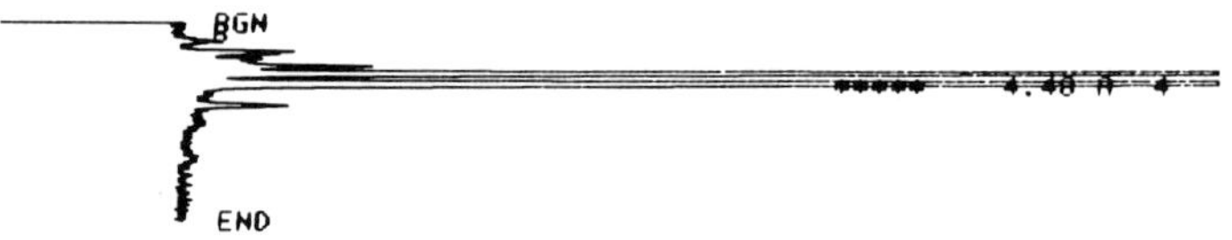


```
INST  1  METH    2  FILE   15

RUN       3  URINE SPIKED WITH OXAZEPAM  EX N 4  11 : 14.2    5 / 16 / 86

SENSITIUITIES  1023 255
```

TIME	AREA	BC	RRT	RF	C	NAME
4.30	88.0435	T	1.000	1.000	1.0000	2-CLORO-9-ACRIDANONE:

Fig. 4. HPLC chromatographic pattern obtained by method 1 (details are in the text).

potassium carbonate, that gave a product with the same retention
time of the acridone synthetized by diazepam.

Consequently this procedure cannot be distinguished between
different benzodiazepines and their major metabolites. However
there is some justification for monitoring total benzodiazepines
in blood because the major metabolites are biologically active
too, and they contribute to drug's effect.

INST 1 METH 2 FILE 6

RUN 2 URINE SPIKED WITH OXAZEPAM EX N 2

SENSITIVITIES 1023 255

BGN

4.37 A 4

END

INST 1 METH 2 FILE 6

RUN 2 URINE SPIKED WITH OXAZEPAM EX N 2 10 : 16.2 5 / 16 / 86

SENSITIVITIES 1023 255

TIME	AREA	BC	RRT	RF		NAME
2.19	53.1072	T	0.511	1.000	1.0000	2-CLORO-9-ACRIDANONE

INST 1 METH 1 FILE 7

RUN 1 BLANK URINE EXTRACT N 1

SENSITIVITIES 1023 255

BGN

4.23 A 4

END

INST 1 METH 1 FILE 7

RUN 1 BLANK URINE EXTRACT N 1 9 : 5.5 5 / 16 / 86

SENSITIVITIES 1023 255

TIME	AREA	BC	RRT	RF	NAME
1.68	6.2956	T	0.168	1.000	1.9976 '
2.01	22.3795	T	0.201	1.000	7.1009 '
2.19	53.4169	T	0.219	1.000	16.9489 '
2.61	18.0608	T	0.261	1.000	5.7306 '
2.88	36.8473	T	0.288	1.000	11.6915 '
3.57	56.0000	T	0.357	1.000	17.7685 '
3.95	16.5875	T	0.395	1.000	5.2651 '
4.41	7.5323	T	0.441	1.000	1.1208 '

Fig. 5. HPLC chromatographic pattern obtained by method 2
(details are in the text).

9-acridone derivatives can be resolved by thin layer chromatography and high performance liquid chromatography.

TLC conditions are: HPTLC pre-coated plates RP18 as
stationary phase; acetonitrile:1 mmol/l acetate buffer pH 4.0
(7:3) as mobile phase; detection was performed under UV lamp.

HPLC conditions are: 5 μm Bondapak C8 12 cm length column
(Waters, Milford, Ma, USA); acetonitrile:1 mmol/l acetate buffer

```
INST  1  METH    1  FILE   11

RUN      1  BLANK URINE EXTRACT N 5

SENSITIVITIES  1023 255
```

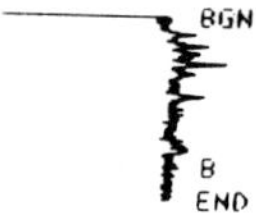


```
INST  1  METH    1  FILE   11

RUN      1  BLANK URINE EXTRACT N 5 11 : 55.4    5 / 16 / 86

SENSITIVITIES  1023 255
```

TIME	AREA	BC	RRT	RF		NAME
1.11	0.2365	T	0.111	1.000	1.4177	'
1.28	1.2308	T	0.128	1.000	5.2965	'
1.58	0.5644	T	0.158	1.000	2.4287	'
1.95	0.3505	T	0.195	1.000	1.5882	'
2.21	1.7106	T	0.221	1.000	2.3610	'
2.71	1.8454	T	0.271	1.000	2.9410	'
3.03	0.6626	T	0.303	1.000	2.8516	'
3.48	2.4064	T	0.348	1.000	10.3549	'
3.80	0.2800	T	0.380	1.000	1.2450	'
4.02	0.4798	T	0.402	1.000	2.0649	'
4.25	0.8058	T	0.425	1.000	3.4676	'
4.33	0.8120	T	0.433	1.000	3.4941	'
4.66	0.2846	T	0.466	1.000	1.2249	'
5.13	0.6402	T	0.513	1.000	2.7550	'
5.37	0.4345	T	0.537	1.000	1.8699	'
5.87	1.9164	T	0.587	1.000	8.2464	'
6.27	0.2897	T	0.627	1.000	1.2469	'
6.38	0.4898	T	0.638	1.000	2.1079	'
6.77	0.3523	T	0.677	1.000	1.5161	'
6.90	0.2873	T	0.690	1.000	1.2363	'
7.17	0.2730	T	0.717	1.000	1.1749	'
7.45	0.2388	T	0.745	1.000	1.0172	'

```
INST  1  METH    2  FILE   16

RUN      1  URINE SPIKED WITH OXAZEPAM EX N 6

SENSITIVITIES  1023 255
```

```
INST  1  METH    2  FILE   16

RUN      1  URINE SPIKED WITH OXAZEPAM EX N 6 11 : 50.0    5 / 16 / 86

SENSITIVITIES  1023 255

     TIME      AREA  BC    RRT       RF           C  NAME
```

Fig. 6. HPLC chromatographic pattern obtained by method 3
(details are in the text).

```
INST  1  METH   1  FILE   6

RUN      1  9-ACRIDANONE

SENSITIVITIES  1023 255

            BGN
          ┌──┐
          │  │                 ***** 3.50 A  4
          │  │
          │  B
          │  END

INST  1  METH   1  FILE   6

RUN      1  9-ACRIDANONE  6 : 25.3   0 /  0 /  0

SENSITIVITIES  1023 255

     TIME      AREA   BC     RRT      KF          C    NAME

     0.59    0.1965  T     0.059    1.000      1.1777 '
     1.45    1.1544  T     0.145    1.000      5.9168 '
     1.64    0.1938  T     0.164    1.000      1.1617 '
     1.85    0.3083  T     0.185    1.000      1.8475 '
     2.50    0.2886  T     0.250    1.000      1.7294 '
     3.26   10.2153  T     0.326    1.000     51.2068 '
     4.07    0.1723  T     0.407    1.000      1.0325 '
     4.61    0.2147  T     0.461    1.000      1.2869 '
     4.70    0.2666  T     0.470    1.000      1.5979 '
     4.85    0.1881  T     0.485    1.000      1.1271 '
     4.94    0.1931  T     0.494    1.000      1.1571 '
     5.22    0.2125  T     0.522    1.000      1.2732 '
     7.24    0.1825  U     0.724    1.000      1.0937 '
     7.35    0.1906  U     0.785    1.000      1.1425 '
```

Fig. 7. Chromatographic pattern of 2-aminobenzophenone
derivative.

pH 4.0 (1:1) as eluent system; fluorimetric detector was set at
260 nm for the excitation and 430 nm for the emission wavelength.

About biological applications of this reaction and subsequent
liquid chromatographic analysis, we tested three different
procedures:

1)   hydrolysis of biological material, benzophenone extraction by
     diethylether, evaporation of organic phase, reconstitution
     with cyclisation medium, extraction by ethyl acetate;
2)   as the previous procedure omitting the first extraction step;
3)   pre-extraction of benzodiazepines by disposable reversed
     phase microcolumns Baker 10-SPE Octadecyl (Baker, Deventer,
     Holland), by adding the sample (1 ml) to the column, washing
     with water and eluting with methanol. The column eluate was
     then hydrolyzed and processed as before.

Chromatographic patterns obtained from urine specimens spiked
with therapeutical amounts of oxazepam are shown in Fig. 4-6.

Fig. 4 is relative to procedure n. 1. In the upper part the
chromatographic run of blank urine is reported and in the lower
part the same urine after addiction of 0.1 µg/ml of oxazepam. The
chromatogram results in this case quite clean: the presence of an
unknown peak only does not affect acridone elution.

The chromatograms shown in Fig. 5 were obtained by using the
direct procedure. In this way the chromatogram results particularly dirty, however the identification of the acridone related to

```
INST  1  METH    1  FILE    8

RUN       1  OXAZEPAM ACRIDANONE 38,44 NG ML

SENSITIVITIES  1023 255
```

```
            BGN
        *****    4.20 A  4
            END
```

```
INST  1  METH    1  FILE    8

RUN       1  OXAZEPAM ACRIDANONE 38,44 NG ML  0 : 55.8   0 / 0 / 0

SENSITIVITIES  1023 255

   TIME       AREA  BC    RRT      RF          C   NAME

   1.27     0.3501 T    0.127   1.000       1.1046 '
   1.43     0.4326 T    0.143   1.000       1.3649 '
   4.18    26.7942 T    0.418   1.000      84.5269 '


INST  1  METH    1  FILE    9

RUN       1  OXAZEPAM ACRIDANONE 9,62 NG ML

SENSITIVITIES  1023 255
```

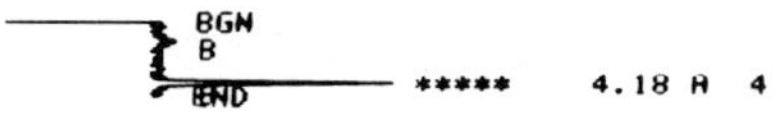

```
            BGN
            B
            END   *****    4.18 A  4
```

```
INST  1  METH    1  FILE    9

RUN       1  OXAZEPAM ACRIDANONE 9,62 NG ML  1 :  4.4   0 / 0 / 0

SENSITIVITIES  1023 255

   TIME       AREA  BC    RRT      RF          C   NAME

   0.13     0.1098 U    0.013   1.000       1.1323 '
   0.33     0.1239 T    0.033   1.000       1.2779 '
   0.54     0.1624 U    0.054   1.000       1.6755 '
   1.33     0.4691 T    0.133   1.000       4.8373 '
   1.42     0.1488 T    0.142   1.000       1.5352 !
   1.63     0.1015      0.163   1.000       1.0469 '
   2.14     0.1338 T    0.214   1.000       1.3804 '
   2.25     0.1284 T    0.225   1.000       1.3243 '
   2.41     0.1104 T    0.241   1.000       1.1384 '
   2.53     0.1465 T    0.253   1.000       1.5113 !
   2.75     0.1270 T    0.275   1.000       1.3098 !
   2.83     0.1692 U    0.283   1.000       1.7448 '
   3.07     0.1083 T    0.307   1.000       1.1176 !
   3.30     0.2443 T    0.330   1.000       2.5194 '
   3.70     0.1667 U    0.370   1.000       1.7198 '
   3.83     0.1152 T    0.383   1.000       1.1879 '
   4.16     6.5462 T    0.416   1.000      67.5044 '
```

Fig. 8. Detection limit for oxazepam related acridones
according to method 3 (for details see text).

oxazepam is still possible; since this acridone is the first
eluted, other acridanones related to different benzodiazepines are
well resolved too. We suggest this procedure when saving in
analytical time may be precious.

As shown in Fig. 6, the pre-extraction step gives very clean
extracts and the identification of acridone peaks is easier in
respect to the other procedures. Moreover in this case the use of

2-aminobenzophenone is suggested as internal standard because this substance, which undergoes the same reaction, is eluted at 3.3 min retention time (Fig. 7).

Finally we studied the detection limit for oxazepam related acridones (chlordiazepoxide, clorazepate, desmethyldiazepam, oxazepam) adding a known amount of this benzodiazepine to blank urine (Fig. 8). The analytical procedure was made according to method n. 3 (preliminary extraction of benzodiazepines from 1 ml of urine) and the cyclisation reaction gave sufficient amount of acridone to perform an easy detection of 5 ng/ml, with a variation coefficient of 4.2%.

The fluorescence detector response was a linear function of concentration from 5 to 500 ng/ml.

CONCLUSION

Our studies confirm the possibility of benzodiazepine analysis by related 9-acridanones. The reaction has been performed on thirteen different drugs and in all the cases an intense fluorescent compound resulted. The suggested final liquid chromatographic separation enhance the specificity, avoiding problems related to the presence of endogenous and/or exogenous fluorescent compounds.

The purposed method may be successfully used as a screening tool with sufficient sensitivity (ng/ml) to detect benzodiazepines in biological specimens at therapeutical levels.

Acknowledgments

We are thankful to the Mass Fragmentography Center of C.N.R. (Rome), for the cooperation given to this work.

REFERENCES

1. H. Schutz, "Benzodiazepines - A Handbook", Springer, Berlin, Heidelberg, New York (1982).
2. J.C. Valentour, J.R. Monforte, B. Lorenzo and I. Sunshine, Fluorimetric screening method for detecting benzodiazepines in blood and urine, Clin. Chem., 21:1976 (1975).
3. M. Divoll and D.J. Greenblatt, Plasma concentrations of temazepam, a 3-hydroxybenzodiazepine, determined by electron capture gas-liquid chromatography, J. Chromatogr. 222:125 (1981).
4. L. Kangas, Determination of nitrazepam and its main metabolites in urine by gas-liquid chromatography: use of electron-capture and nitrogen-selective detectors, J. Chromatogr., 172:273 (1979).
5. R.I. Fryer, J. Earley and L.H. Sternbach, A new synthesis of 9-acridones, J. Chem. Soc. 4979 (1963).
6. J.A.F. De Silva and N. Strojny, Determination of flurazepam and its major biotranformation products in blood and urine by spectrophotofluorometry and spectrophotometry, J. Pharm. Sci. 60:1303 (1971).
7. R.A.Y. Jones, "Physical and Mechanistic Organic Chemistry", Cambridge University Press, Cambridge (1979).

HPLC SIMULTANEOUS DETERMINATION OF 1,4 BENZODIAZEPINES IN
BIOLOGICAL FLUIDS

A. Scotto di Tella, C. Di Nunzio, and P. Ricci*

Institute of Legal Medicine
First Medical School, University of Naples
Naples
* Department of Public Health and Cell Biology
Second University of Rome
Rome, Italy

INTRODUCTION

The most frequently encountered drugs in emergency toxicology
in our country are benzodiazepines, alone or associated to other
psychoactive drugs and/or alcohol[1].

The usually applied TLC procedures, however, although reasonably rapid, provide only approximate results of limited clinical
usefulness. Many other analytical techniques[2-4] (i.e. EMIT®, RIA,
GLC, GC/MS) have been applied for the quantitation of this group
of drugs, but generally they do not allow simultaneous determinations of the most frequently used benzodiazepines. Since this
condition is particularly important in emergency toxicology, where
the time factor is the most binding one, we developed a simple
analytical method through HPLC, that allows the simultaneous assay
of fifteen benzodiazepines, and here we report its main features.

METHODS AND RESULTS

Apparatus

The liquid chromatograph used was a Beckman (Palo Alto, Ca,
Usa) system mod.420 with a Hewlett-Packard HP 3390 integrator
(Palo Alto, Ca, USA). The stainless steel separation column
(250 x 4.6 mm;) was filled with Altex (Palo Alto, Ca, USA) Ultrasphere C18, 5μm. The injector was an Altex mod.210 with a 20 μl
loop. The UV detector (Beckman mod. 165) was set at 240 nm. The
eluent mixture was methanol:0.011 mol/l KH_2PO_4 pH 7.5 containing
6% CH_3CN (60:40) from time 0 to time 18 min; methanol:0.011 mol/l
KH_2PO_4 pH 7.5 with 6% CH_3CN (70:30) from time 19 min to time 30 min
The flow rate was 1.1 ml/min; all the experiments were carried out
at room temperature.

Scheme 1. Extraction Procedure of 1,4 Benzodiazepines from Serum and Urine

---

0.5 ml of ethanolic standard solution

- drying at 35°C and 15 mm Hg

Residue + 0.5 ml of serum and urine

- mix with vortex 2 min and store 30 min at room temperature
- extract with ethilene chloride:methilene chloride: ethyl acetate (1:1:8) 3 x 2 ml, buffered with 0.1 ml of a solution containing 0.1 g/ml of Na_2CO_3:$NaHCO_3$ (1:1)
- centrifuge 5 min at 8,000 rpm

Aqueous phase

Organic phase

- dehydrate with Na_2SO_4

Dehydrated organic phase

- dry at 35°C and 15 mm Hg

Residue + 0.5 ml of methanol

- mix with Vortex 2 min
- filtrate through 0.20 μm membranes (OE66 Schleicher & Schull)

HPLC

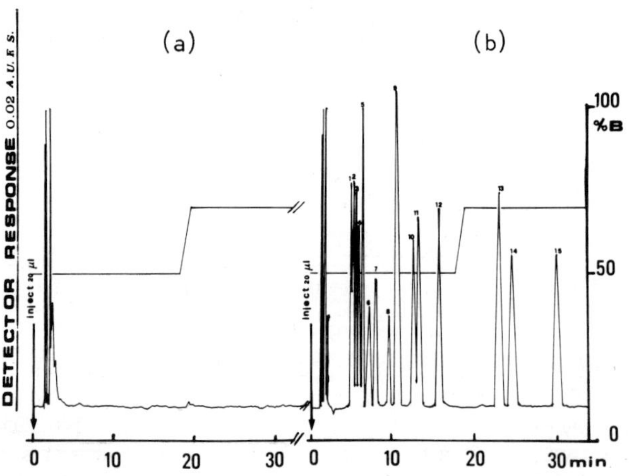

Fig. 1. Representative chromatograms of blank human serum (a) and of human serum to which benzodiazepines have been added (b). No major interferences of normal serum components can be appreciated.

Table 2. Recovery (± S.D.) of 1,4 Benzodiazepines added to Human
          Serum and Urine (μg/ml)

	SERUM				
	20	10	5	2.5	1.25
Bromazepam	99±1.2	100±1.1	99±1.3	98±1.2	98±1.1
Clonazepam	99±1.2	99±1.2	98±1.2	98±1.4	97±1.5
Nitrazepam	98±1.2	98±1.2	98±1.3	98±1.6	97±1.7
Flunitrazepam	100±1.1	98±1.2	98±1.2	98±1.2	97±1.4
Estazolam	99±1.2	98±1.3	98±1.4	98±1.3	97±1.5
Lorazepam	99±1.2	98±1.2	98±1.2	97±1.5	97±1.8
Oxazepam	99±1.1	98±1.2	98±1.2	98±1.3	98±1.2
Chlordiazepam	98±2.1	97±2.2	97±2.5	97±1.4	97±1.5
Nordiazepam	98±1.2	98±1.2	98±1.1	98±1.1	97±1.3
Diazepam	99±1.2	98±1.1	98±1.3	98±1.2	98±1.3
Pinazepam	99±1.2	98±1.1	98±1.1	97±1.2	97±1.2
Camazepam	98±1.1	98±1.4	98±1.3	97±1.5	97±1.4
Prazepam	100±1.1	99±1.2	99±1.1	98±1.1	98±1.2
Flurazepam	100±1.1	99±1.2	99±1.1	99±1.2	99±1.2
Medazepam	98±1.2	98±1.2	98±1.3	97±1.6	97±1.5

	URINE				
Bromazepam	100±1.0	100±0.8	99±1.2	99±1.1	99±1.3
Clonazepam	100±0.9	99±1.2	99±1.2	98±1.3	98±1.2
Nitrazepam	99±1.2	99±1.1	99±1.2	98±1.2	98±1.0
Flunitrazepam	100±1.1	99±1.2	99±1.3	98±1.4	98±1.2
Estazolam	100±0.8	99±1.2	99±1.1	98±1.2	98±1.3
Lorazepam	99±1.1	99±1.3	98±1.2	98±1.2	98±1.4
Oxazepam	99±1.1	99±1.3	99±1.3	98±1.1	98±1.2
Chlordiazepam	98±1.3	98±1.3	97±1.6	97±1.3	97±1.4
Nordiazepam	99±1.1	99±1.1	98±1.3	98±1.3	97±1.6
Diazepam	99±1.1	99±1.2	99±1.2	99±1.1	99±1.1
Pinazepam	99±1.2	98±1.3	98±1.3	98±1.4	98±1.3
Camazepam	100±1.1	100±0.4	99±1.2	99±1.1	99±1.2
Prazepam	100±0.8	99±1.0	99±1.1	99±1.4	98±1.2
Flurazepam	99±1.2	100±0.8	98±1.1	99±1.2	99±1.2
Medazepam	99±1.2	98±1.3	98±1.2	98±1.2	97±1.7

Estazolam recovery as I.S. (10 μg/ml) = 98.39±0.85

<u>Recoveries</u>

The recoveries of drugs from serum and urine obtained at
different concentrations of the benzodiazepines investigated are
listed in Table 2.

<u>Accuracy Data</u>

The regression slope values from serum and urine are summarized in Table 3.

DISCUSSION

The reported extraction procedure allows to minimize the
interferences due to endogenous substances, but other possible
interferences are still under investigation.

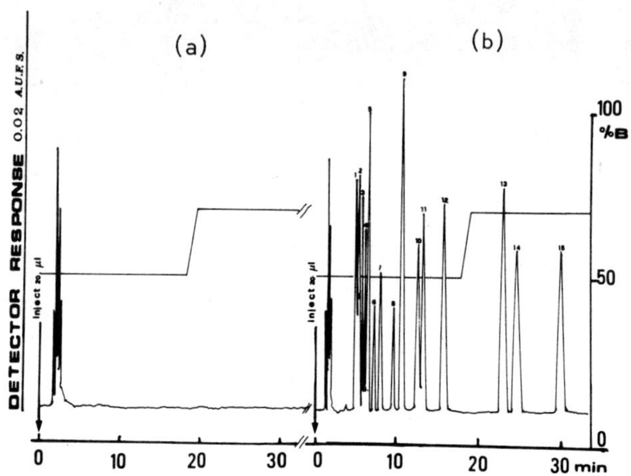

Fig. 2. Representative chromatograms of blank human urine (a) and
of human urine to which benzodiazepines have been added
(b). No major interferences of normal urine components
can be appreciated.

Extraction Procedure

The extraction procedure is shown in Scheme 1.

Retention times

The retention times are listed in Table 1. Typical chromatograms obtained from serum and urine are shown in Fig. 1 and 2.

Table 1. Retention Time, Capacity Factor (k) and Resolution
Factor of the separated 1,4 Benzodiazepines

	Drug	Retention Time (min ± S.D.)	k	Resolution Factor
1	Bromazepam	5.78 ± 0.04	0.96	0.09
2	Clonazepam	5.96 ± 0.08	1.05	0.10
3	Nitrazepam	6.07 ± 0.08	1.09	0.11
4	Flunitrazepam	6.16 ± 1.09	1.12	0.12
5	Estazolam	7.12 ± 0.07	1.45	0.15
6	Lorazepam	7.67 ± 0.04	1.64	0.17
7	Oxazepam	8.12 ± 0.05	1.80	0.18
8	Chlordiazep-oxide	10.30 ± 0.02	2.55	0.26
9	Nordiazepam	11.50 ± 0.07	2.96	0.30
10	Diazepam	13.25 ± 0.09	3.71	0.38
11	Pinazepam	13.95 ± 0.06	3.77	0.39
12	Camazepam	16.67 ± 0.09	4.75	0.49
13	Prazepam	23.40 ± 0.06	7.07	0.73
14	Flurazepam	24.72 ± 0.08	7.52	0.78
15	Medazepam	31.00 ± 0.09	9.69	1.00

Table 3. Linear Regression Equations of the Determination of
          1,4 Benzodiazepines added to Human Serum and Urine

Drug	Serum	Urine
Bromazepam	$7.02 \cdot 10^{-2}$ x $-4.42 \cdot 10^{-3}$	$6.98 \cdot 10^{-2}$ x $+5.00 \cdot 10^{-4}$
Clonazepam	$7.02 \cdot 10^{-2}$ x $-4.83 \cdot 10^{-3}$	$7.06 \cdot 10^{-2}$ x $+2.25 \cdot 10^{-3}$
Nitrazepam	$6.50 \cdot 10^{-2}$ x $-2.71 \cdot 10^{-3}$	$6.50 \cdot 10^{-2}$ x $-7.08 \cdot 10^{-4}$
Flunitrazepam	$5.50 \cdot 10^{-2}$ x $+1.04 \cdot 10^{-3}$	$5.49 \cdot 10^{-2}$ x $+2.09 \cdot 10^{-3}$
Estazolam	$9.60 \cdot 10^{-2}$ x $+1.72 \cdot 10^{-2}$	$9.75 \cdot 10^{-2}$ x $+4.79 \cdot 10^{-2}$
Lorazepam	$5.21 \cdot 10^{-2}$ x $-1.54 \cdot 10^{-3}$	$5.19 \cdot 10^{-2}$ x $+9.83 \cdot 10^{-3}$
Oxazepam	$5.99 \cdot 10^{-2}$ x $-1.83 \cdot 10^{-3}$	$5.98 \cdot 10^{-2}$ x $+1.00 \cdot 10^{-3}$
Chlordiazep-oxide	$4.95 \cdot 10^{-2}$ x $-1.33 \cdot 10^{-2}$	$5.01 \cdot 10^{-2}$ x $+4.37 \cdot 10^{-3}$
Nordiazepam	$7.02 \cdot 10^{-2}$ x $-5.42 \cdot 10^{-4}$	$7.05 \cdot 10^{-2}$ x $-7.29 \cdot 10^{-3}$
Diazepam	$4.97 \cdot 10^{-2}$ x $-5.04 \cdot 10^{-4}$	$4.98 \cdot 10^{-2}$ x $+3.12 \cdot 10^{-3}$
Pinazepam	$5.49 \cdot 10^{-2}$ x $+1.71 \cdot 10^{-3}$	$5.48 \cdot 10^{-2}$ x $+2.92 \cdot 10^{-3}$
Camazepam	$6.00 \cdot 10^{-2}$ x $-2.29 \cdot 10^{-3}$	$6.00 \cdot 10^{-2}$ x $+1.10 \cdot 10^{-3}$
Prazepam	$6.51 \cdot 10^{-2}$ x $-3.08 \cdot 10^{-3}$	$6.52 \cdot 10^{-2}$ x $-4.37 \cdot 10^{-3}$
Flurazepam	$5.18 \cdot 10^{-2}$ x $+1.29 \cdot 10^{-3}$	$5.19 \cdot 10^{-2}$ x $+2.08 \cdot 10^{-3}$
Medazepam	$7.20 \cdot 10^{-2}$ x $-1.54 \cdot 10^{-3}$	$7.20 \cdot 10^{-2}$ x $+1.50 \cdot 10^{-3}$

The correlation coefficient of each equation was > 0.998. The
method we described allows a very rapid assessment of the most
commonly used 1,4 benzodiazepines in both human serum and urine.
The consistent quantitative recovery of the added benzodiazepines, the sensitivity and the reproducibility of the results
make this method highly reliable for quantitative analytical
purposes.

The described procedure has already been utilized for emergency toxicological analyses and in some forensic cases; the results have always been in agreement with TLC and GLC confirmatory
analyses.

REFERENCES

1. A. Fiori, Studio del fenomeno della mortalità correlata
    all'abuso di sostanze stupefacenti e psicotrope, Boll. Farm.
    Alcolismo 2-3:169 (1985).
2. R.C. Baselt, "Analitycal Procedure for Therapeutic Drug
    Monitoring and Emergency Toxicology", Biomedical Publ., Canton
    (1980).
3. H. Schutz, "Benzodiazepine. A Handbook", Springer-Verlag,
    New York (1982).
4. D.M. Hailey, Chromatography of the 1,4 benzodiazepines,
    J. Chromatogr. 98:527 (1974).

SALIVA/PLASMA RATIOS FOR FORENSIC MONITORING OF DRUG ASSUMPTION
IN OPIATE ADDICTS

F. Mari, E. Bertol, R. Biagioli, and R. Chiarugi

Forensic Medicine Department
University of Florence
Florence, Italy

INTRODUCTION

In the recent years the determinations of drugs in biological
fluids were not only carried out in serum, plasma and urine, but
also in other specimens such as saliva, milk, spinal fluid etc.
The saliva drug levels are of special interest, because its
concentration values correlate with plasma levels. Saliva offers
particular advantages in various application areas, such as:

1. bioavailability and pharmacokinetic studies;
2. therapeutic drug monitoring;
3. forensic and epidemiological monitoring of drugs of
   abuse.

The main advantages are the following:

- non-invasive method of samples collection;
- no discomfort to patients or volunteers;
- no need of particularly trained personnel in sampling.

For the applications in the areas of points 1 and 2 numerous
investigators have studied the relationship between saliva and
plasma concentrations for a large number of drugs, such as
benzodiazepines[1-7], antipyretics and analgesics[8-11], antiarrhyth-
mics[12], anticonvulsants[13-17], xanthine derivatives[18-20], barbiturates[21]
and other drugs[22-24].

On the contrary there are only a few studies on saliva/plasma
ratios (S/P) in drugs abuse[25-29]. The purpose of the present paper
is the study of saliva and plasma total opiate levels in volunteer
subjects after administration of opiates, in order to obtain
experimental correlation data between the two biological fluids.

SAMPLE COLLECTION AND PREPARATION

The plasma and saliva samples collected belonged to three
groups of volunteers:

1st: 5 subjects who received an oral dose of 30 mg of codeine
     phosphate;
2nd: 5 subjects who received an i.m. dose of 30 mg of
     morphine hydrochloride;
3rd: 5 subjects who self-administered an unknown dose of
     heroin.

Both blood and saliva samples were collected 1 h and 2 hrs
after drug administration. The plasma was separated from blood
samples (7-10 ml) by centrifugation; the saliva samples (3-5 ml)
were collected after the volunteers had stimulated the saliva
flow, chewing strips of Parafilm®. The saliva samples were then
centrifuged and the pH value was measured. All samples were stored
immediately at 4°C until they were analysed.

REAGENTS

Codeine phosphate, morphine hydrochloride and nalorphine
hydrobromide were F.U. (Italian Official Pharmacopea) purity.
Hydrochloric acid, ammonium hydroxide, chloroform and isopropanol
were of analytical grade, obtained from Merck (Darmstadt, FRG).
MethElute® was obtained from Pierce (Rockford, Il, USA).

INSTRUMENTATION

A Carlo Erba (Milan, Italy) series 4150 HRGC, equipped with a
SPB-5 CFS capillary column (15 m x 0.25 μm I.D.) and a flame
ionization detector, was used for the chromatographic analyses.
The conditions were: oven 250°C; detector and injector 300°C;
carrier gas (hydrogen) flow rate 3.5 ml/min.

METHODS

Two calibration curves were prepared for codeine and morphine
respectively. The calibrators consisted of blank plasma samples
spiked with codeine phosphate or morphine hydrochloride to give
0.15, 0.30, 0.60, 1.20 μg/ml concentration levels, to which 1.0 ml
of nalorphine hydrobromide aqueous solution (at 0.80 μg/ml concentration) was added as an internal standard.

Plasma and saliva samples (1.0 to 2.0 ml) containing the
internal standard solution (1.0 ml) were hydrolysed with concentrated HCl (1.0 ml) in an autoclave for 30 min. After cooling, the
samples were alkalinised with concentrated ammonium hydroxide to
obtain a pH value of 8.8-9.0 and then extracted with 2 x 5 ml of
chloroform:isopropanol (3:1). The combined extracts were evaporated to dryness at 45°C under a nitrogen stream. An amount of
5-10 μl of MethElute was added to the residue and 1 μl was injected into the column. The methylated derivatives were measured
by capillary gas chromatography.

RESULTS AND DISCUSSION

The calibration curves were constructed by plotting the peak
height ratio (codeine/nalorphine) versus codeine or morphine
concentrations of the calibrators (Fig. 1).

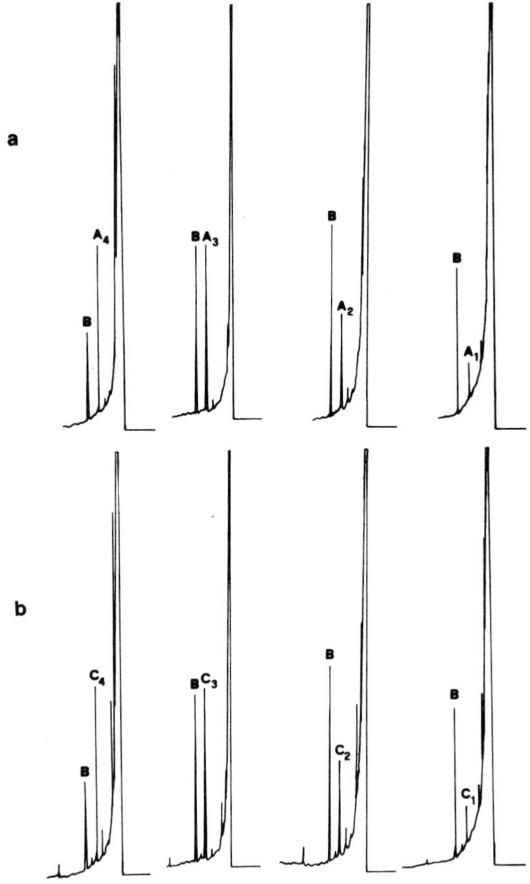

Fig. 1. Chromatograms of codeine and morphine calibrators.
       a) codeine calibrators: A₁=0.15, A₂=0.30, A₃=0.60,
          A₄=1.20 µg/ml codeine conc.; B=internal standard.
       b) morphine calibrators: C₁=0.15, C₂=0.30, C₃=0.60,
          C₄=1.20 µg/ml methylated morphine conc.; B=internal
          standard.

For both drugs a linear relationship was obtained (Fig. 2).
The regression equations for the two calibration curves were:

$$Y = 1.659\ X - 0.004\ (r = 0.999)\ \text{for morphine and}$$
$$Y = 1.515\ X + 0.066\ (r = 0.998)\ \text{for codeine.}$$

The analytical recovery was found to be excellent, ranging
95-102% for both the compounds, obtaining a CV% of 7.5 and 6.8 on
10 analyses of a calibrator at known concentration of codeine and
morphine respectively.

The concentration levels of codeine or morphine (as the major
heroin metabolite also) in all saliva and plasma samples at the
two collection times are reported in Table 1; the S/P ratios and
their mean values with ± S.D. for each drug group are also
reported.

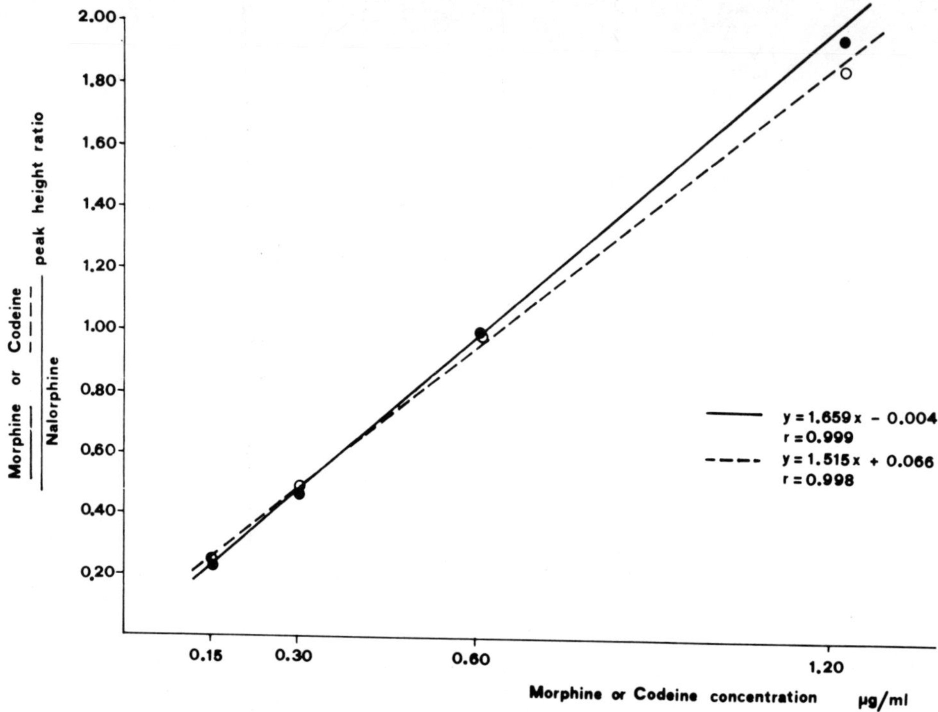

Fig. 2. Calibration curves of morphine and codeine vs. nalorphine
(internal standard).

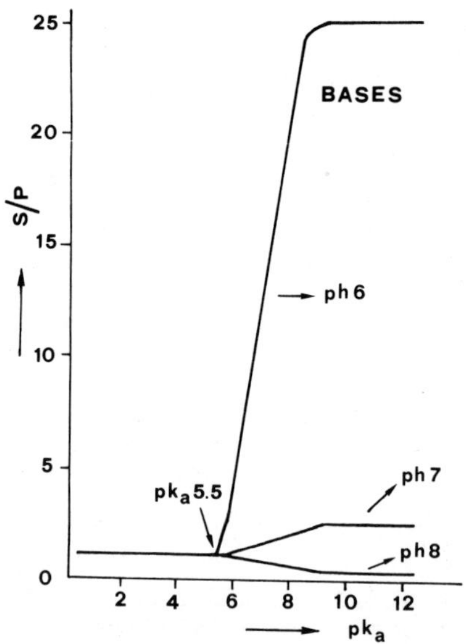

Fig. 3. Variation of the saliva/plasma concentration ratio (S/P)
with drug pKa values at three different salivary pH
values (From R.C. Baselt, ed.: Advances in analytical
toxicology, vol. 1, Biomed. Pub. 1984).

Table 1. Concentration Levels (μg/ml) in Saliva and Plasma after
          Administration of Drugs

| Administered Drug | Subjects | Concentrations | | | | Saliva/Plasma Ratio | |
| | | Saliva | | Plasma | | | |
		1h	2h	1h	2h	1h	2h
Codeine (30 mg os)	1	0.96	0.91	0.37	0.41	2.59	2.21
	2	0.28	0.25	0.15	0.13	1.86	1.92
	3	0.65	0.47	0.38	0.26	1.71	1.80
	4	0.45	0.38	0.20	0.19	2.25	2.00
	5	0.58	0.46	0.21	0.20	2.76	2.30
					mean	2.23	2.05
					±S.D.	0.45	0.21
Morphine (30 mg i.m.)	1	0.65	0.95	0.28	0.38	2.32	2.50
	2	0.72	0.65	0.31	0.28	2.32	2.32
	3	0.95	0.70	0.42	0.50	2.26	1.40
	4	1.16	0.97	0.59	0.51	1.96	1.90
	5	0.89	1.04	0.39	0.60	2.28	1.73
					mean	2.22	1.97
					±S.D.	0.15	0.44
Heroin (unkn. e.v.)	1	0.30	0.25	0.14	0.12	2.14	2.08
	2	0.85	0.70	0.41	0.39	2.07	1.79
	3	0.28	0.22	0.15	0.13	1.86	1.70
	4	0.98	0.80	0.39	0.38	2.51	2.10
	5	1.25	0.75	0.67	0.38	1.86	1.97
					mean	2.09	1.93
					±S.D.	0.27	0.18

The generated results demonstrate that the saliva/plasma
concentration ratio obtained for the 3 groups of the studied drugs
at the 2 times following the drug administration was found to be
fairly constant. In fact this ratio was not greatly variable at
the mean of the values obtained after each drug administration:
2.23, 2.22, 2.09 at the first hour and 2.05, 1.97, 1.93 at the
second hour.

Furthermore the ratio values in our experiments resulted in
agreement with the predictable S/P ratios for basic drugs, considering the pKa of the drugs and the pH value of the saliva (ranging
in our cases from 6.0 to 7.0) and plasma (pH = 7.4), according to
the graph of Fig. 3[30].

The advantage of the use of saliva in respect to plasma is
evident: it is a non-invasive and an easy method of sample
collection and, most importantly, higher concentrations of opiates
(or their metabolites) are found in saliva. Consequently, in cases
of very low concentrations in the plasma, saliva should be the
specimen of choice because of the opiate concentrations in this
fluid well correlated with the plasma concentrations.

REFERENCES

1. H.G. Giles, D.H. Zilm, R.G. Frecker, S.M. Macleod and E.M. Sellers, Saliva and plasma concentrations of diazepam after a single oral dose, Brit. J. Clin. Pharm. 4:711 (1977).
2. H. Ogata, S. Horii, T. Shibazaki, N. Aoyagi, N. Kaniwa and A. Ejima, Salivary excretion of barbital and diazepam from humans, Eisei Shikensho Hokuku 96:27 (1978).
3. R. Dixon, R. Lucek, J. Early and C. Perry, Chlordiazepoxide: a new, more specific and sensitive radioimmunoassay, J. Pharm. Sci. 68:261 (1979).
4. L. Kangas, H. Allonen, R. Lammintausta, M. Salonen and A. Pekkarinen, Pharmacokinetics of nitrazepam in saliva and serum after a single oral dose, Acta Pharm. Tox. 45:20 (1979).
5. C. Halstrom, M.H. Lader and S.H. Curry, Diazepam and N-desmethyldiazepam concentrations in saliva, plasma and CSF, Brit. J. Clin. Pharm. 9:333 (1980).
6. R. Lucek and R. Dixon, Chlordiazepoxide concentrations in saliva and plasma measured by radioimmunoassay, Res. Comm. Chem. Path. Pharm.27:397 (1980).
7. J.J.T. De Geier, B.J. Hart, P.F. Wilderink and F.A. Nelemans, Comparison of plasma and saliva levels of diazepam, Brit. J. Clin. Pharm. 10:151 (1980).
8. E.S. Vesell, G.T. Passananti, P.A. Glenwright and B.H. Dvorchik, .Studies on the disposition of antipyrine, aminopyrine and phenacetin using plasma, saliva and urine, Clin. Pharm. Ther. 18:259 (1975).
9. R.M. Welch, R.L. De Angelis, M. Wigsfield and T.W. Former, Elimination of antipyrine from saliva as a measure of metabolism in man, Clin. Pharm. Ther. 18:249 (1975).
10. H.S. Fraser, J.C. Mucklow, S. Murray and D.S. Davies, Assessment of antipyrine kinetics by measurement in saliva, Brit. J. Clin. Pharm. 3:321 (1976).
11. P.J. Meffin, R.L. Williams, T.F. Blaschke and M. Rowland, Application of salivary concentration data to pharmacokinetic studies with antipyrine, J. Pharm. Sci. 66:135 (1977).
12. A. Barchowsky, W.W. Stargel, D.G. Shand and P.A. Routledge, Saliva concentrations of lidocaine and its metabolites in man Ther. Drug Mon. 4:335 (1982).
13. C.E. Cook, E. Amerson, W.K. Poole, P. Lesser and L. O'Tuama, Phenytoin and phenobarbital concentrations in saliva and plasma measured by radioimmunoassay, Clin. Pharm. Ther. 18:742 (1975).
14. T. Nishikawa, M. Saito and H. Muira, Determination of antiepileptic drugs by enzyme immunoassay and comparison of the drug concentrations in saliva and in plasma, Kitazato Igaku 7:363 (1977).
15. J.C. Mucklow, C.J. Bacon and A.M. Hierons, Monitoring of phenobarbitone and phenytoin therapy in small children by salivary samples,Ther. Drug Mon. 3:275 (1981).
16. I.M. Friedman, I.F. Litt and R. Henson, Saliva phenobarbital and phenytoin concentrations in epileptic adolescents, J. Pediat. 98:654 (1981).
17. R.F. Goldsmith and R.A Ouvier, Salivary anticonvulsant levels in children: a comparison of methods, Ther. Drug. Mon. 3:151 (1981).
18. N.N. Khanna, H.S. Bada and S.M. Somani, Use of salivary concentrations in the prediction of serum caffeine and theophylline concentrations in premature infants, J. Pediat. 96:494 (1980).

19   H.W. Kelly, W.M. Hadley, S.A. Murphy and B.G. Skipper,
     Monitoring children on sustained released therapy by salivary
     theophylline levels, Am. J. Dis. Child. 135:137 (1981).
20. S.M. Lena, P. Hutchins, C.B.S. Wood and P. Turner, Salivary
     theophylline estimation in the management of asthma in
     children, Postgrad. Med. J. 56:85 (1980).
21. T. Inaba and W. Kalow, Salivary excretion of amobarbital in
     man, Clin. Pharm. Ther. 18:558 (1975).
22. K. Feller and G. Le Petit, On the distribution of drugs in
     saliva and blood plasma, Int. J. Clin. Pharm. 15:468 (1977).
23. M.G. Horning, L. Brown, J. Nowlin, K. Lertratanangkoon,
     P. Kellaway and T.E. Zion, Saliva in therapeutic drug
     monitoring.Clin. Chem. 23:157 (1977).
24. J.C. Mucklow, M.R. Bending, G.C. Kahn and C.T. Dollery, Drug
     concentrations in saliva, Clin. Pharm. Ther. 24:563 (1978).
25. S.H. Wan, S.B. Martin and D.L. Arzarnoff, Kinetics, salivary
     excretion of amphetamine isomers and effect of urinary pH,
     Clin. Pharm. Ther. 23:585 (1978).
26. A.W. Jones, Inter and intra-individual variations in the
     saliva blood alcohol ratio during ethanol metabolism in man,
     Clin. Chem. 25:1394 (1979).
27. G. Boettger and K. Feller, Determination of the ethanol
     concentration in blood and saliva, Zent. Pharm. Pharmakother.
     Lab.118:328 (1979).
28. N.P.E. Vermeulan, M.W.E. Teunissen and D.D. Breimer,
     Pharmacokinetics of pemoline in plasma, saliva and urine
     following oral administration, Brit. J. Clin. Pharm. 8:459
     (1979).
29. M.E. Sharp, S.M. Wallace, K.W. Hindmarsh and H.W. Peel,
     Monitoring saliva concentrations of methaqualone, codeine,
     secobarbital, diphenhydramine and diazepam after single oral
     doses, J. Anal. Toxicol.7:11 (1983).
30. B. Caddy, Saliva as a specimen for drug analysis, In:
     "Advances in Analytical Toxicology", R.C. Baselt, ed., Biomed.
     Publ., Foster City, Ca, USA (1984).

DETERMINATION OF MORPHINE IN HAIR BY IMMUNOCHEMICAL AND GAS

CHROMATOGRAPHIC MASS SPECTROMETRIC TECHNIQUES

F. Centini*, C. Offidani, A. Carnevale, M. Chiarotti,
and I. Barni Comparini*

Institute of Forensic Medicine
Catholic University, Rome
* Institute of Forensic Medicine
University of Siena
Siena, Italy

INTRODUCTION

Hair analysis promises to be a suitable complement to serum
and urine analysis as diagnostic tool. This is due to the
accumulation of drugs, other organic chemicals and metals in hair.
Hence, much of the original interest in analysis of hair involved
its application to forensic science.

Among abused drugs, morphine has been demonstrated to
accumulate in hair[1,2] and its determination was suggested as
diagnostic tool expecially for investigating past chronic intake
of heroin[3,4].

Radioimmunoassay (RIA) is generally used in thid kind of
analysis, however, in the forensic field confirmatory tests are
required because the possible specific and non specific
interferences[5].

Gas chromatography/mass spectrometry (GC/MS) represents a
method with high specificity and sensitivity and it has emerged as
a reference method[5,6].

In our study we used two different immunochemical methods for
morphine assay in hair and GC/MS for confirmatory analysis, to
investigate their reability in terms of specificity and
sensitivity.

EXPERIMENTAL AND DISCUSSION

We analyzed 42 hair samples of heroin abusers under detoxication program. Ten samples from healthy subjects were used as
negative control specimens.

Enzyme immunoassay (EMIT®) and radioimmunoassay (RIA) were chosen
for immunological analysis. Immunoenzymatic analyses were

Table 1. Morphine Detection in the Hair

Sample treatment

- cleaning procedure by organic solvent (CH_2Cl_2)
- hydrolysis with NaOH 1 mol/l, 30 min at 80°C
- neutralization by HCl 10 mol/l, then 0.1 mol/l phosphate
  buffer pH 7.4 (for RIA a) or pH 5.5 (for RIA b)
- centrifugation

	liquid/liquid extraction by $CHCl_3$/ter-buthanol (9/1) after alkalinization
direct RIA and EMIT	GC/MS confirmatory analysis

---

carried out with a commercial kit (Syva, Palo Alto, Ca, USA) for morphine detection in urine, having a sensitivity of 3 µg/l for morphine. Two radioimmunoassay with different detection limits were also used:

1) RIA-a consisting of a commercial kit "Abuscreen" for morphine (Roche, Basel, Switzerland) with a detection limit of 50 ng/ml.
2) RIA-b a solid phase I125 RIA (Coat a Count) provided by D.P.C.(Los Angeles, Ca, USA), with a sensitivity of 1 ng/ml.

Table 2. Sample Preparation for GC/MS Analysis

Liquid-liquid extraction:	chloroform/ter-buthanol (9/1) after alkalinization to pH 9.6. Evaporate to dryness by N_2
Acetylation:	pyridine/acetic anhydride (2/1). Incubate for 1 h at 60°C. Add methanol and evaporate to dryness by N_2. Reconstitute the residue with 20 µl of methanol. Inject 1 µl
Mass spectrometer:	operating in electron-impact mode
GC conditions:	- fused silica capillary column 25 m OV1, film thickness 0.4 µm - carrier gas He at 3 ml/min - injector split-splitless (20 s splitless time) - temperature program: 1 min at 160°C, then to 250°C (5°/min), 30 min at 250

About 100 mg hair samples were cut into small sections and washed twice by methilene chloride, then hydrolyzed by sodium hydroxide 1 mol/l (1 ml) for 30 min at 80°C. Samples were neutralized by stoichiometric amounts of HCl 10 mol/l, then phosphate buffer was added to obtain a suitable pH for immunoassay detection as suggested by manufacturers.

Known amounts of neutralized solutions were made alkaline by bicarbonate then extracted according to Felby's procedure for confirmatory GC/MS analyses (Table 1). Before GC/MS, the extracts were derivatized by pyridine-acetic anhydride.

Gas chromatographic conditions are reported in Table 2.

A Finnigan 1020 GC/MS (San Jose, Ca, USA) apparatus was used for mass spectrometry determinations. The spectrograms were obtained operating in total ion current (scanning range 40-400 amu). Selected ion monitoring (m/e 369, 327, 268) was performed when higher sensitivity was required[8,9].

In these conditions a detection limit of about 50 pg of injected heroin was obtained corresponding to a theoretical amount of 0.01 ng of morphine per mg of hair.

Information about drug abusers and results obtained with EMIT and RIA are reported in Table 3.

EMIT assay gave positive results only in 52% of the samples with morphine levels ranging from 3 to 37 ng/mg. The low percent of positive results were due to the sensitivity of this method. As a matter of fact only samples containing more than 3 ng/mg of morphine were seen as positive.

On the contrary RIA-a detected morphine in 93% of samples ranging from 0.5 to 23.6 ng/mg. Good linear relationship between positive results was obtained by the two techniques as shown in

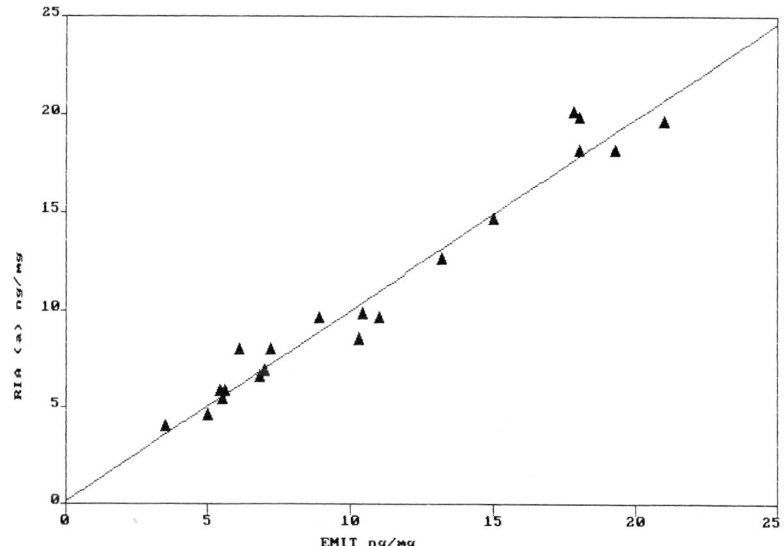

Fig. 1. Correlation between EMIT and RIA data.

Fig. 1. No positive data were obtained in control group with both methods.

Samples from addicts containing morphine below 0.5 ng/mg and negative controls were also tested with RIA-b.

Table 3. Information on the Studied Heroin Addicts

Case	Months*	Dose g/day	A.L.M. +/-	Emit U. +/-	Emit H. ng/mg	RIA H. (a) ng/mg
1	> 12	0.7	+	+	-	2.6
2	> 12	1.5	+	+	7.0	6.8
3	> 12	2.0	n.d.	n.d.	7.2	7.9
4	> 12	2.5	n.d.	n.d.	15.0	14.5
5	> 12	2.0	n.d.	n.d.	18.0	18.0
6	> 12	2.0	n.d.	n.d.	8.9	9.5
7	> 12	3.0	-	-	17.8	20.0
8	> 12	1.5	-	-	13.2	12.4
9	> 12	1.5	-	-	6.1	7.9
10	> 12	n.d.	-	-	5.0	4.5
11	5	1.0	n.d.	-	-	1.2
12	> 12	3.0	-	-	-	2.4
13	3	n.d.	n.d.	n.d.	-	0.0
14	6	1.0	n.d.	n.d.	5.6	5.7
15	> 12	2.0	+	-	-	2.3
16	> 12	2.5	n.d.	n.d.	10.4	9.7
17	4	0.5	+	-	-	0.8
18	> 12	0.8	+	+	-	1.7
19	10	1.0	+	+	-	0.9
20	2	0.5	n.d.	n.d.	-	0.8
21	> 12	1.5	+	+	-	2.3
22	5	0.2	+	+	-	0.6
23	> 12	2.5	+	+	18.0	19.7
24	8	0.5	+	+	6.8	6.5
25	> 12	1.0	+	+	5.5	5.3
26	> 12	n.d.	+	+	37.0	23.6
27	2	n.d.	-	-	-	0.0
28	> 12	3.0	+	+	11.0	9.5
29	> 12	1.0	+	+	3.5	3.9
30	2	0.8	+	+	3.0	0.6
31	6	0.3	-	-	-	0.6
32	> 12	0.5	+	+	-	1.6
33	> 12	3.0	+	+	19.3	18.0
34	8	0.5	-	-	-	1.2
35	12	0.5	+	+	5.4	5.7
36	> 12	2.0	+	+	10.3	8.4
37	> 12	0.1	+	-	-	1.7
38	> 12	3.0	+	+	-	0.8
39	> 12	0.5	+	-	-	2.1
40	12	2.5	+	+	21.0	19.5
41	1	0.5	-	-	-	0.0
42	3	0.2	-	-	-	0.5

* months of heroin addiction; A.L.M.: assumption referred to the last month; EMIT U.: assay on urine samples; EMIT H.: assay on hair samples; RIA H.: determination carried out on hair samples by the Roche kit (a); n.d.: not determined.

Table 4. Morphine in hair (ng/mg) by RIA

	a (Roche)	b (D.P.C.)
1	2.6	2.93
10	4.5	4.62
11	1.2	1.41
12	2.4	2.61
13	0.0	0.15
15	2.3	2.27
17	0.8	0.93
18	1.7	1.84
19	0.9	0.19
20	0.8	0.74
21	2.3	2.16
22	0.6	0.47
27	0.0	0.06
29	3.9	4.13
30	0.6	0.85
31	0.6	0.72
32	1.6	1.67
34	1.2	1.20
37	1.7	1.88
38	0.8	0.70
39	2.1	2.51
41	0.0	0.14
42	0.5	0.52

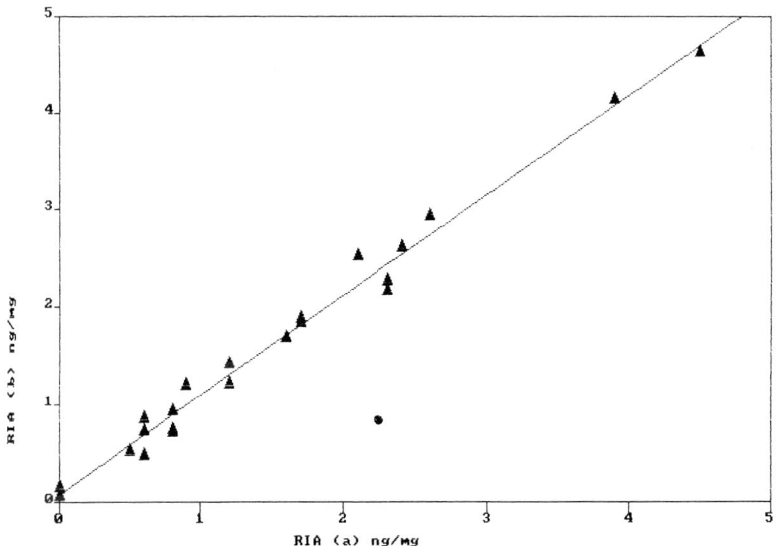

Fig. 2. Morphine in hair measured by two different RIAs.

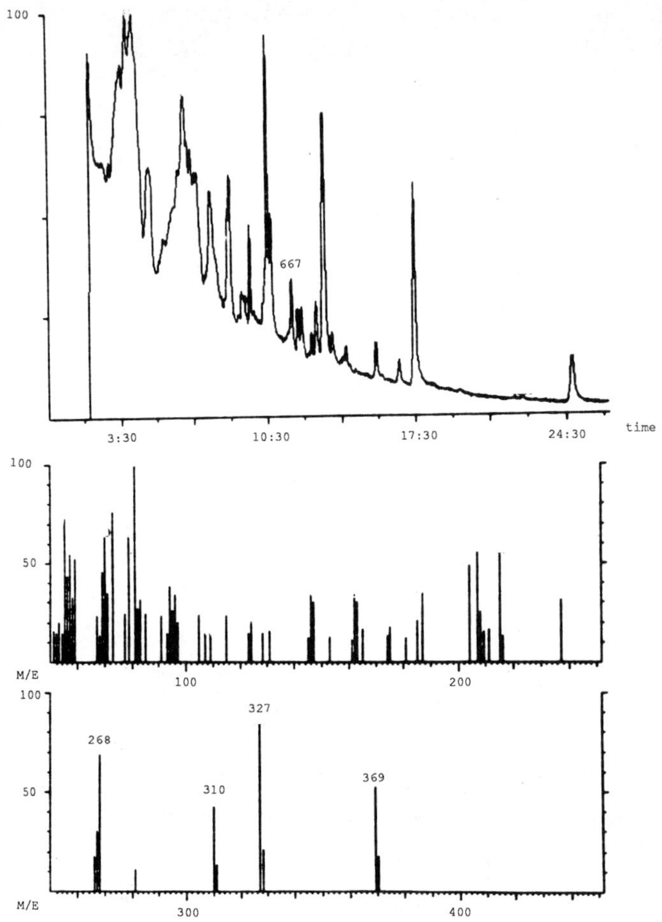

Fig. 3. Chromatogram and mass fragmentogram at scanning time 667 of a positive heroin sample.

In our experimental conditions,the detection limit of RIA-b (0.01-0.05 ng/mg) was fifty times lower that that of RIA-a and consequently all the samples of the heroin addicts resulted positive, even if morphine concentration was less than 0.5 ng/mg. The results obtained by both RIA methods are reported in Table 4.

Unexpectedly two positive results, showing morphine in trace amount (0.05 and 0.06 ng/mg), were also observed in the control group by RIA-b. Good linear relationship between both RIA data was also observed (Fig. 2).

Confirmatory GC/MS analyses were well correlated with RIA results. Only in one case the presence of morphine was not confirmed.

In the other positive samples with morphine concentration higher than 0.1 ng/mg mass fragmentography supported radioimmunoassay data. A positive heroin sample with specific mass fragmentogram at scanning time 667 is shown in Fig. 3.

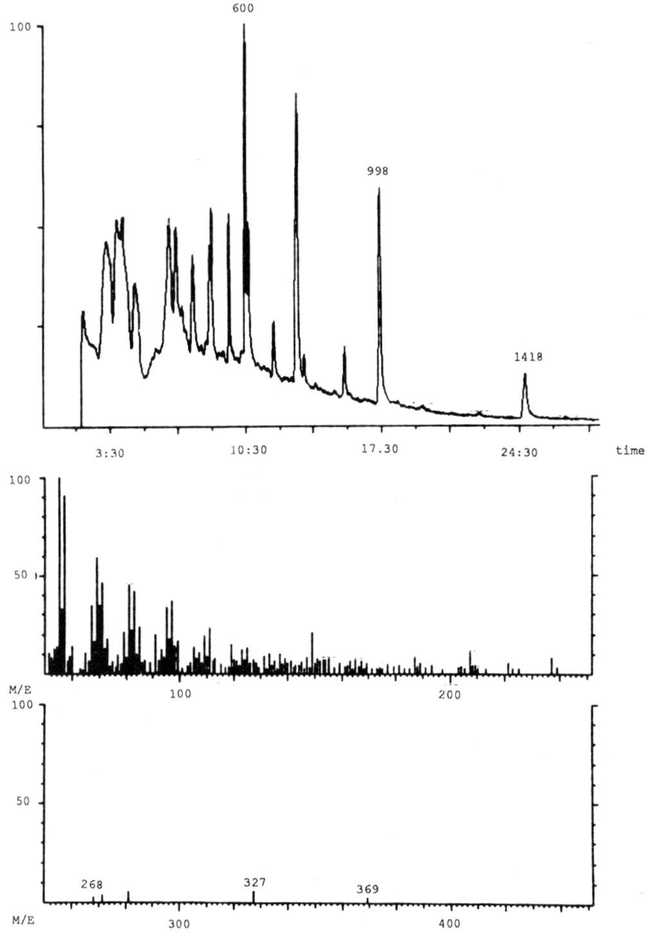

Fig. 4. GC/MS assay of a RIA positive sample containing morphine
in trace amount (<0.1 ng/mg).

    However when morphine concentration is below 0.1 ng/mg some
identification problems could arise. In these cases to enhance the
sensitivity it is necessary operating in selected ion monitoring
(SIM) mode, focusing on ions at m/e 369, 327, 268. Under these
conditions all the RIA results were confirmed both in addicts and
in control group.

    However, by replicate GC/MS analyses in total ion current,
positive results with morphine levels near 0.05 ng/mg, found in
the control group, were not confirmed.

    As a matter of fact fragmentation did not show peaks at
heroin scanning time and were not observed on the mass-spectra
significant ion as 310, 215; their absence renders the data
unsuitable to surely support the presence of heroin in these
samples (Fig. 4).

CONCLUSION

The EMIT method is not sensitive enough for morphine
detection in the hair; actually, a too high number of samples from
addicts result negative, since morphine levels are bolow 3 ng/mg.

RIA technique has been confirmed as a suitable tool for this
analysis, when coupled with a more specific method as GC/MS.
Both RIA and GC/MS data were in agreement with clinical-
toxicological findings of abusers, even if no positive correlation
was established between heroin intake and morphine amount in the
hair.
The high sensitivity obtained by some commercially available RIA
kits renders difficult GC/MS confirmatory tests, especially when
loss in sensitivity can arise after extraction and chromatographic
processes. Consequently unambiguous demonstration of morphine in
the hair near 0.05 ng/mg is difficult and it could require more
sophisticated techniques. However, it must be noticed that in our
causistry, very low levels of morphine in hair are confined to
about 1% of the cases.

REFERENCES

1. A.M. Baumgartner, P.F. Jones, W.A. Baumgartner and C.T.Black,
    Radioimmunoassay of the hair for determining opiate-abuse
    histories, J. Nucl. Med. 20:748 (1979).
2. D. Valente, M. Cassani, M. Pigliapochi and G. Vanzetti, Hair
    as the sample in assessing morphine and cocaine addiction,
    Clin. Chem. 27:11 (1981).
3. K. Puschel, P. Tomasch and W. Arnold, Opiate levels in hair,
    Forens. Sci. Int. 21:181 (1983).
4. A. Carnevale, M. Marchetti and A. Fiori, La ricerca
    radioimmunologica della morfina nei capelli, Boll. Farmacod.
    Alcoolismo, 6:579 (1983).
5. F. Tagliaro, S. Lafisca, S. Maschio, A. Parolin, G. Lubli and
    M. Marigo, Quantitative determination of morphine in hair: a
    comparison between RIA and HPLC methods, 13th Congress
    International Academy of Forensic and Social Medicine,
    Budapest 1985.
6. B.D. Paul, L.D. Mell, J.M. Mitchell, J. Irving and A.J. Novak,
    Simultaneous identification and quantitation of codeine and
    morphine in urine by capillary gas chromatographiy and mass
    spectroscopy, J. Anal. Tox., 9:222 (1985).
7. A.W. Jones, Determination of morphine in biological samples by
    gas chromatography/mass spectrometry, J. Chromatogr. 309:73
    (1984).
8. H. Audier, M. Fetizon, D. Ginsburg, A. Mandelbaum and
    T. Rull., Mass spectrometry of the morphine alkaloids,
    Tetrahedron Letters, 1:13 (1965).
9. D.M.S. Wheeler, T.H. Kinstle and K.L. Rinehart, Mass spectral
    studies of alkaloids related to morphine, J. Am. Chem. Soc.
    89:17 (1967).

DETERMINATION OF MORPHINE AND OTHER OPIOIDS IN THE HAIR OF HEROIN ADDICTS BY RIA, HPLC AND COLLISIONAL SPECTROSCOPY

F. Tagliaro, P. Traldi*, B. Pelli*, S. Maschio,
C. Neri, and M. Marigo

Institute of Forensic Medicine
University of Verona, Verona
* C.N.R. Mass Spectrometry Service
Research Estate of Padua, Padua
Italy

INTRODUCTION

The determination of morphine in the hair, according to several Authors' reports, could become a useful tool for identifying past, chronic heroin intake.

Since 1979, some researchers[1] have pointed out that this compound, which is the major biological metabolite of heroin, passes from circulating fluids into the hair, and here remains, bound to the matrix throughout hair lifetime.

In case of chronic heroin intake this kinetics allows a slow, continuous enrichment with morphine of the hair, theoretically related to this habit. On the contrary, hair morphine content should not be considered closely dependent on the dose and the time of the last heroin injection.

This feature makes the hair unique in comparison with other tissues and body fluids, commonly used as samples for toxicological investigations; in fact they are strongly influenced by the recent intake of opioids, and therefore reflect only a very short term addiction history.

Unfortunately, morphine levels in the hair are low (ng/mg), and therefore very sensitive assays are needed.

On this ground, the use of radioimmunoassay (RIA) has been proposed by most of Authors[1-4], because of its high sensitivity coupled with operative simplicity. Yet well known cross-reactions of the commonly used antisera with other opiates such as codeine, that concentrate in the hair too[4], hamper the practical use of this technique.

In order to overcome this problem, we applied a new commercially available RIA, that uses an antiserum not cross-reacting with codeine. This method gains high sensitivity (1-5 ng/ml),

through an easy sample preparation and simplified analitycal
procedures.

Nevertheless, because of the verified existence of matrix
interferences that can affect RIAs, in Forensic Toxicology it
is widely accepted that RIA results must be validated by chromatographic methods[5].

In the recent years, HPLC has been proposed for the investigation of a lot of compounds of toxicological interest in biological matrices. Particularly, a number of morphine assays using a
normal or reversed-phase separations and UV, electrochemical or
fluorescence detection have been reported[6-8]. Quite recently, in
our laboratory a very sensitive HPLC assay of morphine has been
developed, by using dansyl derivatisation, straight phase
chromatography and fluorimetric detection[9].

With the aim of confirming the data obtained by RIA, this
method has been applied to the analysis of morphine in the hair
matrix, with promising results[10].

Unfortunately, the high specificity of the proposed HPLC assay is not coupled with a large analytical spectrum, and therefore
heroin metabolites other than morphine cannot be detected. In order to investigate by more powerful techniques, another method has
been developed through collisional spectroscopy, using a reverse
geometry, double focusing mass spectrometer[11].

In the present paper these methods are compared in view of an
unambiguous assessment of heroin abuse histories.

SUBJECTS, SAMPLES AND SAMPLE PREPARATION

Human hair samples (weighing 100-200 mg) were collected without any pre-treatment by cutting near the scalp from a group of 40
addicts to heroin, aging 18 to 30, admitted at a methadone detoxication program. Other larger samples were collected from addicts
who died in Verona from intravenous narcotism.

At the beginning and at intervals during the detoxication
program, urine samples were assayed by immunoenzymatic methods

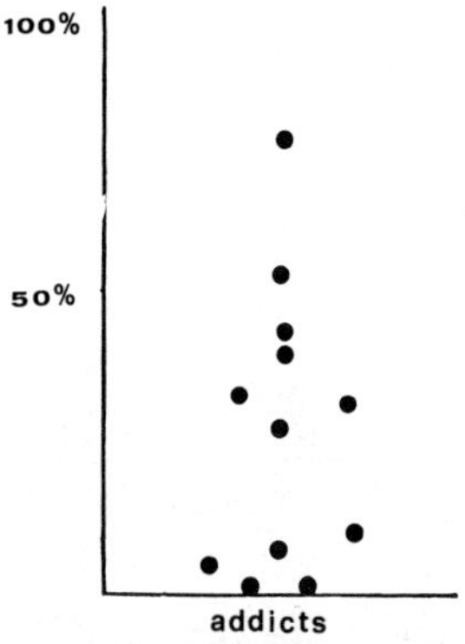

Fig. 1. Percentage of morphine removable by washing on the whole
morphine hair content, evaluated in 12 addicts.

(EMIT-DAU®, Syva, Palo Alto, Ca, USA) and by HPTLC (Merck, Darmstadt, FRG).

In corpses more complete toxicological investigations on biological fluids and tissues were carried out by EMIT, HPTLC and gas-liquid chromatography with FID/NPD.

A matched age group of 20 controls, without any clinical and laboratory evidence of heroin abuse history, was also studied.

In order to remove any external contamination, hair samples were washed with 10 ml of ethyl ether and then with 12 ml of 0.01 mol/l HCl on a porous glass filter.

The complete removal of loosely bound morphine, that likely comes from external contamination, was verified by the following experiment. Four hair samples of heroin addicts were washed with ethyl ether and with 6 seriate aliquots of 2 ml 0.01 mol/l HCl; then morphine was measured in all washing solutions by RIA, confirming that after the fourth aliquot of diluted HCl no more alkaloid was washed away.

The percentage of morphine removable by washing in 12 cases resulted in the range from 0 to 87% of the whole drug extracted from the hair (Fig. 1). These data do not agree with Baumgartner[1], who reported negligible amounts of alkaloid in the wash solutions. Therefore this discordance can be explained by the lighter washings used by this Author, who employed detergent solutions instead of ethyl ether/diluted HCl.

Morphine was then extracted from the hair matrix by incubation of samples in 0.1 mol/l HCl at 45°C for 12 h.

Lacking hair with known morphine content, it was impossible for us to calculate the recovery of the described extraction procedure. Nevertheless, we observed that a stronger treatment with 0.6 mol/l NaOH at 120°C for 30 min, causing the complete dissolution of the matrix, allowed only a little increase of the extracted morphine, meanly of 17%.

MATERIALS, METHODS, RESULTS AND DISCUSSION

Radioimmunoassay

A commercially available RIA (DPC, Los Angeles, Ca, USA) was directly applied on extraction mixtures, previously neutralized and buffered at pH 7.

The adopted RIA kit used a specific anti-morphine antiserum absorbed onto the walls of the assay tubes, with an extremely low cross-reactivity with other narcotics such as dextromethorphan meperidine, methadone, codeine and hydrocodone. After 15 min incubation at room temperature, bound/free separation was carried out by decanting. A typical standard curve is shown in Fig. 2.

Precision was comparable to other RIAs, with a within-run CV = 4.1% and an interassay CV = 5.3%, at a morphine level of 15 ng/ml.

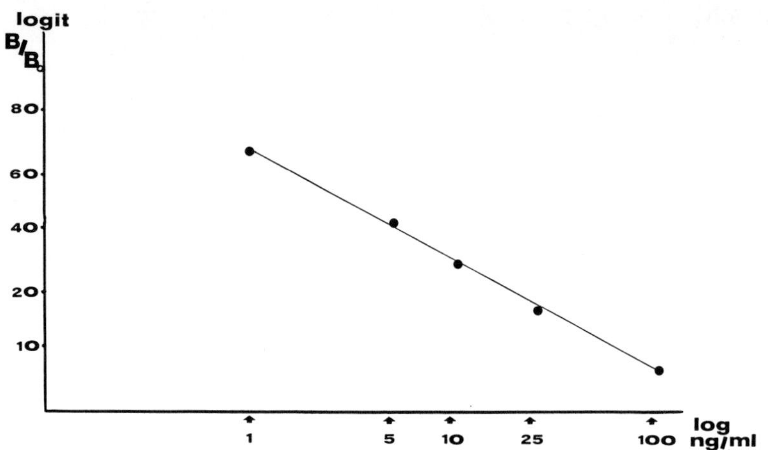

Fig. 2. A typical standard curve of the used solid-phase RIA of
morphine.

Unfortunately, according to Carnevale[4], we found that RIA can
be affected by non specific interferences causing random false
positive results up to 0.27 ng/mg, with an unacceptable frequency
(5-20%).

<u>High Performance Liquid Chromatography</u>

On the basis of a previous work by Frei[12] concerning pharmaceuticals, an original, very sensitive HPLC determination of
morphine in biological fluids was developed in our laboratory: it
uses pre-column dansyl derivatisation, straight phase chromatography and fluorimetric detection[9]. With minor modifications,
this method was applied to the determination of morphine in the
hair matrix.

The analytical procedure has been summarized in Fig. 3.

After neutralization, hair extraction mixtures were poured
into commercially available, ready-to-use tubes for liquid-liquid
extraction of alkaloids (Toxi-Tubes A, Analytical Systems, Laguna
Hills, Ca, USA), containing buffering salts (sodium carbonate and
bicarbonate) and a mixture of organic solvents (heptane,
dichloromethane, dichloroethane). After mixing by vortex for 2
min, the organic phases were transferred into glass conical tubes
and the aqueous phases were re-extracted. Then the organic phases
were pooled and brought to dryness under a stream of Nitrogen. The
dried extracts were dissolved in a small volume (50 µl) of distilled water and mixed with 50 µl of 10 g/l dansyl chloride in
acetone and with an equal volume of 0.1 mol/l sodium carbonate.
The mixtures were incubated for 90 min at room temperature, with
the exclusion of light.

Derivatives were extracted with 1 ml toluene on a vortex,
with a recovery better than 85%. Then the solvent was evaporated
under Nitrogen stream and the extracts were redissolved in a
proper volume of chromatography eluant and injected.

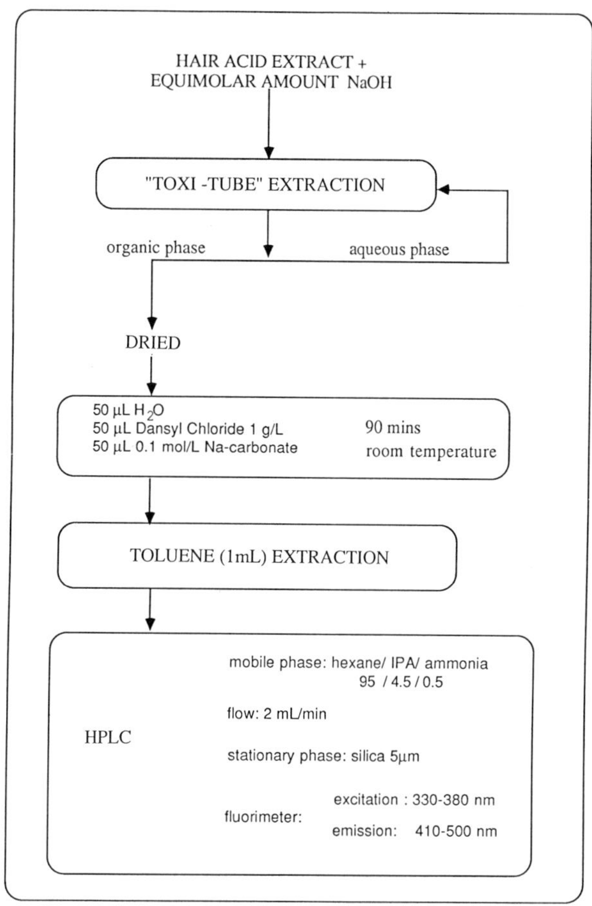

Fig. 3. "Flow chart" of extraction, derivatisation and chromatography procedures of the proposed HPLC assay of morphine in the hair.

The overall recovery of the sample preparation procedures, calculated by using standard morphine in the range 5 to 1,000 ng, was meanly 58.8% with a CV = 3.8% (n=10).

The used chromatography apparatus consisted of two Gilson 302 pumps, a 5 µm silica packed column (Spherisorb S5W 150 x 4.4 mm I.D., Phase Separations, Queensferry, U.K.) and a Gilson 121 filter fluorimeter (fitted with a 330-380 nm excitation filter and a 410-500 nm emission filter). As a mobile phase a mixture of hexane:isopropanol:ammonia 95:4.5:0.5 was used at a flow rate of 2 ml/min.

Typical chromatograms of a negative and a positive hair sample are shown respectively in Fig. 4a and 4b.

Lacking a suitable internal standard (nalorphine retention time was too short compared to morphine), quantitation was carried out by external standardization. The standard curve was found linear over all the used range of injected morphine (0.3-1 to 100 ng).

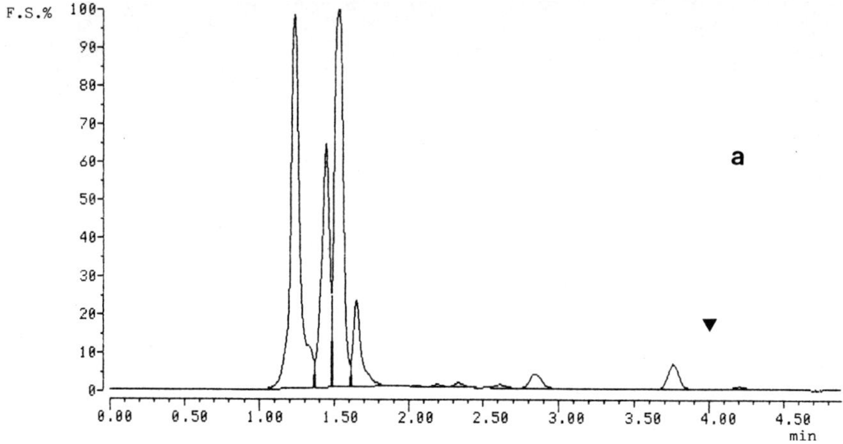

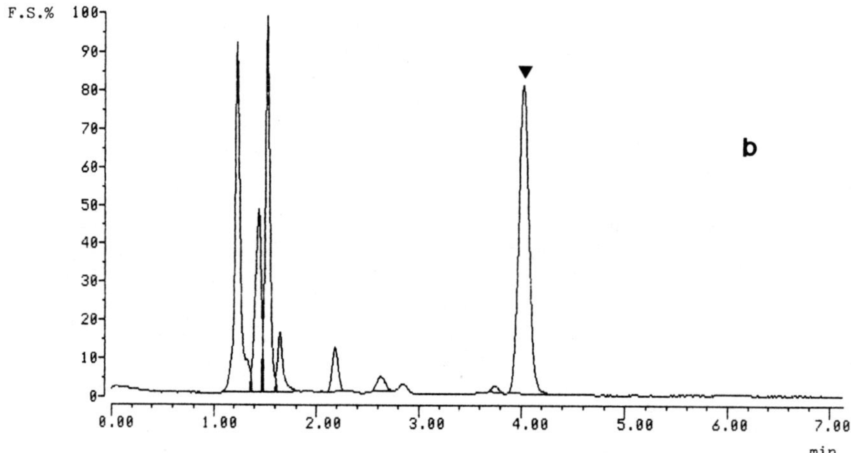

Fig. 4. (a) typical chromatogram of a negative hair sample (150 mg); (b) chromatogram of a positive sample with a morphine content of 0.5 ng/mg (▼ indicates the retention time of morphine).

Within-run and day-to-day variation coefficients resulted quite acceptable being 5.6% and 7.8% respectively.

No interferences were observed from codeine, that was not derivatised, and from more than 70 other opioid and non-opioid drugs[9].

All the 40 addicts investigated showed measurable morphine levels in the range 0.08 to 15.7 ng/mg.

On the other hand, 20 control samples did not show any peaks interfering with morphine. Four subjects on methadone treatment, who claimed to having been heroin abstinent for three months before hair sampling, in agreement with negative toxicological assays of plasma and urine, resulted slightly positive in the range 0.08-0.4 ng/mg. On the contrary, high levels were found in cases with recent histories of heavy heroin abuse.

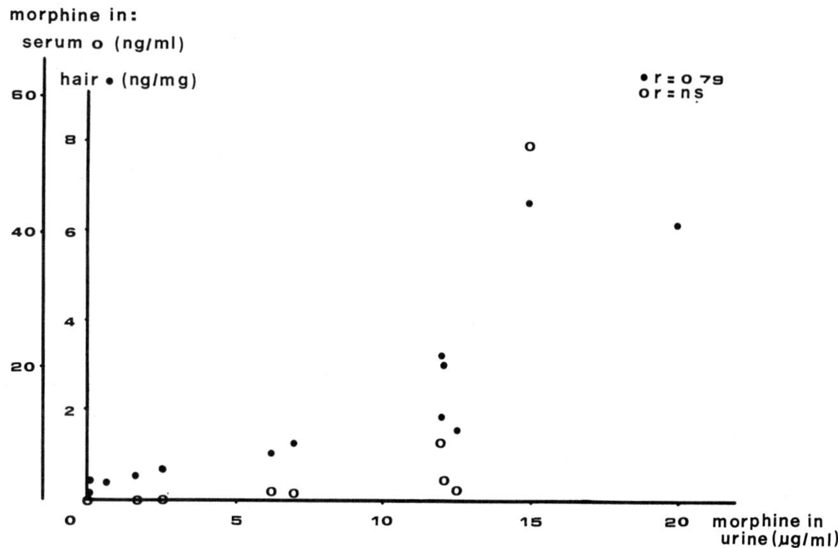

Fig. 5. Morphine levels in hair (o) or serum (•) plotted vs. mean
morphine levels in urines, assayed randomly (from 8 to 16
times per subject) in the same addicts during the last
four months before hair and serum sampling.

On the basis of these observations, the hair and serum morphine levels of 12 addicts have been tentatively correlated with
the urine mean morphine concentrations, that resulted from seriate
random tests made during the last four months before hair sampling
(Fig. 5). A significant correlation was found between hair and
urine (as mean of multiple determinations) levels, but not between
serum and urine.

Though too scanty, these data seem in agreement with the
hypothesis that hair morphine levels, but not serum ones, roughly
reflect the addition habit during some months before sam-pling.

Collisional Spectroscopy

Looking at the interest on this field of research and in
particular at the need of comparing the results obtained with
different analytical techniques in order to gain an unambiguous
ascertainment of heroin abuse history, we dealt with the problem
also through a new and fascinating approach, such as collisional
spectroscopy.

All mass spectrometric measurements were obtained with a
ZAB2F instrument (VG, Altrincham, U.K.) operating in electron
impact (EI) conditions (70 eV, 200 µA). Samples were introduced in
direct electron impact (DEI) conditions, with a source temperature
of 200°C. Collisionally induced decomposition (CID) B/E = const
linked scans and mass analyzed ion kinetic energy spectra (MIKES)
were obtained by colliding 8 keV ions with air in the first and
second field-free regions respectively. The pressure in the colli-
sion cells was such to reduce the main beam intensity to 50% of
its usual value.

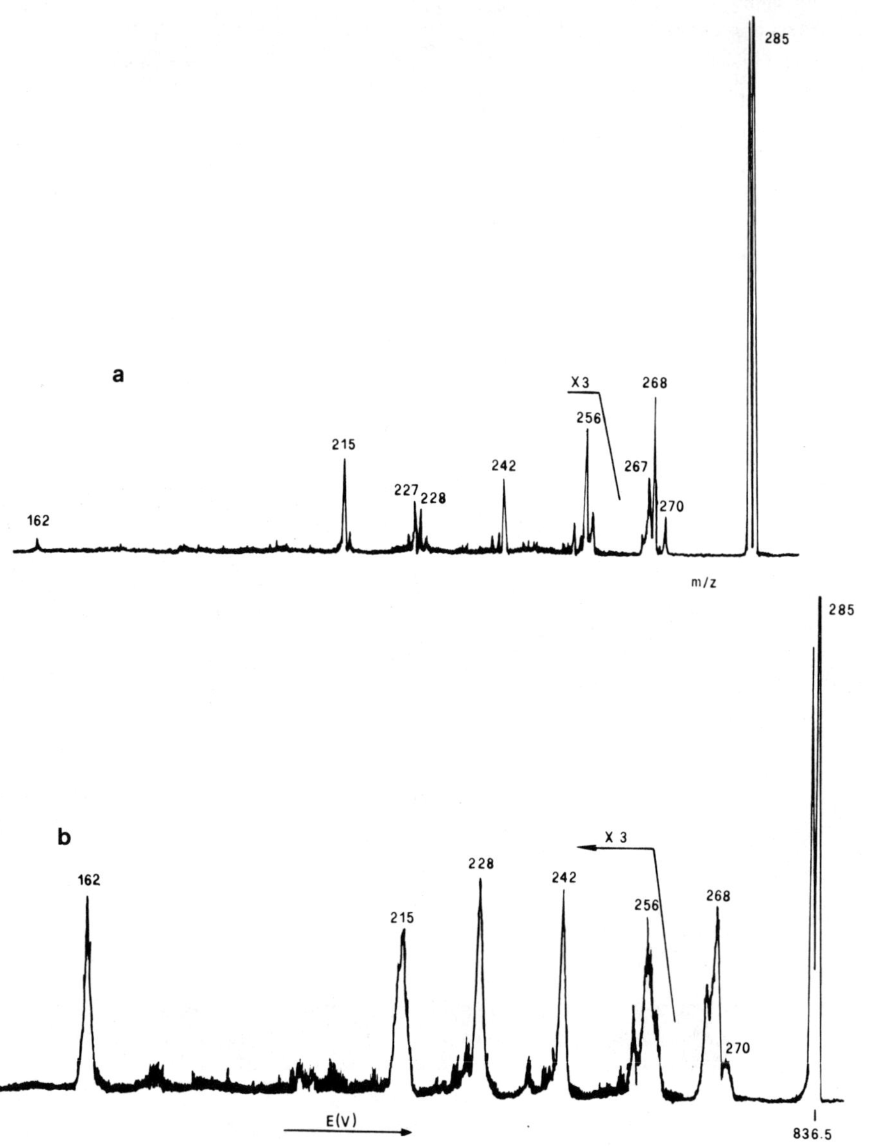

Fig. 6. CID B/E (a) and CID MIKE (b) spectra of M⁺ of morphine
(m/z 285) obtained under EI conditions (by kind
permission of Wiley and Sons, New York, Chichester).

Fig. 6 shows the CID B/E and CID MIKE spectra of M+· of morphine. Except for minor differences, they are overimposable. The proposed related collisionally activated decomposition pattern is shown in Fig. 7. The most of fragmentation processes are related to the cleavage of the nitrogen containing ring, leading to ions at m/z 256, 242, 228 and 215. Losses of CH_3·, OH· and H_2O give rise to ions at m/z 270, 268 and 267 respectively. Finally, an interesting fragmentation process leads to ions at m/z 162, which result particularly abundant in the CAD MIKE spectrum only and can be considered highly diagnostic.

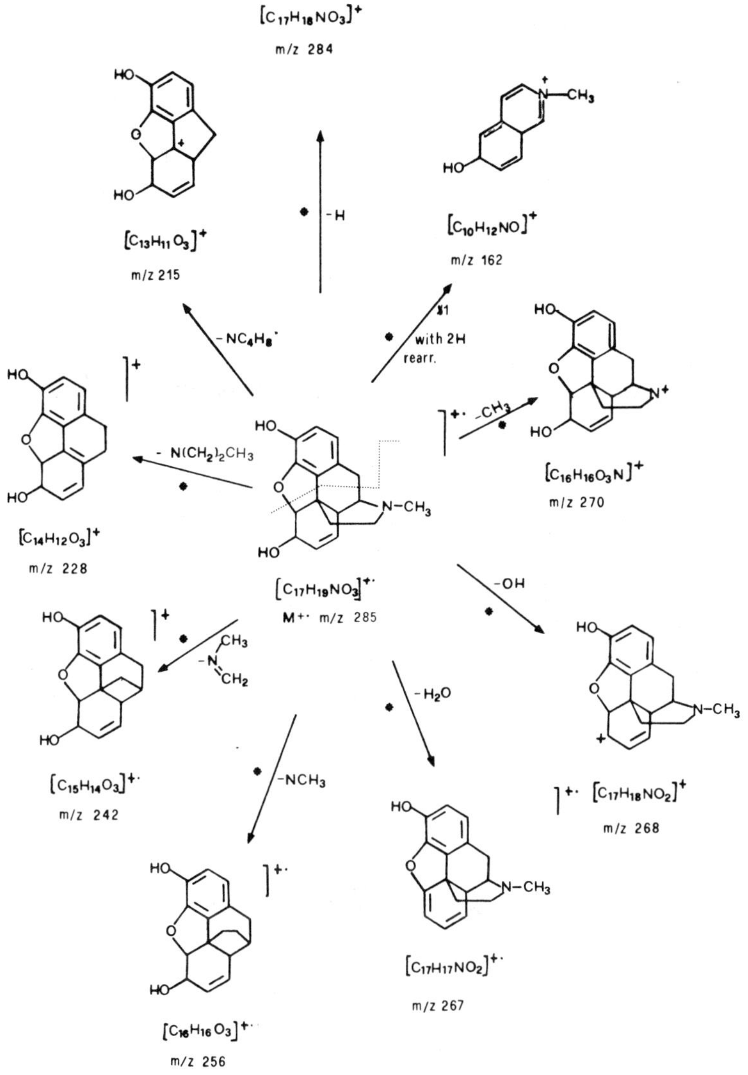

[C17H18NO3]+
m/z 284
-H
[C13H11O3]+
m/z 215
[C10H12NO]+
m/z 162
-NC4H8·
with 2H
rearr.
[C14H12O3]+
m/z 228
- N(CH2)2CH3
-CH3
[C16H16O3N]+
m/z 270
N-CH3
[C17H19NO3]+·
M+· m/z 285
-OH
[C15H14O3]+·
m/z 242
-H2O
-NCH3
[C17H18NO2]+
m/z 268
[C17H17NO2]+·
m/z 267
[C16H16O3]+·
m/z 256

Fig. 7. Proposed collisionally activated decomposition pattern of morphine (by kind permission of Wiley and Sons, New York, Chichester).

The minimum detectable amount of morphine standard, with a signal-to-noise ratio of 5/1 for the peak at m/z 162, was estimated better than 10 fg. This limit increased to 50-500 fg when morphine was dissolved in the natural matrix.

By introducing samples of blank hair into the EI ion source, very complex spectra were generated, with peaks for every mass value. Nevertheless the CID MIKE spectrum of ions at m/z 285 resulted free of collisionally induced fragments related to morphine fragmentation pattern (Fig. 8a).

On the other hand, when a sample from a positive hair was introduced into the EI source, it gave rise to a quite complex mass spectrum, but the CID MIKE spectrum of ions at m/z 285 was clearly

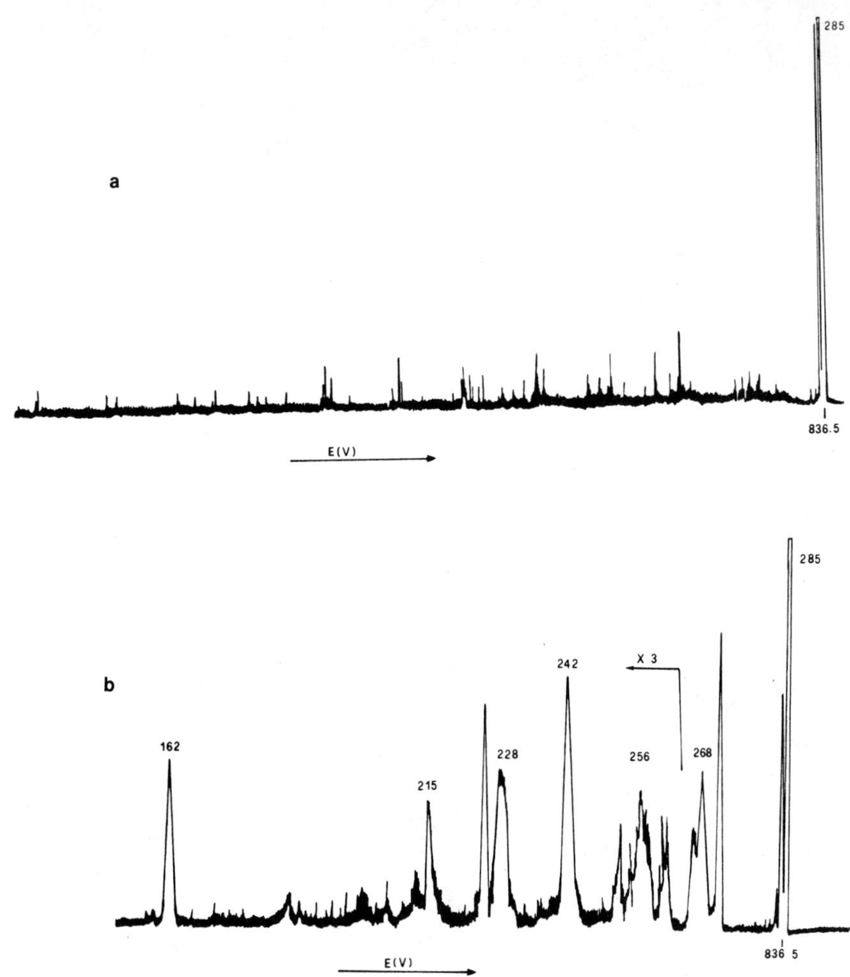

Fig. 8. CID MIKE spectra of ionic species at m/z 285 originating
from EI of: (a) 1 μg of blank hair extract; (b) 1 μg of
positive hair extract (by kind permission of Wiley
and Sons, New York, Chichester).

overimposable to that of morphine, proving the presence of such a
heroin metabolite (Fig. 8b).

In this case, quantitative determinations obtained by HPLC
gave a result of 0.3 ng of morphine/mg of hair.

Concordant results were obtained from other 6 cases, with
morphine levels in the range 0.12 to 3.3 ng/mg.

The wide analytical spectrum of collisional spectroscopy led
to other interesting results, consisting in the identification of
monoacetylmorphine and, surprisingly, heroin itself in the hair
matrix. Actually, looking at the collisional spectra of ions at
m/z 327 and 369, arising from the hair samples of heroin addicts,
they resulted mainly overimposable to those obtained by collisional spectroscopy of molecular ions of monoacetylmorphine and
heroin respectively (Fig. 9,10).

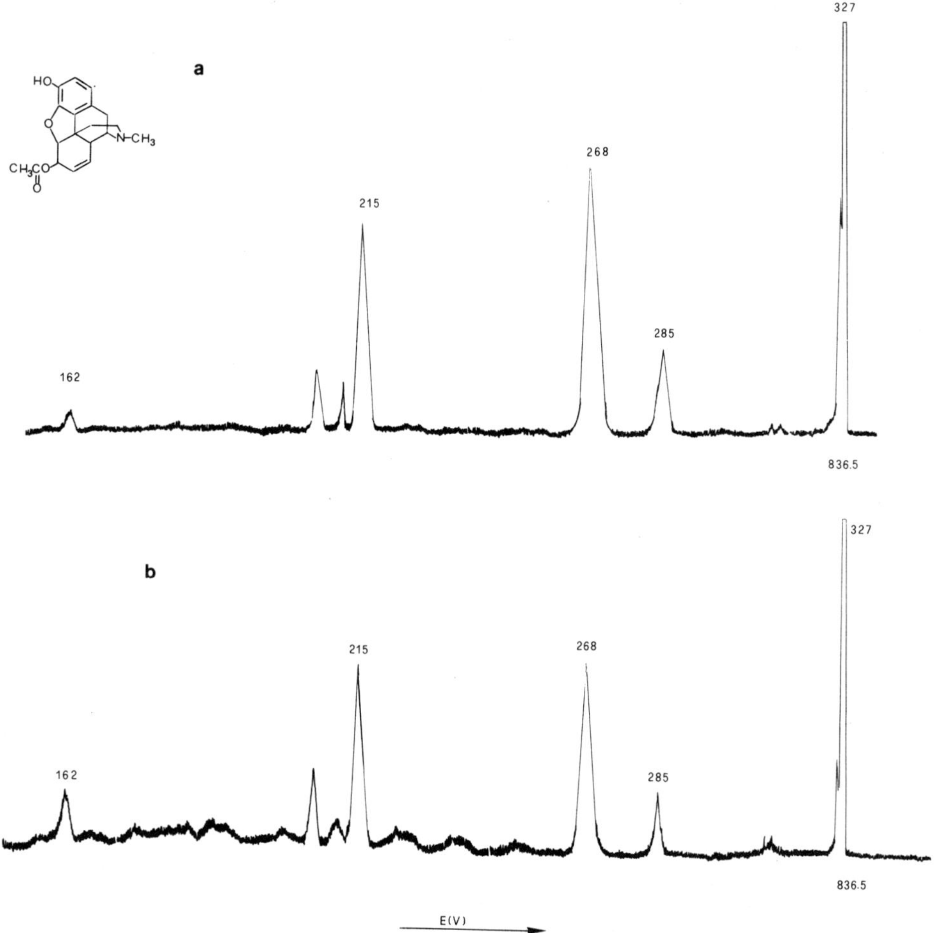

Fig. 9. CID MIKE spectra of ionic species at m/z 327 originating from EI of: (a) standard mono-acetyl-morphine and (b) hair sample from heroin addict.

If the presence of the former compound could be reasonably expected, being the first metabolite arising from heroin deacetylation, the finding of heroin in the hair matrix was a real surprise. Actually the half-life of this compound in the biological fluids is so brief that one can hardly hypothesize its straight passing from the blood into the hair in significant amounts.

This fact could be ascribed to external contamination, but the extensive washings carried out, that, as experimentally observed were able to remove the loosely bound morphine, should have avoided any risk of exogenous interferences.

As an alternative hypothesis, it could be suggested that the heroin amounts found in the hair were of endogenous origin, through a till now undemonstrated morphine re-acetylation pathway. In this case, heroin should be present also in samples from patients treated with morphine or from subjects addicted to morphine. Works are in progress to verify this point.

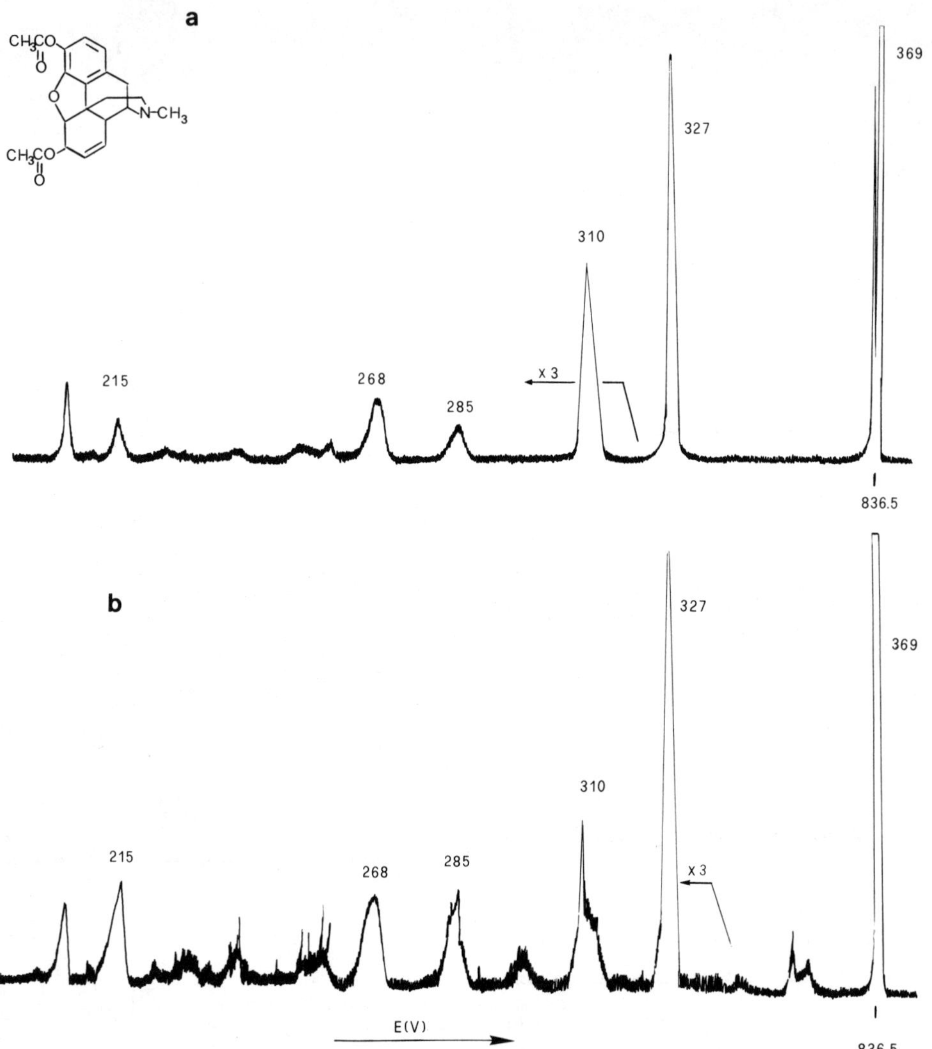

Fig. 10. CID MIKE spectra of ionic species at m/z 369 originating
from EI of: (a) standard heroin and (b) hair sample from
heroin addict.

REFERENCES

1. A.M. Baumgartner, P.F. Jones, W.A. Baumgartner and C.T. Black,
   Radioimmunoassay of hair for determining opiate-abuse
   histories J. Nucl. Med. 20:748 (1979).
2. D. Valente, M. Cassani, M. Pigliapochi and G. Vanzetti, Hair
   as the sample in assessing morphine and cocaine addiction,
   Clin. Chem. 27:1952 (1981).
3. A. Carnevale, M. Marchetti and A. Fiori, La ricerca radio-
   immunologica della morfina nei capelli, Boll. Farmacod.
   Alcoolismo 6:579 (1983).
4. K. Puschel, P. Thomasch and W. Arnold, Opiate levels in hair,
   Forens. Sci. Int. 21:181 (1983).

5. F. Black, in:"Methods of Morphine Estimation in Biological Fluids and the Concept of Free Morphine", J.F.B. Stuart, ed, The Royal Society of Medicine, London (1983).
6. I. Jane and F. Taylor, Characterization and quantitation of morphine in urine using high-pressure liquid chromatography with fluorescence detection, J. Chromatogr. 109:37 (1975).
7. J.E. Wallace, S.C. Harris and M.W. Peek, Determination of morphine by liquid chromatography with electrochemical detection, Anal. Chem. 52:1328 (1980).
8. B.L. Posey and S.N. Kimble, Simulteneous determination of codeine and morphine in urine and blood by HPLC, J. Anal. Toxicol. 7:241 (1983).
9. F. Tagliaro, A. Frigerio, R. Dorizzi, G. Lubli and M. Marigo, Liquid chromatography with pre-column dansyl derivatization and fluorimetric detection applied to the assay of morphine in biological samples, J. Chromatogr. 330:323 (1985).
10. M. Marigo, F. Tagliaro, C. Poiesi, S. Lafisca and C. Neri, Determination of morphine in the hair of heroin addicts by High Performance Liquid Chromatography with fluorimetric detection, J. Anal. Toxicol. 10:158 (1986).
11. B. Pelli, P. Traldi, F. Tagliaro, G. Lubli and M. Marigo, Collisional spectroscopy for inequivocal and rapid determination of morphine at ppb level in the hair of heroin addicts, Biomed. Mass Spectrom. 14:63 (1987).
12. R.W. Frei, W. Santi and M. Thomas, Liquid chromatography of dansyl derivatives of some alkaloids and the application to the analysis of pharmaceuticals, J. Chromatogr. 116:365 (1976).

COCAINE, BENZOYLECGONINE AND ECGONINE METHYL ESTER DETERMINATIONS
IN POST MORTEM HUMAN URINE AND BLOOD BY GAS CHROMATOGRAPHY AND
CAPILLARY GAS CHROMATOGRAPHY AFTER "EXTRELUT®" EXTRACTION

R. Froldi, V. Gambaro*, and A. Groppi**

Institute of Forensic Medicine
University of Macerata, Macerata
* University of Milan, Milan
** University of Pavia, Pavia
Italy

Blood and urine level determinations of cacaine and its
metabolites are important for the correct toxicological evaluation
of "death due to cocaine". In fact the significance of the
biological fluid concentration in fatal cases can be assessed from
human clinical studies[1-3]. On the other hand the urine and blood
concentrations provide toxicological data that may assist forensic
toxicologist and pathologist in the interpretation of the time and
also the manner of death. The concentration of cocaine and its metabolites in body tissues is less important for the death investigation from the medico legal point of view[4-9].

Benzoylecgonine (BE) and ecgonine methyl ester (EME) are
considered to be the principal metabolites of cocaine (COC) in man
and animals. BE is a nitrogenous polar, amphoteric compound highly
soluble in water containing a carboxylic functional group. It is
not efficiently extracted by simple organic solvents and gas
chromatographic assays require derivatization. EME is produced by
action of liver and blood cholinesterases and can be chromatographed directly without derivatization[3,10-12].

This report presents a method for the quantitation of COC, BE
and EME in post mortem urine and blood by gas chromatography (GLC)
and capillary gas chromatography (CGC). The procedure involves extraction at pH 9 using a column filled with "Extrelut". The blood
is previously protein denatured with acetonitrile. As internal
standards are used n-propylecgonine for EME and n-propylbenzoylecgonine for COC and BE. Chloroform:isopropanol (3:1) is used as
the extraction solvent. The conversion of BE to the isopropylbenzoylecgonine is carried out with trimethylanilinium hydroxide
(TMAOH) and 2-iodopropane. This method has been applied to post
mortem blood and urine "blank" spiked with COC, BE and EME at
concentrations of 2 and 4 µg/ml. These values were chosen according to the usual mean blood concentrations in cases of death due
to cocaine. Previously the analysis were performed by gas liquid
chromatography with a NP detector: an Apiezon L basic packed

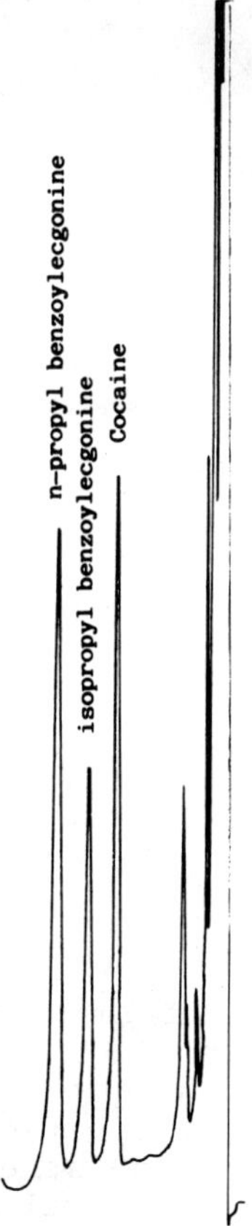

Fig. 1. Sample: spiked blood. COC and BE determinations on a
OV-1 packed column with a NP detector.

column was used to detect EME and an OV-1 packed column to detect
COC and BE (as isopropylBE). Subsequently capillary gas chromatography with a FID was used to detect COC and BE (as isopropylBE)
in splitless mode.

Recently the method has been applied also to post mortem
blood and urine in a case of death due to cocaine using a
capillary gas chromatography system with a mass spectrometric
detector (GC/MS).

EXPERIMENTAL

Materials

All chemicals were chosen for their purity and used without
further purification. Buffer solution was a pH 9 buffer prepared
with Titrisol (Merck, Darmstadt, FRG). Derivatizing reagents were
trimethylanilinium hydroxide 0.2 mol/l in methanol (Pierce,
Rockford, USA) and 2-iodopropane (Aldrich, Steinheim, FRG).
Extraction columns were obtained from a 10 ml polyethylene
hypodermic syringe filled with 2 g. of Extrelut (Merck). Extraction solvents were chloroform:isopropanol (3:1). Internal

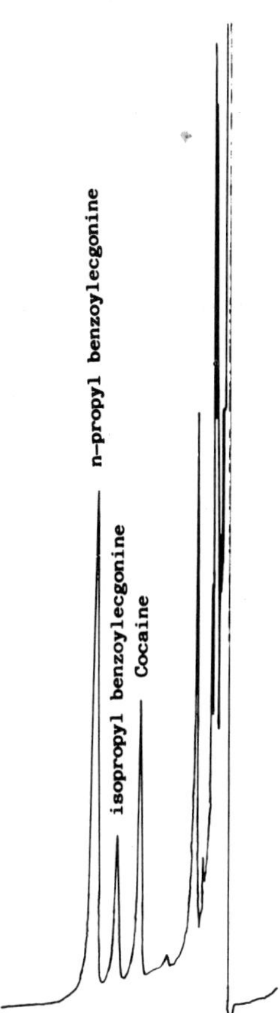

Fig. 2. Sample: spiked urine. COC and BE determinations on a
OV-1 packed column with a NP detector.

standards were n-propylbenzoylecgonine (20 ng/μl in methanol) and
n-propylecgonine (20 ng/μl in methanol).

<u>Instrumentation</u>

Gas liquid chromatography: a Perkin Elmer Sigma 3B (Norwalk,
Ct, USA) equipped with a NP Detector was used.

To detect EME the conditions were:

column: glass 2 m x 2 mm I.D.- Apiezon L 10% + 2% KOH Supelcoport
(Supelco, Bellefonte, Pa, USA) 80-100 mesh;
temperatures: oven 170°C - injector 225°C - detector 225°C;
carrier gas: nitrogen 1.50 kg/cm^2.

To detect BE and COC the conditions were:

column: glass 2 m x 2 mm I.D. OV-1 3% on Supelcoport 100-120 mesh;
temperatures: oven 230°C - injector 260°C - detector 260°C;
carrier gas: nitrogen 1.25 kg/cm^2.

Capillary gas chromatography: a Fractovap model 4161 gas chromatograph (C. Erba, Milan, Italy) equipped with a FID was used to detect COC and BE. The conditions were:

injector: splitless (split valve closed for 30 seconds);
column: SE-52 glass capillary column 15 m x 0.32 mm I.D., 0.1 μm;
oven program: 60°C - 40°C/min - 200°C - 10°C/min - 270°C;
carrier gas: hydrogen 0.3 kg/cm^2.

<u>Analytical Procedure and Results</u>

The biological fluids were added with internal standard solution. Then 1 ml urine (or 1 ml blood previously protein denatured with 1 ml acetronitrile) and 2 ml pH 9 buffer were placed into a centrifuge tube. After brief vortexing the solution was transferred onto the top of the extraction column and allowed to soak for 10 min. Three portions of 2 ml of extraction solvent were passed through the column, collected in a conical glass tube and evaporated to dryness under a stream of nitrogen, at room temperature. The residue was dissolved in 1 ml acetonitrile and

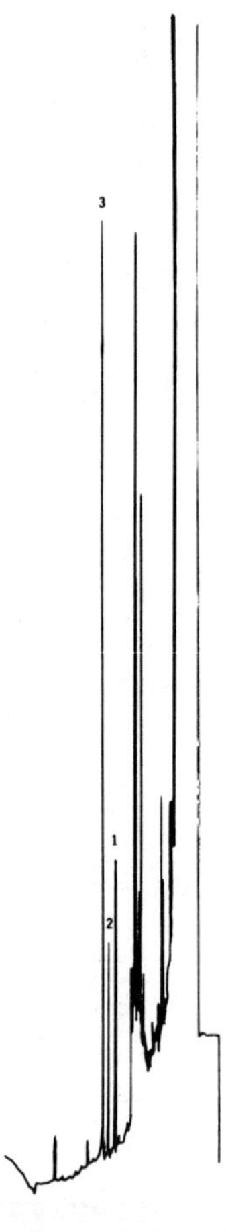

Fig. 3. Sample: spiked urine. COC and BE determinations on a
capillary column SE-52 with a FID.
1. Cocaine
2. isopropylbenzoylecgonine
3. n-propylbenzoylecgonine

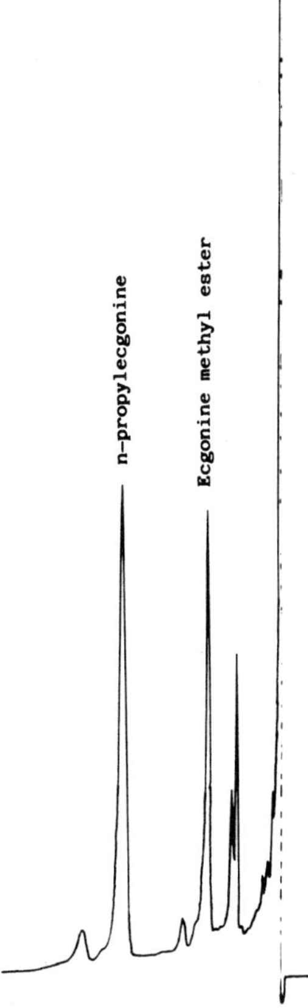

Fig. 4. Sample: spiked blood. EME determination on a APIEZON L basic packed column with a NP detector.

divided in two equal aliquots. One for determination of COC and BE and the other for determination of EME.

The first portion of 0.5 ml was transferred into a vial with 50 µl of TMAOH. After 30 s, 15 µl of 2-iodopropane was added and allowed to stand for 10 min. in an 80°C water bath. After coolingthe reaction products were extracted with 8 ml of n-hexane. After vortexing for 1 min. the n-hexane was placed in a conical tube and back extracted with 0.5 ml of 0.05 mol/l sulphuric acid. The aque-ous layer made basic with solid sodium bicarbonate was reextracted with 4 ml of n-hexane. The organic layer was concentrated until about 100 µl and 1 µl was injected into the gas chromatograph to detect COC and BE. Typical chromatograms are shown in Fig. 1-3.

The other portion of acetonitrile was diluted with 4 ml of diethyl ether and extracted with 0.5 ml of 0.05 mol/l sulphuric acid. After vortexing the aqueous layer made basic with solid sodium bicarbonate was reextracted with 4 ml of ethyl ether. The organic layer was concentrated until about 100 µl and 1 µl was injected into the gas chromatograph to detect EME. Typical chromatograms are shown in Fig. 4-5. A scheme of the proposed procedure is shown in Table 1.

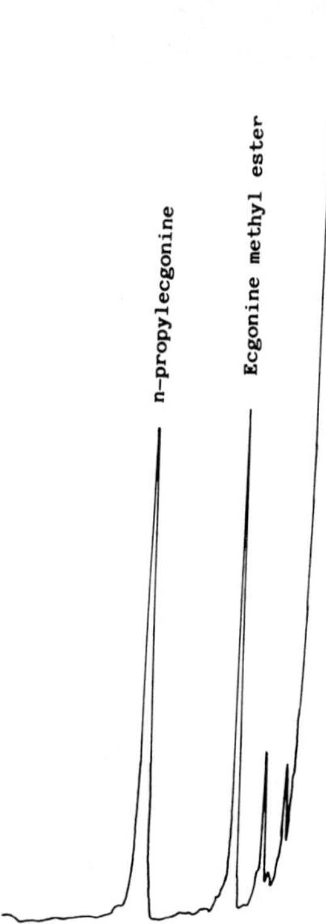

Fig. 5. Sample: spiked urine. EME determination on a APIEZON L basic packed column with a NP detector.

Table 2 lists the coefficients of variation per cent of the peak area ratios between COC and BE or EME and the internal standards for post mortem urine and blood spiked with COC, BE and EME and extracted with the proposed procedure. The C.V. was calculated on the basis of 8 determinations. R is the mean of the peak area ratios. The values were obtained by GLC and also by CGC only to detect COC and BE in urine samples (ratios of height).

A CASE OF DEATH DUE TO COCAINE

Though good results were obtained also in capillary gas chromatography, recently blood and urine samples of a case of cocaine death has been analyzed using a capillary gas chromatograph with a mass-spectrometer as the detector. GC/MS was used to detect COC, BE and also EME with the same capillary column. The instrument was a HP 5890 model equipped with a mass selective detector model 5970 (Hewlett-Packard, Palo Alto, Ca, USA).

The conditions were:

injector: splitless (split valve closed for 30 seconds);
column: fused silica cross-linked methylsilicone 12 m x 0.20 mm I.D., 0,33 µm;
oven program: 60°C - 40°C/min - 160°C - 8°C/min - 300°C;
carrier gas: hydrogen 0.3 kg/cm^2;

scan acquisition: solvent delay 4.50 min; eM volts O relative;
resulting voltage 1,400; start time 4.50 min; low mass 50.0; high
mass 400; scan threshold 20; a/d samples (2^N) 2; scan per
second 1.2.

The method was performed on 2 aliquots of 0.5 ml of blood and
2 aliquots of 0.1 ml of urine (one aliquot for COC and BE, the

Table 1. Scheme of Analytical Procedure

---

1 ml urine (1 ml blood + 1 ml acetonitrile)
+ n-propylecgonine, n-propylbenzoylecgonine
(I. S.) + 2 ml buffer pH 9

transfer onto the top of the Extrelut
column. Add chloroform:isopropanol (3:1)

evaporate to dryness. Add 1 ml CH_3CN

(COC + BE)	(EME)
0.5 ml CH_3CN + 50 µl TMAOH + 15 µl 2 iodopropane	0,5 ml CH_3CN + diethyl ether
allow to stand at 80°C for 10 min	
extract with n-hexane	extract with 0.05 mol/l H_2SO_4
extract with 0.05 mol/l H_2SO_4	make basic the aqueous layer with $NaHCO_3$ extract with diethyl ether
make basic the aqueous layer with $NaHCO_3$ extract with n-hexane	
GLC ASSAY	GLC ASSAY

Table 2. Gas Liquid Chromatography

Sample: spiked blood

Cocaine (4 µg/ml) n = 8	Benzoylecgonine (4 µg/ml) n = 8	Ecgonine methyl ester (4 µg/ml) n = 8
$R = \dfrac{A.\ COC}{A.\ I.S.} = 1.0716$	$R = \dfrac{A.\ BE}{A.\ I.S.} = 0.8148$	$R = \dfrac{A.\ EME}{A.\ I.S.} = 0.9610$
S.D. = 0.0432	S.D. = 0.1413	S.D. = 0.0696
C.V.% = 4.0	C.V.% = 17.3	C.V.% = 7.2

Sample: spiked urine

Cocaine (2 µg/ml) n = 8	Benzoylecgonine (2 µg/ml) n = 8	Ecgonine methyl ester (4 µg/ml) n = 8
$R = \dfrac{A.\ COC}{A.\ I.S.} = 0.5050$	$R = \dfrac{A.\ BE}{A.\ I.S.} = 0.3932$	$R = \dfrac{A.\ EME}{A.\ I.S.} = 0.8440$
S.D. = 0.0096	S.D. = 0.0104	S.D. = 0.0415
C.V.% = 1.9	C.V.% = 2.6	C.V.% = 4.9

Capillary Gas Chromatography

Sample: spiked urine

Cocaine (2 µg/ml) n = 8	Benzoylecgonine (2 µg/ml) n = 8
$R = \dfrac{H.\ COC}{H.\ I.S.} = 0.1351$	$R = \dfrac{H.\ BE}{H.\ I.S.} = 0.2696$
S.D. = 0.0103 C.V.% = 7.7	S.D. = 0.0033 C.V.% = 1.2

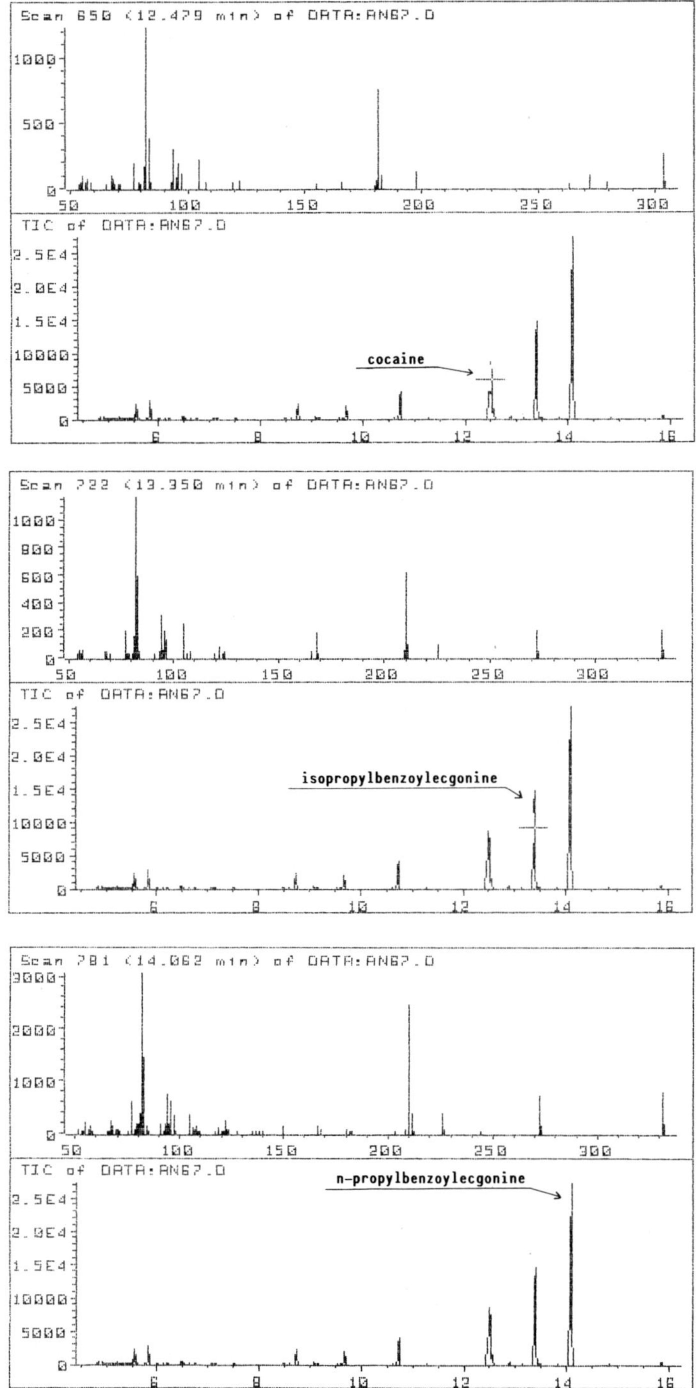

Fig. 6. A case of death due to cocaine. COC and BE determinations
in blood by GC/MS.

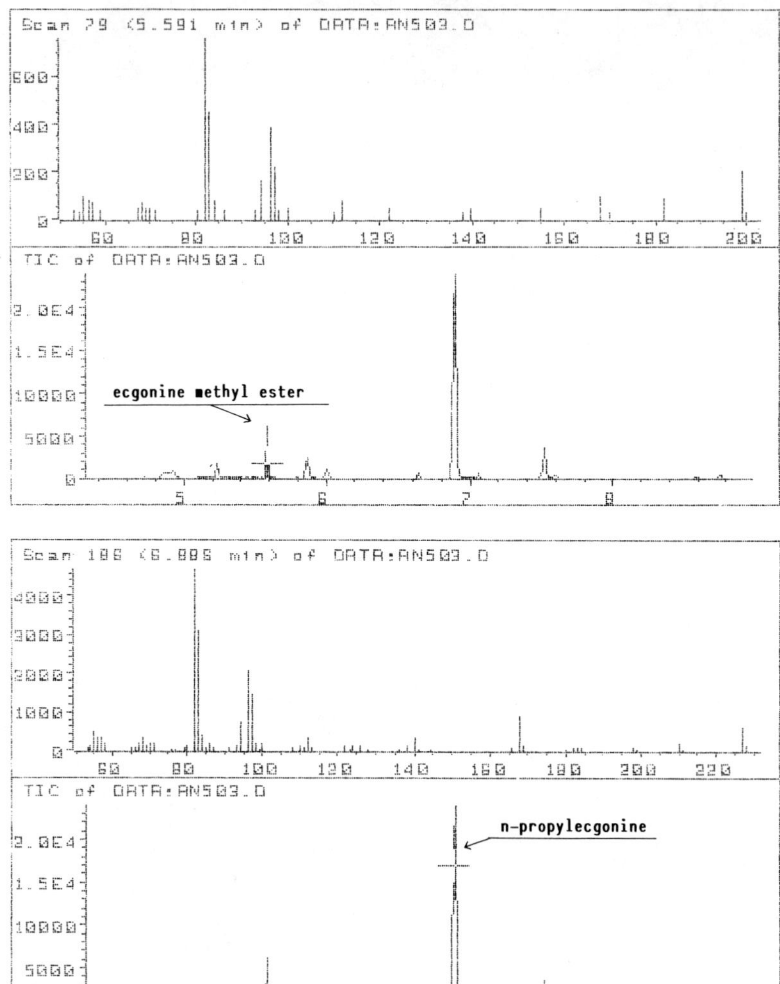

Fig. 7. A case of death due to cocaine. EME determination in blood by GC/MS.

other for EME) (Fig. 6-9) which were spiked separately with suitable standards. The results are shown in Table 3.

CONCLUSION

The method described is both simple and rapid especially with regard to extraction. In fact the elution through the "Extrelut" columns is sufficient to complete the procedure. Since the proposed method has been studied to detect cocaine and its metabolites in post mortem biological fluids, the protein denaturation with acetonitrile is very important for recovery of cocaine and its metabolites in post mortem blood and allowing to use "Extrelut". The procedure could be improved, especially regarding to the instrumental step. Actually we believe that reliable determinations of cocaine, benzoylecgonine and ecgonine methyl

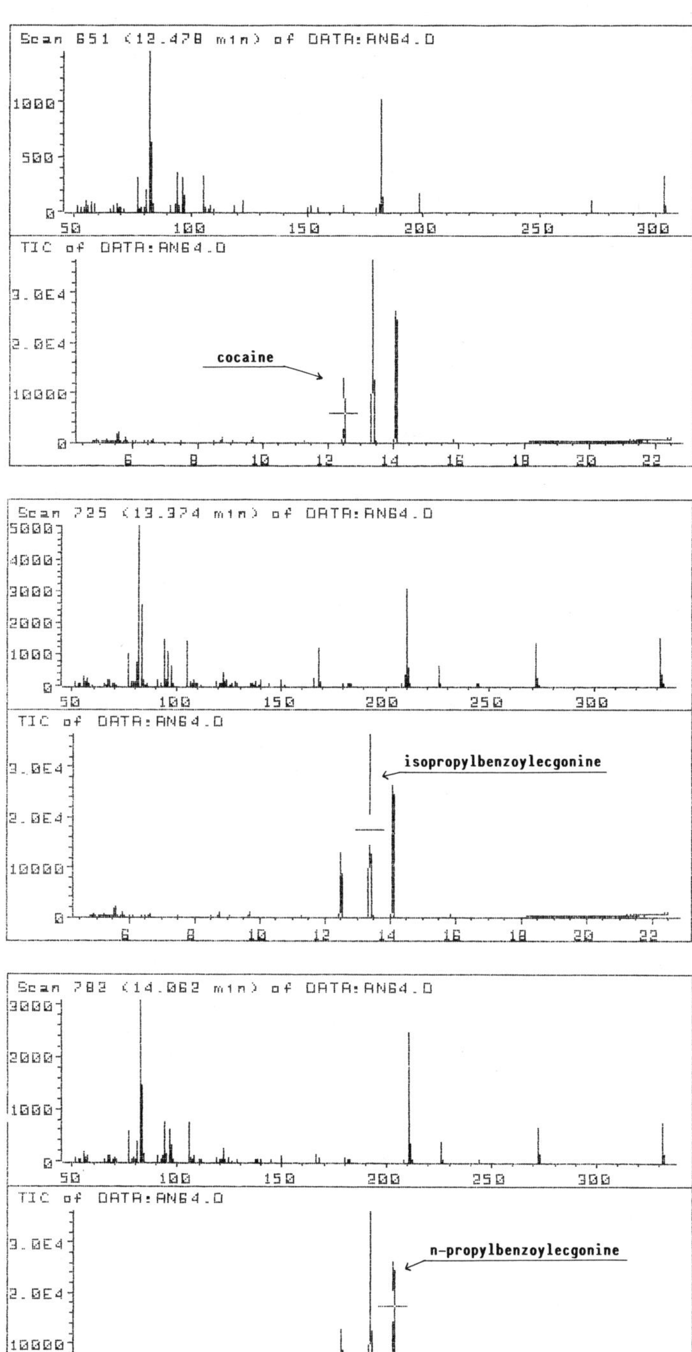

Fig. 8. A case of death due to cocaine. COC and BE determinations in urine by GC/MS.

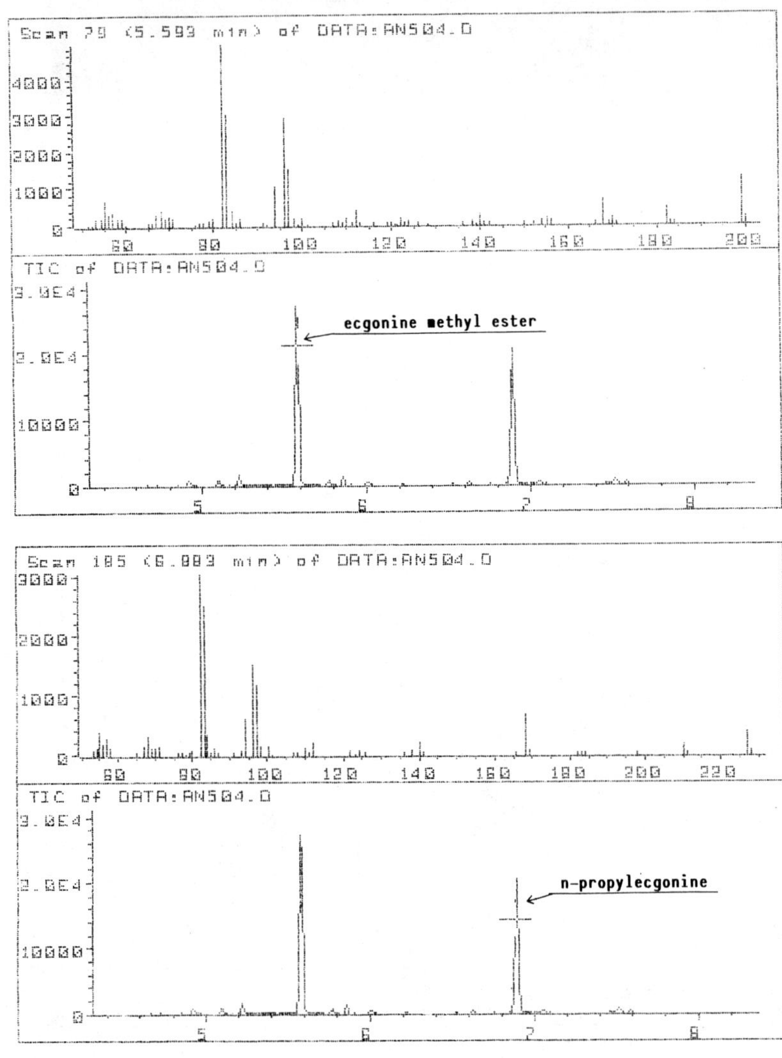

Fig. 9. A case of death due to cocaine. EME determination in urine by GC/MS.

Table 3. Cocaine (COC), Benzoylecgonine (BE) and Ecgonine Methyl Ester (EME) Concentrations (μg/ml) in a Case of Death caused by Cocaine

	COC	BE	EME
Blood	3.0	11.3	23.5
Urine	113.0	863.0	325.0

ester could be obtained through capillary gas chromatography with
NP detector, using the same capillary column.

REFERENCES

1. J.E. Wallace, H.E. Hamilton, D.E. King, D.J. Bason, H.A.
   Schwertner and S.C. Harris, Gas liquid chromatographic
   determination of cocaine and benzoylecgonine in urine, Anal.
   Chem. 48:34 (1976).
2. H.E. Hamilton, J.E. Wallace, E.L. Shimek, P. Land, S.C. Harris
   and J.G. Christenson, Cocaine and benzoylecgonine excretion in
   humans, J. Forens. Sci. 22:697 (1977).
3. M.J. Kogan, K.G. Verebey, A.C. De Pace, R.B. Resnik and
   S.J. Mùle, Quantitative determination of benzoylecgonine and
   cocaine in human biofluids by gas liquid chromatography,
   Anal. Chem. 49:1965 (1977).
4. B.S. Finkle and K.L. McCloskey, The forensic toxicology of
   cocaine (1971-1976), J. Forens. Sci. 23:173 (1978).
5. J.C. Valentour, V. Aggarwal, M.P. McGee and S.W. Goza, Cocaine
   and benzoylecgonine determinations in postmortem samples by
   gas chromatography, J. Anal. Toxicol. 2:134 (1978).
6. H.H. McCurdy, Quantitation of cocaine and benzoylecgonine
   after Jetube extraction and derivatization, J. Anal. Toxicol.
   4:82 (1980).
7. J.E. Lindgren, Guide to the analysis of cocaine and its
   metabolites in biological material, J. Ethnoph. 3:337 (1981).
8. C.V. Wetli and R.E. Mittleman, The "body packer syndrome".
   Toxicity following ingestion of illicit drugs packaged for
   transportation, J. Forens. Sci. 26:492 (1981).
9. R.E. Mittleman and C.V. Wetli, Death caused by recreational
   cocaine use, JAMA 252:1889 (1984).
10. J.J. Ambre, T. Ruo, G.R. Smith, D. Backes and C.M. Smit,
    Ecgonine methyl ester, a major metabolite of cocaine, J. Anal.
    Toxicol. 6:26 (1982).
11. J.J. Ambre, M. Fischman and T. Ruo, Urinary excretion of
    ecgonine methyl ester, a major metabolite of cocaine in
    humans, J. Anal. Toxicol. 8:23 (1984).
12. K. Matsubara, M. Kagawa and Y. Fukui, In vivo and in vitro
    studies on cocaine metabolism: ecgonine methyl ester as a
    major metabolite of cocaine, Forens. Sci. Int. 26:169 (1984).

A RAPID AND SENSITIVE HPLC METHOD FOR DETERMINATION OF ALPHA

AMANITIN IN URINE

R. Fenoil, R. Alfieri, and G. Weisz

Laboratory of Clinical Chemistry
"Martini Nuovo" Hospital - USL 1-23
Turin, Italy

INTRODUCTION

Amatoxins contained in Amanita Phalloides and other Amanita species are responsible for almost all fatal human intoxications by mushrooms that occur in Italy and in most of the European countries (about 200 cases in northern Italy every year, 10-15% of mortality)[1].

The work of Wieland and Faulstich[2] showed that these toxic compounds are chemically classified as bicyclooctapeptides with M.W. of 900 Daltons and named alpha, beta, gamma and epsilon amanitin, the first and the second being in larger quantity in the mushrooms. The mechanism of molecular action studied by Fiume and Wieland[3] consists in a strong inhibiting effect on the DNA dependent RNA polymerase B in the nuclei of eucaryotic cells and their consequent necrosys. The latecy period of the gastrointestinal symptoms (about 6 to 12 hours) makes the toxin removal more difficult; after this period serum amatoxins concentration is generally very low, as the toxins have been already fixed in the tissues (expecially in liver and bowel). It is so advisable an immediate and heavy removing therapy (forced diuresis, hemoperfusion), to be therefore avoided in not ascertained cases of intoxications. The aim of a confirmation, even after the start of therapy, has developed quick and sensitive RIA techniques for serum and urine[4,5]. The use of a RIA method by our group revealed however a few problems of no correlation between clinical and analytical data, so we decided to develope an HPLC technique which can be used as a confirmation and comparison with RIA results.

Since in acute intoxications urinary α-amanitin concentrations can reach values up to 100 ng/ml, with a mean concentration level 10 times higher than in plasma[6], we developed a relatively simple and rapid procedure to purify and concentrate urine before the HPLC analysis.

MATERIALS AND METHODS

Standards and reagents were obtained from commercial sources:
α-amanitin was purchased from Boehringer Biochemia Robin (Milan,
Italy); methanol, acetic acid, dichloro and trichloromethane, all
HPLC grade, from Merck (Darmstadt, FRG). Disposable columns packed
with silica RP C-18 (3 ml)(Baker, Deventer, Holland) were used for
the first purification step, discarding those lots with an
α-amanitin recovery lower than 85%. For the second purification
step we used disposable columns packed with silica gel (Baker,
3 ml).

The isocratic liquid chromatograph was composed of the
following units: a Model 112 solvent delivery pump (Beckman, Palo
Alto, Ca, USA) and a Model 210 sample injector (Beckman) equipped
with a 20 µl loop. Detection was by optical density with a Model
160 photometer (Beckman) equipped with a mercury lamp and a 313 nm
filter. Separations were achieved on a Hibar Lichrospher column
(Merck), 125 x 4 mm, filled with C-18 bonded silica particles
(5 µm). The mobile phase used for the assay of urines was a
mixture of methanol:water (25:75), with a flow rate of 1.1 ml/min.
Graphs were obtained with an attenuation setting corresponding to
0.01 AUFS at 5 mV input.

The purification and concentration of urine specimens were
obtained as follows: 5 ml of centrifuged urine were concentrated
on silica RP C-18 disposable columns which were washed with 3x1 ml
of distilled water and eluted with 3 x 400 µl of a mixture of
methanol:water (40:60). The pooled methanolic phases were evap-

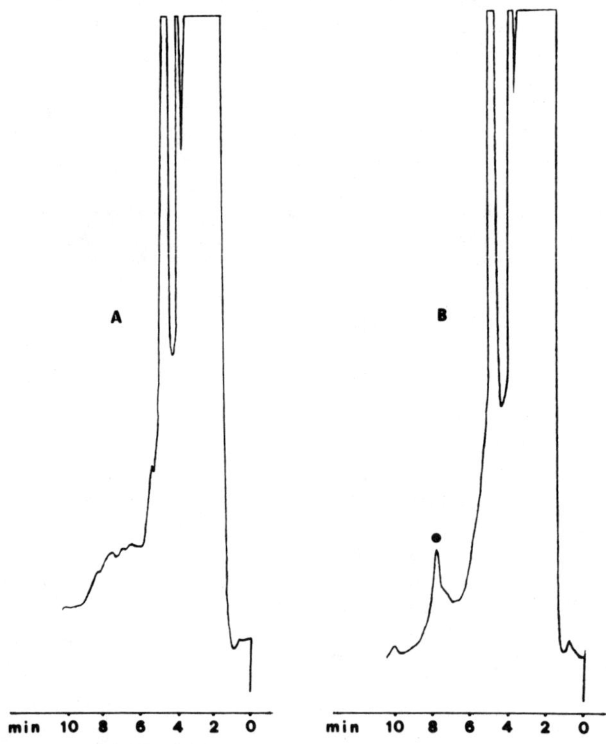

Fig. 1. HPLC chromatograms of normal human urine (A) and normal
human urine spiked (B) with 20 ng/ml of α-amanitin (●).

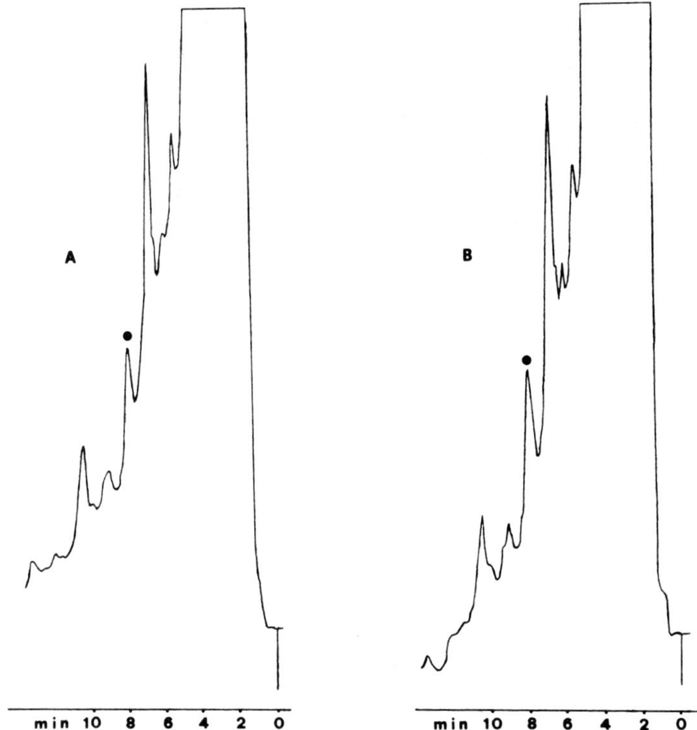

Fig. 2. HPLC chromatograms of urine (A) of a patient suspected of Amanita Phalloides poisoning (confirmed by RIA) and the same urine spiked (B) with 20 ng/ml of amanitin.

orated under stream of N_2 at 50°-60°C and the residues, dissolved with 200 µl of a mixture of trichloromethane:methanol:acetic acid (80:50:4.5), were purified on disposable columns packed with silica gel and preconditioned with 4 ml of the same solvent. The columns were washed with 2 ml of the same solvent followed by 0.5 ml of a mixture of dichloromethane:methanol (70:30), dried by suction and finally eluted with 2 x 1.2 ml of a mixture of dichlororomethane:methanol (50:50). The pooled eluates, dried under stream of N_2 at 50°-60°C, were taken up in 100 µl of the HPLC mobile phase and analyzed by HPLC.

RESULTS

The sensibility obtained spiking normal urine with α-amanitin (Fig. 1) is about 10 ng/ml which allows the detection of acute intoxication over a period of 48-72 h and is 10 times better than other HPLC methods presented for α-amanitin determination in urine[7] while the overall purification step takes no more than 3 hours, which makes this method suitable even for emergency analysis in those toxicological laboratories without beta or gamma counters.

The recovery was measured on 12 replicates and the mean obtained was of 67% with a standard deviation of ± 5.9, while linearity was tested up to 150 ng/ml.

The normal urines were analyzed without any false positive result and we are now comparing the RIA and HPLC data for patients surely intoxicated with Amanita Phalloides (Fig. 2).

Serum analysis can be accomplished by this method omitting the second step on silica gel column: in this case the sensibility is lowered to 5 ng/ml owing to the better recovery, mantaining the same HPLC conditions.

REFERENCES

1. D. Costantino, Amanita Phalloides related nephropaty, Control. Nephrol. 10:84 (1978).
2. T. Wieland and H. Faulstich, Amatoxins, phallotoxins, phallolysins and antamide, the biologically active components of poisonous Amanita mushrooms, Crit. Rev. Biochem. 5:195 (1978).
3. L. Fiume and T. Wieland, Amanitins chemistry and action, Febs. Lett. 8:1 (1970).
4. H. Faulstich, S. Zobeley and H. Trischmann, A rapid immunoassay, using a nylon support, for amatoxins from Amanita mushrooms, Toxicon 20:913 (1982).
5. R. Andres, W. Frei, K. Gautschi and D.J. Von der Schmitt, Radioimmunoassay for amatoxins by a rapid ^{125}I-based system, Clin. Chem. 9:1751 (1986).
6. C. Busi, L. Fiume and D. Costantino, Radioimmunoassay of amatoxins, in:"Amanita Toxins and Poisoning", H. Faulstich, B. Kommerel and T. Wieland, eds., Verlag Gerhard Witzstrock, Baden-Baden (1980).
7. F. Jehl, C. Gallion, P. Birckel, A. Jaeger, F. Flesch and R. Mink, Determination of alpha amanitin and beta amanitin in human biological fluids by high-performance liquid chromatography, Anal. Biochem. 149:35 (1985).

FATAL KETAMINE ABUSE: REPORT OF A CASE AND ANALYTICAL

DETERMINATION BY GAS LIQUID CHROMATOGRAPHY/MASS SPECTROMETRY

F. Centini, M. Gabbrielli, V. Fineschi,
and I. Barni Comparini

Institute of Forensic Medicine
University of Siena
Siena, Italy

INTRODUCTION

Ketamine (2-o-chlorophenyl - 2 methylaminocyclohexanone)
is the only drug capable of producing a condition known by
anaesthesiologists as "dissociative anaesthesia" with amnesia and
deep analgesia[1].

According to Saidman[2], such properties are due to the
mechanism of the action of the drug, which presents an elective
affinity for some opiate receptors. An interaction between
Ketamine and opiate receptors was recently reported by Bion[3,4] and
De Simoni[5].

The analgesic effect occurs very rapidly after the intravenous injection and lasts for a relatively short period of time.
After the anaesthesia period, effects on the central nervous
system remain for a very short time, such as delirium, marked
excitation, hallucinations and dysphasia[6]. However, no risk of
permanent damage for mental integrity occurs[7].

Ketamine biotransformation occurs in the drug metabolizing
system of hepatic endoplasmic reticulum[8]. The first step consists
in demethylation to norKetamine which is then hydroxilated in two
different positions of the molecule to form other metabolites.
These can undergo conjugation and be excreted as such or, more
likely, after being converted to other products by dehydration.
Notwithstanding all psychic effect, no cases of Ketamine poisoning
or dependence have been reported up to now.

Thus it seems of interest to report a recent case of Ketamine
dependency in a subject that died from an overdose.

CASE REPORT

In August 1985, a 34 year old, hospital attendant was found
dead in an advanced state of decomposition at home where he lived
alone. He had a syringe still inserted in his vein. Fifty two

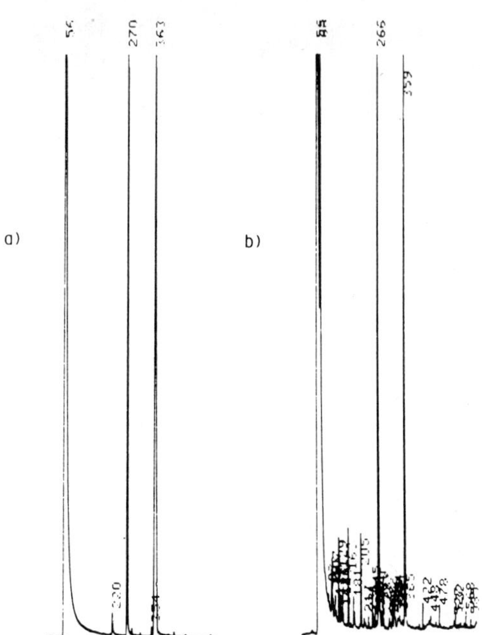

Fig. 1. a) Gas-chromatogram of ketamine standard.
b) Gas-chromatogram of kidney extract.

bottles of Ketalar (Parke-Davis 50 mg/ml) were found near the bed
or elsewhere in the house. The cadaver was covered in maggotts.
The autopsy did not show specific pathological alterations. Only
blood infiltrations in the elbow fold were evident, demostrating
numerous acupunctures. The histological examinations showed
diffused pulmonary edema, even if the cadaver was in an advanced
state of decomposition.

ANALYSIS

Because of the advanced state of decomposition it was not
possible to draw blood, urine and bile. The analyses were
therefore accomplished on tissues and were carried out by means of
gas-chromatography (GC) and GC/mass spectrometry (MS).

Samples (10 g) of liver, kidney, spleen, lung and brain were
homogenized in a borate buffer. The homogenate was extracted with
dichloromethane by shaking in a mechanical shaker for 30 mins.
After phase separation by centrifugation, the extract was purified
by the addition of an acid solution. The solvent was evaporated
under vacuum and the residue was redissolved in a small amount of
methanol. Quantitative analysis was performed by means of high
resolution gas-chromatography (HRGC). A 4160 gas-chromatograph
(C.Erba, Milan, Italy), equipped with a capillary column, was
used. The stationary phase was OV1 (film 0.40-0.45 μm).

The same samples were also analyzed by GC/MS. Since only
Ketamine could be detected, the samples were derivated to detect
possible Ketamine metabolites. Derivatisation was performed with
heptafluorobutyric anhydride in the presence of pyridine at 80°C

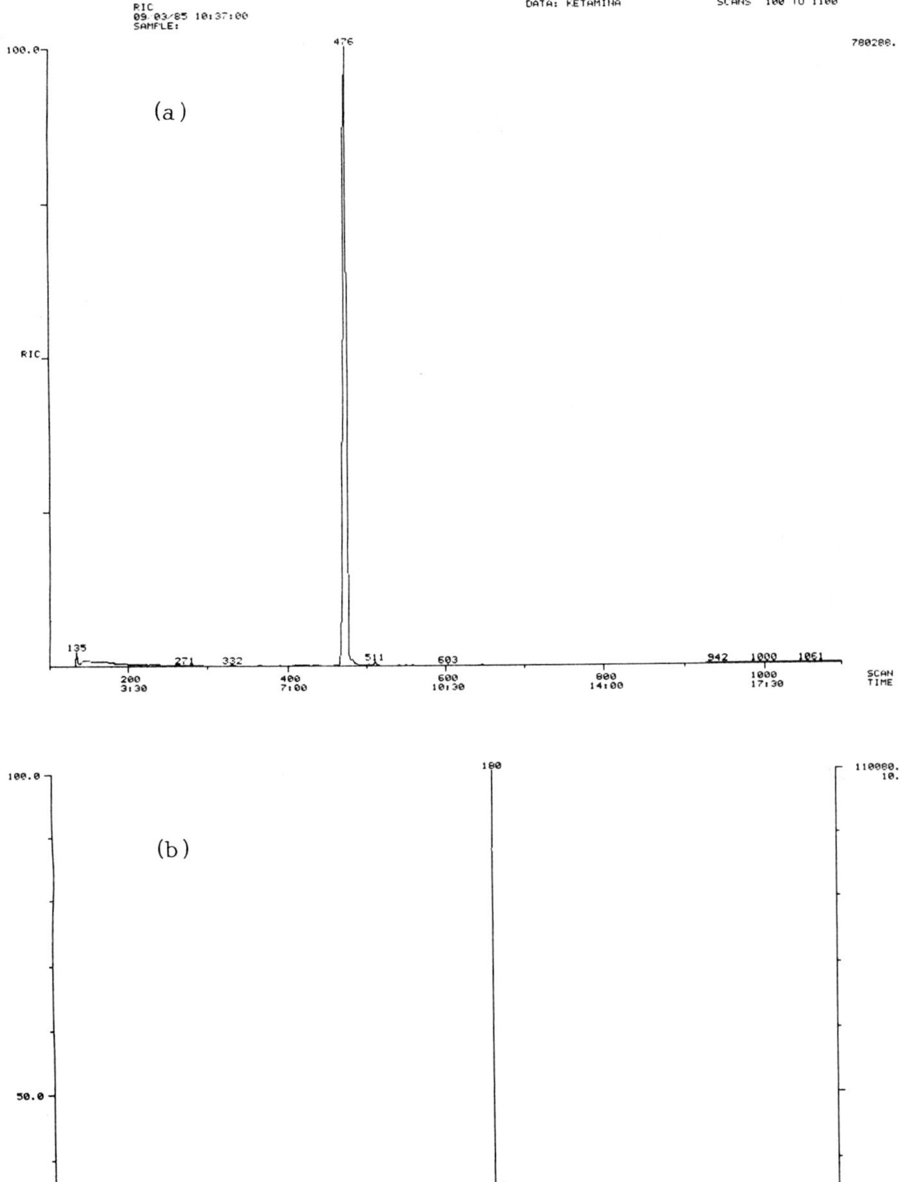

Fig. 2. a) Total ion chromatogram of ketamine standard.
b) Mass spectrum of ketamine standard.

for 2 h. The solvents were evaporated under N2 and the residue was
dissolved with 0.1 ml of dichloromethane. The samples were
analyzed again by GC/MS with a Finnigan 1020 GC/MS (Finnigan, San
Jose, Ca, USA) system operated in the electron impact mode. Fused
silica capillary columns OV1 (25 m) were used.

RESULTS

    Ketamine was detected in all the tissues examined. Fig. 1
shows the analysis of the kidney extract. As can be seen, Ketamine
was present in a relatively high level. Fig. 2 shows the total ion

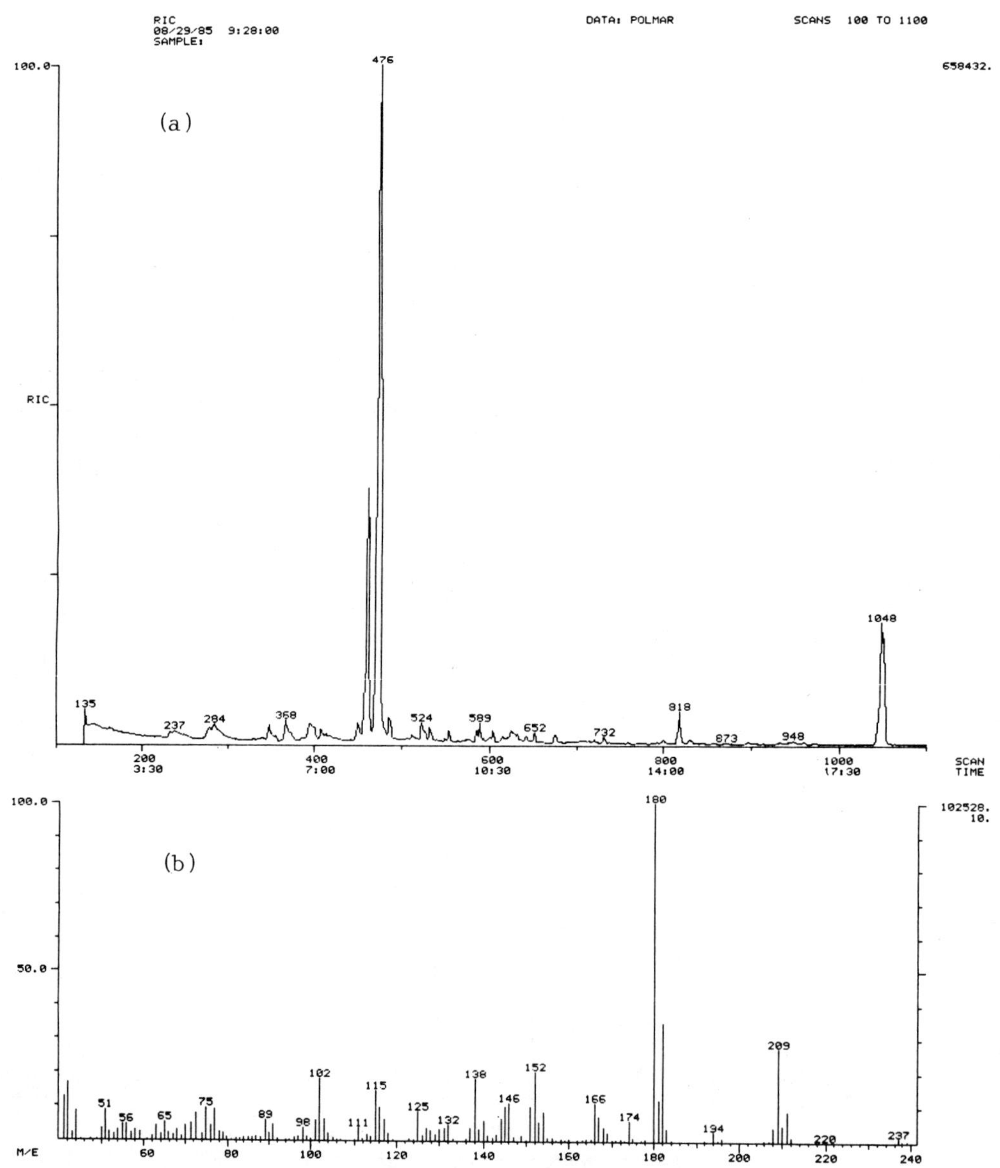

Fig. 3. a) Total ion chromatogram of lung extract.
        b) Mass spectrum of ketamine.

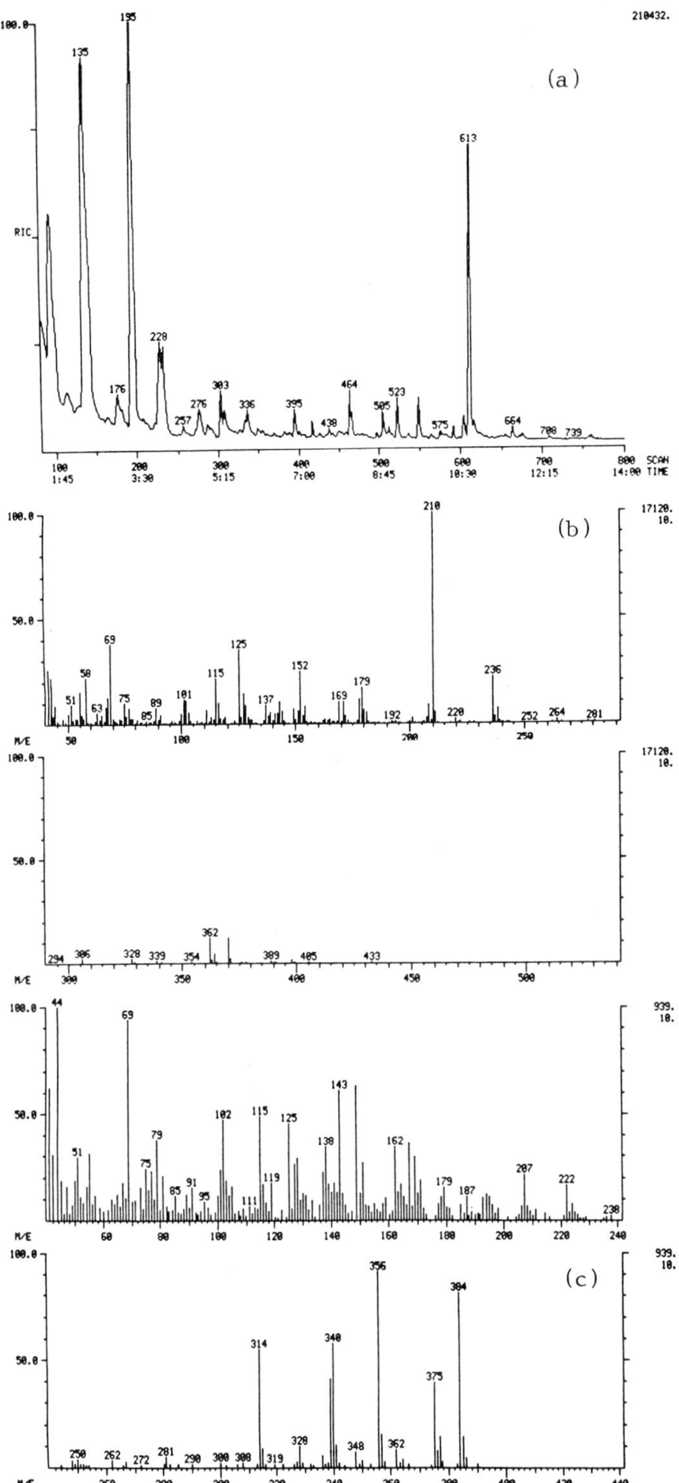

Fig. 4. a) Total ion chromatogram of liver extract after
derivatization.
b) Mass spectrum of HFBA derivative of ketamine.
c) Mass spectrum of HFBA derivative of nor-ketamine.

Table 1.   Quantitative Determinations of Ketamine and norKetamine

	GLC Ketamine	GC/MS Ketamine	norKetamine
Liver	11,8 mg/Kg	12.2 mg/Kg	0.15 mg/Kg
Brain	3.5 mg/Kg	3.7 mg/Kg	–
Lung	15.6 mg/Kg	16.4 mg/Kg	0.17 mg/Kg
Kidney	15.3 mg/Kg	15.5 mg/Kg	0.70 mg/Kg
Spleen	13.6 mg/Kg	14.1 mg/Kg	0.12 mg/Kg

chromatogram of standard Ketamine (a) as well as the pertinent mass spectrum (b). Fig. 3 shows the total ion chromatogram of the lung extract (a) and the mass spectrum of the peak corresponding to Ketamine (b). Fig. 4 shows the total ion chromatogram of liver extract after derivatisation with heptafluorobutyric anhydride (a) and the mass spectra of the peaks corresponding to either Ketamine (b) or norKetamine (c).

The quantitative estimations of Ketamine and norKetamine obtained by either GC or GC/MS (after derivatisation) are reported in Table 1.

As can be seen, the amount of Ketamine determined in the variuos tissues by the two analyses show a strict correspondence.

The high levels of Ketamine found in the tissues indicate that the subject died after a Ketamine overdose. No other drugs were found. The numerous empty tiny bottles of Ketalar found in the house and the numerous acupunctures in the elbow fold demonstrated that the hospital attendant made use of Ketamine as a narcotic drug.

It is certainly a case of Ketamine dependence, a drug which can be easily found by people who attend, for work or study reasons, hospitals because this drug is not watched over in our country.

It is clear that Ketamine is capable of producing, even through more attenuate and short, effect on the central nervous system as phencyclidine (PCP).

REFERENCES

1.   T.N. Calvey and N.E. Williams, "Principles and practice of Pharmacology for Anaesthesists", Blackwell Scientific Publications (1982).
2.   L.J. Saidman, Opiate receptor mediator of ketamine analgesia, in: "The Year Book of Anaesthesia 1983", R.D. Miller, R.R. Kirby, G.W. Ostheimer, L.J. Saidman and R.K. Stoelting,eds., Year Book Medical Publishers, Chicago 25:26 (1982).
3.   J.F. Bion, Infusion analgesia for acute war injuries. A comparison of pentazocine and Ketamine, Anaesthesia 39:560 (1984).
4.   J.F. Bion, Ketamine, Anaesthesia 40:306 (1985).

5. M.G. De Simoni, R. Giglio, I. Comporti and S. Algeri, Cross
    tolerance between Ketamine and morphine to some
    pharmacological and biochemical effects, Neuropharmacology
    24:541 (1985).
6. P.F. White, W.L. Way and A.J. Trevor, Ketamine-Its
    Pharmacology and Therapeutic uses, Anaesthesiology
    56:119 (1982).
7. R.J. Moretti, S.Z. Hassan, L.I. Goodman and H.Y. Meltzer,
    Comparison of Ketamine and Thiopental in healty volunteers:
    effects on mental status, mood and personality, Anaesthesia
    and Analgesia 63:1087 (1984).
8. K.J. Denlinger, Prolonged emergence and failure to regain
    conosciousness, in:"Complications in Anaesthesiology", F.K.
    Orkin, L.H. Cooperman ,eds., Lippincott Co., Philadelphia
    (1983).

HEAD-SPACE AND GAS CHROMATOGRAPHY-MASS SPECTROMETRY TECHNIQUES

IN THE DIAGNOSIS OF LETHAL POISONING WITH ETHYL ESTER OF

FLUOROACETIC ACID

L. Ròzanski and R. Wachowiak*

Institute of Plant Protection
* Academy of Medicine, Department of Forensic Science
Poznan, Poland

INTRODUCTION

While working in a laboratory of the Chemical Faculty of the University of Poznan, a fifth-year female student was lethally poisoned due to the inhalation of vapours and skin absorption of the ethyl ester of monofluoroacetic acid.

This compound was originally meant to be used as a substrate for the synthesis of uracyls. It was obtained during the reaction of substitution of chlorine atom in the ethyl ester of chloro-acetic acid, with the fluorine atom. As a result of incidental spilling of about 100 ml of the reaction mixture, the absorption of the components of this mixture to organism through the respiratory system and the skin took place. As was found later, this reaction flask mixture contained almost equal amounts of the substrate (ethyl ester of chloroacetic acid) and the product (ethyl ester of fluoroacetic acid).

THEORETICAL

The ethyl fluoroacetate is a highly toxic substance, which was well known before WW II as a battle gas. The mode of action of this ester in organisms of warm-blooded animals, as proposed by Peters[1], depends on the blocking the biotransformation process of tricarboxylic aliphatic acids. During the process the citric acid is transformed into the fluorocitric acid as a result of the action of the ethyl ester of fluoroacetic acid. This leads to the inhibition of aconitase activity, and to the blocking of enzymatic Krebs cycle[2], as shown in Fig. 1.

The fatal buildup of the citric acid in the tissues leads to the symptoms which culminate in violent convulsions and death from cardiac failure or respiratory arrest. The time span from poisoning to death normally amounts to about 18 hours.

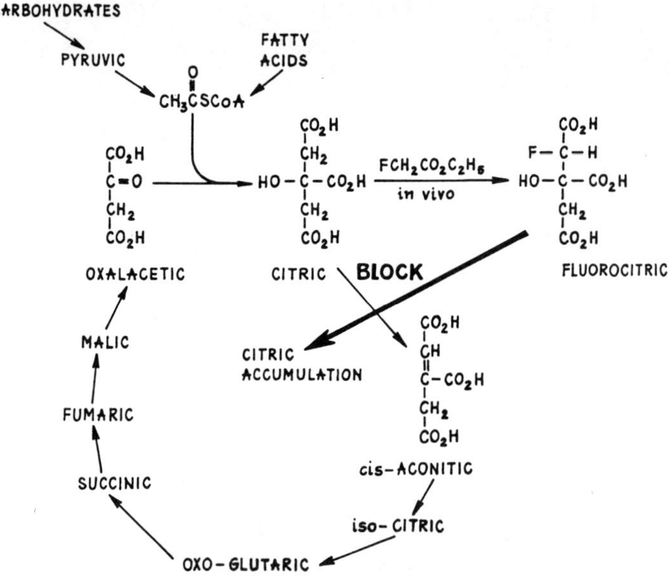

Fig. 1. Mode of action of ethyl ester of monofluoroacetic acid
(according to Griblle, 1972).

The analysis aimed at finding the toxic ester in blood and
pulmonary fluid and was carried out seven days after death.
Because both esters are volatile compounds, we could employ the
head-space technique of analysis.

EXPERIMENTAL

GC/MS Analysis Conditions

The Hewlett-Packard GC/MS 5992 B System equipped with
membrane separator and 6-feet packing column filled with 3% OV-101
on Chromosorb W HP 100-120 mesh, were used. Other conditions of
analysis were: carrier gas - Helium, rate 30 ml/min, temperatures:
injector port - 180°C, oven - 60°C.

1 to 2 ml volumes of vapour liberated from the reaction
mixture and blood samples during their heating to about 70°C in

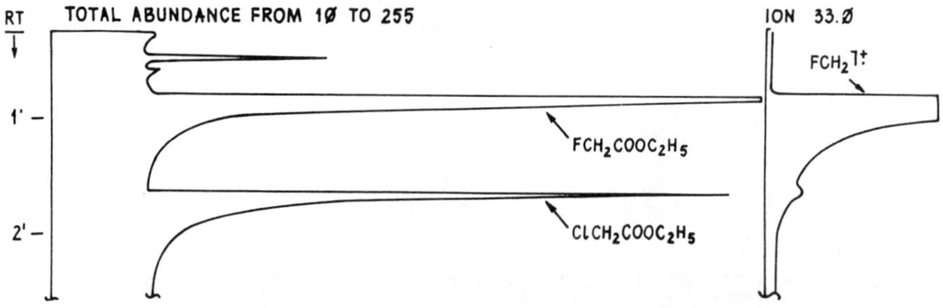

Fig. 2. GC/MS chromatogram of reaction flask mixture.

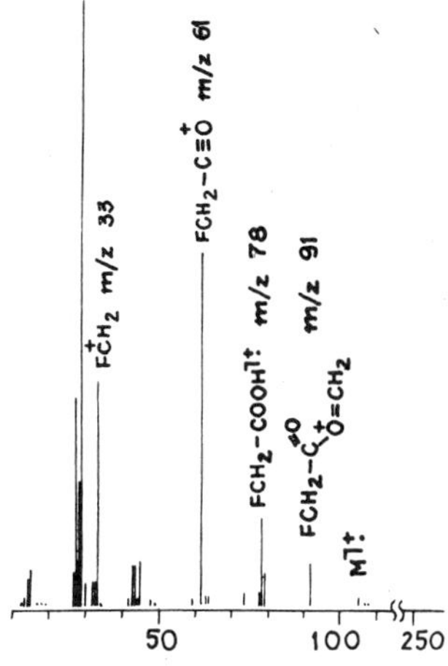

Fig. 3. Mass spectrum of ethyl fluoroacetate.

the sealed reaction vials, were injected into the GC/MS system.

Reaction Flask Mixture Analysis

The total ions chromatogram registered in the above conditions for the vapour of reaction mixture is presented in Fig. 2.

The two highest peaks visible here have retention times of 0.8 and 1.6 min., and belong to the ethyl esters of fluoroacetic and chloroacetic acids, respectively. From the comparison of total abundance values registered for both peaks we conclude that the concentrations of these two esters are almost equal.

The mass spectra of the ethyl fluoroacetate and chloroacetate obtained by this analytical procedure are shown in Fig. 3 and 4.

The proposed way of the fragmentation of toxic ethyl fluoroacetate, in turn is shown in Fig. 5.

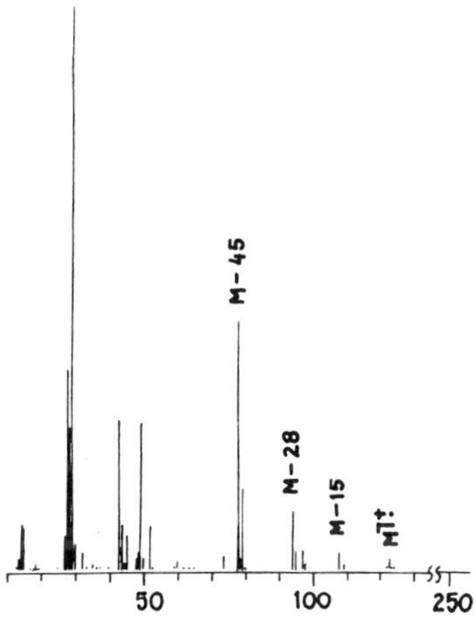

Fig. 4. Mass spectrum of ethyl chloroacetate.

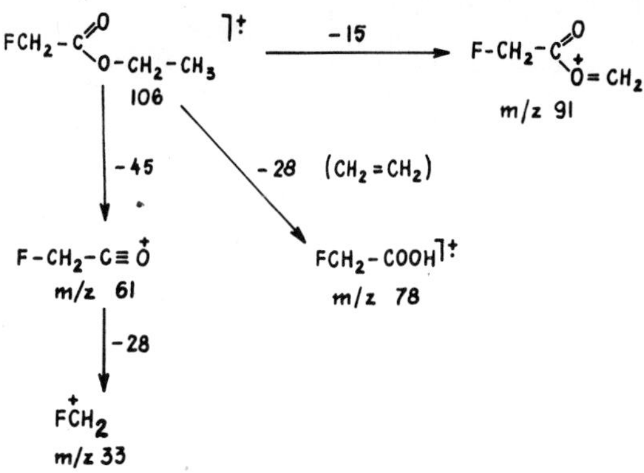

Fig. 5. Way of fragmentation of ethyl ester of fluoroacetic acid.

Blood Sample Analysis

The vapours liberated from the 10 ml of blood sample in a 50 ml flask sealed with gum membrane were injected into GC/MS system. The concentration of the interesting compound in the vapours was too low for recording the good mass spectrum. Fig. 6 shows the total and single ion chromatograms registered for this sample.

There was only slight rise of the total ion current on the left path, at a retention time of 0.8 min. Retention time for ethyl fluoroacetate was the same. But on the right path, a well formed single ion peak could be seen at the same retention time. The monitored ion at a mass of 33 is specific for a fragment of the ethyl ester of fluoroacetic acid molecule, and belongs to fluoromethyl ion.

For a better confirmation of the ethyl flouroacetate presence in the analysed blood sample, we used another programm i.e. Selected Ion Monitoring. In this way the sensitivity of analysis rose substantially.

For the monitoring, the ions at m/z 61, 33, 78 and 91 were chosen from the mass spectrum of ethyl ester of fluoroacetic acid registered earlier. The ions containing the fluorine atom were specially selected. In Fig. 7 are shown well formed single ion chromatographic peaks for the ions at m/z 61 and 33. The peaks for ions 78 and 91 are better visible on the full width of the paper. They are shown in Fig. 8.

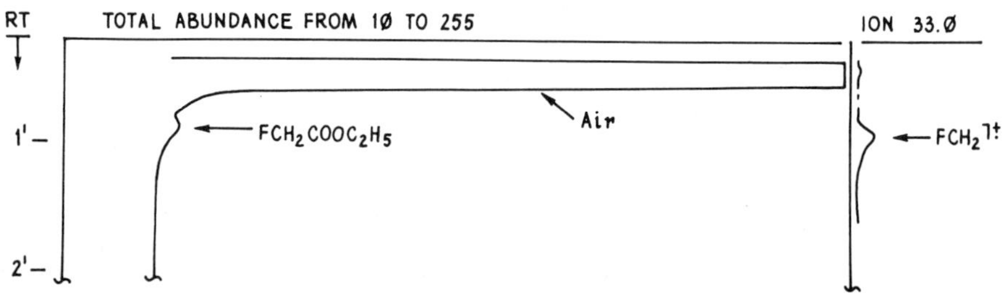

Fig. 6. GC/MS chromatogram of blood sample.

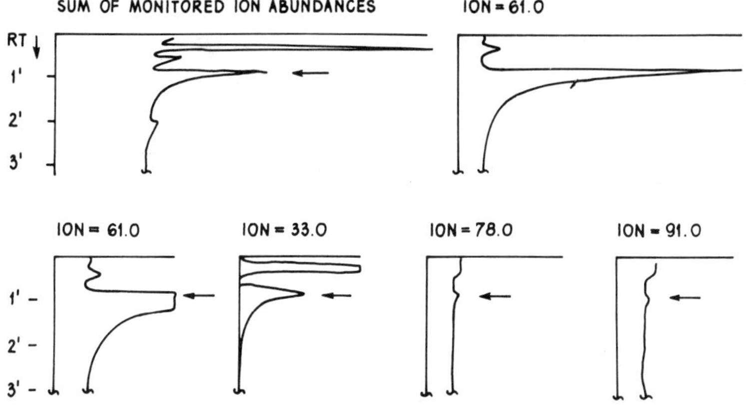

Fig. 7. Single ion chromatograms of blood sample.

DISCUSSION

The results presented above, sufficiently confirmed the presence of unchanged ethyl fluoroacetate in the organism of the poisoned student. We can suppose that most of the absorbed ester was transformed in the organism of the poisoned student. By means of the presented head-space and GC/MS method, we were able to detect the original ethyl ester of fluoroacetic acid only.

In the pulmonary fluid, however, the toxic ester content was too low to be detected by the analytical procedure used here.

As mentioned earlier, the packed column and membrane separator were used. We can suppose that the sensitivity of GC/MS analysis can increase when the capillary column is used.

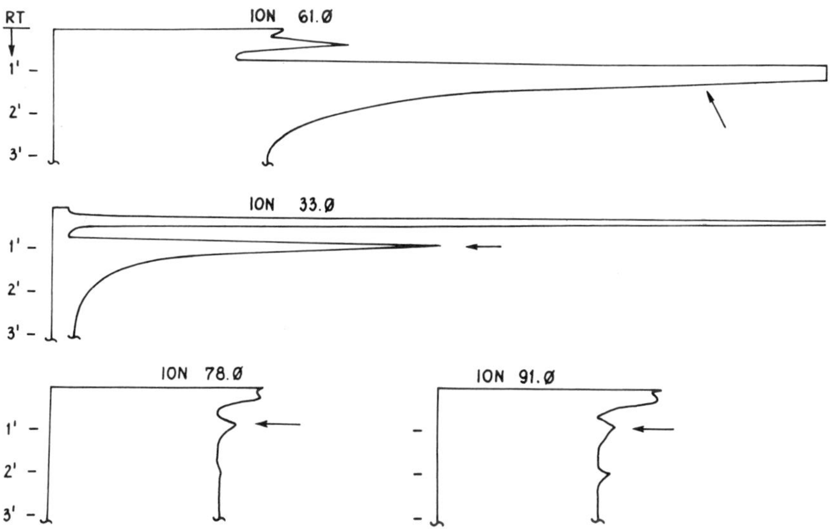

Fig. 8. Single ion chromatograms same as in Fig. 7, printed on the full width of the paper.

Aside from the above results, fluorine was determined in blood, liver and pulmonary fluid by spectrophotometric method with alizarine. The contents found (expressed in $\mu g/g$) were 3.4, 1.7 and 1.1, respectively.

According to Bredemann[3], the concentration of this element in blood is lethal when it reaches the level of 3 $\mu g/g$. Thus in the blood of the poisoned student the lethal limit was not only reached, but it was also exceeded.

REFERENCES

1. R. Peters, "Carbon-Fluorine Compounds Chemistry, Biochemistry and Biological Activities", Ciba Foundation, Elsevier, Amsterdam (1972).
2. G.W. Gribble, Fluoroacetate Toxicity, J. Chem. Educ. 50:460 (1973).
3. G. Bredemann, "Biochemie und Physiologie des Fluors", Akademia Verlag, Berlin (1956).

FACTORS INFLUENCING THE URINARY EXCRETION OF PHENOL AND CRESOLS

F. Sanguinetti, M.C. Sanguinetti, S. Mantovani,
and C. Melandri

"Alberto Sanguinetti" Center
Ravenna, Italy

i. OCCUPATIONAL EXPOSURE, URBAN POLLUTION AND CIGARETTE SMOKING

INTRODUCTION

Following exposure to either benzene, phenol or phenolic
derivatives, phenol is excreted in human urine as the sulphate and
glucuronide. The determination of urinary phenol has been used to
control the level of benzene exposure[1-3]. Likewise, following
exposure to either toluene or cresols, cresols are excreted in
human urine; although the o-cresol is only a minor metabolite of
toluene, the urinary excretion of this compound is considered a
good parameter of toluene exposure[4-6]. Toluene is a widely used
solvent in industry; its main metabolite is benzoic acid, which is
excreted as hippuric acid. However, hippuric acid is an endogenous
metabolite common in human urine and its excretion can be largely
influenced by diet[6]; so the determination of o-cresol in urine has
been suggested as more specific for the biological monitoring of
toluene exposure.

Phenol and cresols occur naturally in human urine as
degradation products of tyrosine by the activity of intestinal
flora[8]. Moreover these compounds are chemicals widely used in many
industrial processes: also their urinary excretion can be derived
from this exposure.

An adequate biological exposure test should not be only
specific, but also gives a good correlation with the occupational
uptake. The urinary excretion of phenol can result not only from
benzene or phenol occupational exposure, but also from intake of
some drugs containing phenolic derivatives, from exposure to
cigarette smoke and other air pollutants[7-9]. We continued our study
concerning the identification of non-occupational factors, which
can interfere in the biological monitoring of exposed workers. We
examined the urinary excretion of phenol and cresols in many
groups of persons, who were exposed to benzene and toluene, trying
to identify some interference factors and to control the real
influence of some factors found by other Authors.

MATERIALS AND METHODS

<u>Apparatus</u>. A Perkin-Elmer Liquid Chromatograph consisting of a Model LC4, a Model Rheodyne 7125 S sample injector with a loop of 6 μl, a Model LC 75 spectrophotometric detector at 240 nm, a Model 204 spectrophotofluorimeter (excitation at 273 nm and emission at 298 nm) and a Model 56 recorder were used.

<u>Chemicals</u>. All chemicals were of analytical grade or HPLC grade and obtained from Merck (Darmstadt, FRG).

<u>Sample Collection and Preparation</u>. Urinary specimens were taken in the morning or after the work-shift as specified, from workers in a petrochemical industry. Hippuric and mandelic acids were injected after dilution: to 0.2 ml of urine was added 1.4 ml of methanol; after centrifugation at 2,000 rpm for 5 min, 3 μl of supernatant were injected into the column. Phenol and cresols were assayed after acid hydrolisis. To 0.2 ml of urine 1.5 ml of H_3PO_4 0.3 mol/l were added: the mixtures were incubated at 95°C for 90 min. After cooling, 0.2 ml of hydrolisate were diluted with 0.8 ml of mobile phase; 6 μl were injected into the column.

<u>Chromatography</u>. The analytical column was an Hypersil 5 ODS. The mobile phase was:

1. KH_2PO_4 0.01 mol/l in acetic acid 0.5%:methanol (100:8) for hippuric and mandelic acids.
2. KH_2PO_4 0.05 mol/l (pH 3.5):acetonitrile (70:30) for phenol and cresols.

Hippuric and mandelic acids were identified and quantified in comparison with reference standards. Phenolic compounds were identified in comparison with standards and the identification was confirmed by the determination of the emission spectra. The urinary metabolites concentrations were corrected for excretion volume by creatinine determination. The samples were stored at -20°C until analysis.

RESULTS AND DISCUSSION

<u>Influence of Smoking</u>

Phenol and cresols are normal constituents of the total particulate matter of tobacco smoke. Phenolic constituents are selectively absorbed by cigarette filters; notwithstanding important quantities of these compounds remain in the mainstream

Table 1.   Urinary Excretion of Phenol and Cresols in Smokers and non-Smokers

Subjects	n	Phenol mg/g creatinine	Cresols mg/g creatinine
non-smokers	65	3.58 ± 1.63	11.78 ± 6.65
smokers	48	3.66 ± 1.48	11.08 ± 6.02

Table 2. Urinary Excretion of Phenol and Cresols: Comparison
          between Town Inhabitants and Country Inhabitants *

Inhabitants	n	Phenol mg/g creatinine	Cresols mg/g creatinine
town	44	2.54 ± 1.32	9.08 ± 4.24
country	22	2.99 ± 1.66	8.27 ± 4.80

* specimens were collected at the morning.

total particulate matter and the whole quantity is dispersed in
the environment by the side-stream[9]. Hansen found that in smokers
the urinary excretion of o-cresol is higher than in non-smokers[7].
In Table 1 we compared the amounts of phenol and cresols excreted
in two groups of workers, employed in a petrochemical industry and
not specifically exposed to benzene and toluene. As smokers we
considered subjects, who smoke daily more than 8 cigarettes; as
non-smokers we classified subjects, who smoke daily less than one
cigarette. All urine samples were collected in the morning.

The results show that smoking does not influence phenol and
cresols excretion in non-exposed people. The different results
obtained by Hansen[7] can be explained by the different experimental
conditions.

Influence of Urban Pollution

In Table 2 and 3 we report results obtained to show the
influence of urban and industrial air pollution on the excretion
of phenol and cresols. In the first study the samples were collected from two groups of ceramic industry workers, inhabitants in
a small town (less than 10,000 inhabitants) or in the country. The
data obtained show no difference between the two groups. In the
second study the whole group of workers of the ceramic industry
were compared to a large group of workers of the petrochemical
industry, working and living in a city (more than 100,000 inhabitants). The differences between the two groups appear highly
significant both for phenol and cresols. These differences can be
attributed mainly to environmental pollution.

Table 3. Urinary Excretion of Phenol and Cresols: Comparison
          between Town Inhabitants and City Inhabitants *

Inhabitants	n	Phenol mg/g creatinine	P	Cresols mg/g.creatinine	P
town	63	2.55 ± 1.27	–	9.63 ± 5.02	–
city	135	3.40 ± 1.57	<.001	11.58 ± 5.69	<.05

* specimens were collected in the morning.

Table 4.   Urinary Excretion of Phenol and Cresols: Correlation
           with Occupational Exposure *

Workers	n	Phenol mg/g creatinine	P	Cresols mg/g creatinine	P
non-exposed	61	2.70 ± 2.1	–	7.80 ± 4.2	–
exposed	53	2.90 ± 4.0	<.1	10.00 ± 4.8	<.05

* specimens were collected at the end-shift.

Influence of Occupational Exposure

In Table 4 we compare data obtained from workers exposed to
benzene (at trace level) and toluene in comparison with a group of
non-exposed subjects employed in the same factory. The urine
samples were collected at the end-shift. The differences between
the two groups do not appear significant for phenol excretion, but
may be significant for cresols excretion.

Conclusions

Our results show that the problem of reference values concerning phenol and cresols urinary excretion in normal subjects
needs further researches.

REFERENCES

1. K. Ogan and E. Katz, Determination of phenol in aqueous
    samples by LC with fluorescence detection, Chromatogr. Newsl.
    7:15 (1979).
2. W.A. Fishbeck, R.R.Langner and R.J. Kociba, Elevated urinary
    phenol levels not related to benzene exposure, Am. J. Ind.
    Hyg. Assoc. 36:920 (1975).
3. C.V. Eadsforth and P.C. Coveney, Measurement of phenol in
    urine using a high performance liquid chromatographic method,
    Analyst, 109:175 (1984).
4. J. Angerer, Occupational chronic exposure to organic solvents
    VII-metabolism of toluene in man, Int. Arch. Occup. Environ.
    Health, 43:638 (1979).
5. K. Hasegawa, S. Shiojima, A. Koizumi and M. Ikeda, Hippuric
    acid and o-cresol in the urine of workers exposed to toluene,
    Int. Arch. Occup. Environ. Health, 52:197 (1983).
6. P. Pfaffli, H. Savolainen, P-L. Kalliomaki and P. Kalliokoski,
    Urinary o-cresol in toluene exposure, Scand. J. Work Environ.
    Health, 5:286 (1979).
7. S.H. Hansen and M. Dossing, Determination of urinary hippuric
    acid and o-cresol, as indices of toluene exposure, by liquid
    chromatography on dinamically modified silica, J. Chromatogr.
    229:141 (1982).
8. L.L. Needham, S.L. Head and R.E. Cline, Determination of
    phenol and cresols in urine by gas chromatography, Anal.
    Letters 17:1555 (1984).
9. B.A. Tomkins, R.A. Jenkins, W.H. Griest, R.R.Reagan and
    S.H.Holladay, Liquid chromatographic determination of phenol

and cresols in total particulate matter of cigarette smoke,
    J. Assc. Off. Anal. Chem. 67:919 (1984).
10. F. della Fiorentina and A. Crippa, Determinazione del fenolo
    nell'urina, Biochim. Clin. 7:761 (1983).
11. F. Sanguinetti, M.C. Sanguinetti and C. Melandri, High
    pressure liquid chromatography determination of urinary
    metabolites of toluene, xylene and styrene, Spectroscopy
    Int. J. 3:408 (1984).

ii. DAILY AND SEASONAL VARIATIONS

INTRODUCTION

The determination of urinary excretion of volatile phenols
has widely been used in the biological monitoring of occupational
exposure to benzene, toluene or phenol. These biological
indicators are usually determined on spot samples of urine taken
at the end of the work-shift to obtain information concerning the
workday exposure. This method of samples collection is considered
reliable in biological monitoring of rapidly metabolized
substances such as benzene and toluene.

However, it is well known that the urinary excretion of these
metabolites is not exclusively dependent from environmental
exposure and can be affected by diet. Normal human urine contains
10 to 50 mg of volatile phenols per day, principally in the form
of phenol and p-cresol. It has been known that they are products
of the bacterial metabolism of tyrosine in the gut[1].

The production of phenol and p-cresol from tyrosine was
investigated in several bacterial species. The aromatic compounds
produced from tyrosine by Clostridia are hydroxyphenylacetic acid,
hydroxyphenyllactic acid, hydroxyphenylpropionic acid, phenol and
p-cresol. The Clostridia appear unable to modify the aromatic
ring[2]. Thus, phenol and p-cresol are absent in the urine of germ-
free animals and their excretion decreases with low protein diet[1].
Instead, a meal rich of protein can increase the volatile phenols
urinary excretion[2]. Other interfering factors are dermal
application of phenol containing preparations or ingestion of
phenol containing drugs[3]. However, the dietary factors can be
considered as most relevant.

The purpose of our investigation is to control the influence
of seasonal and daily variations of volatile phenols excretion in
the reliability of biological monitoring of exposed workers.

MATERIALS AND METHODS

Apparatus

A Perkin Elmer Liquid Chromatograph (Norwalk, Ct, USA)
consisting of a model LC4, a model Rheodyne 7125 S sample injector
with a loop of 6 µl, a model LC 75 spectrophotometric detector at
240 nm, a Model 204 spectrophotofluorimeter (excitation at 273 nm
and emission at 298 nm) and a model 56 recorder was used.

<u>Chemicals</u>

All chemicals were of analytical-reagent grade or HPLC-grade and obtained from Merck (Darmstadt, FRG).

<u>Sample Collection and Preparation</u>

Urinary specimens were taken in the morning or after the work-shift as specified, from workers in a petrochemical industry. Hippuric and mandelic acids were injected after dilution: to 0.2 ml of urine was added 1.4 ml of methanol; after centrifugation at 2,000 r.p.m. for 5 min. 3 μl of supernatant were injected into the column. Phenol and cresols were assayed after acid hydrolisis. To 0.2 ml of urine 1.5 ml of H_3PO_4 0.33 mol/l was added; the mixture was incubated at 95°C for 90 min. After cooling, 0.2 ml of hydrolisate was diluted with 0.8 ml of mobile phase; 6 μl were injected into column.

<u>Chromatography</u>

The analytical column was an Hypersil 5 ODS. The mobile phase was:

1) KH_2PO_4 0.01 mol/l in acetic acid 0.5%:methanol (100:8) for hippuric and mandelic acids.
2) KH_2PO_4 0.05 mol/l (pH 3.5):acetonitrile (70:30) for phenol and cresols.

Hippuric and mandelic acids were identified and quantified in comparison with reference standards. Phenolic compounds were

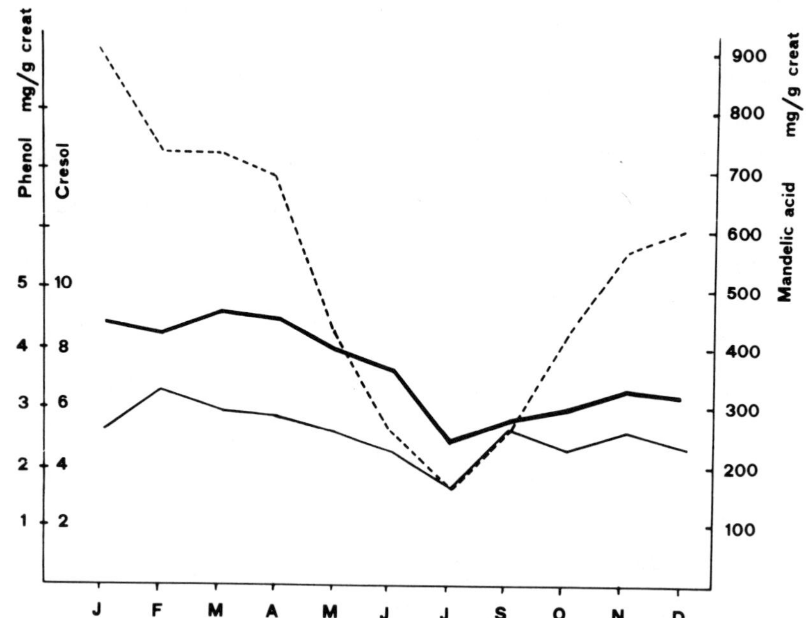

Fig. 1. Seasonal variation of phenol, p-cresol and mandelic acid urinary excretion during 1985. Phenol=thin solid line; p-cresol=thick solid line; mandelic acid=dotted line. The concentrations of metabolites have been corrected by creatinine levels in urines.

Table 1. Urinary Excretion of Phenol, p-Cresol and Hippuric Acid:
          Comparison between Morning Samples and End-Shift Samples
          of the Same Subjects

Metabolite	n	Morning Samples mg/g creatinine	End-Shift Samples mg/g creatinine
phenol	30	3.4 ± 2.1	2.9 ± 2.3
p-cresol	30	10.8 ± 4.2	8.9 ± 4.7
hippuric acid	35	582.0 ± 302.0	321.0 ± 169.0

identified in comparison with standards and the identification was
confirmed by the determination of the emission spectra. The
urinary metabolites concentrations were corrected for the
excretion volume by creatinine determination. The samples were
stored at -20°C until analysis.

RESULTS AND DISCUSSION

     The first study concerns the seasonal variations of the
phenol and p-cresols excretion as appeared in Fig. 1 in comparison
with mandelic acid excretion. As previously reported[6] the urinary
excretion of mandelic acid exhibits a seasonal variation as a
sensible decrease in the summer. Same phenomena are observable in
the phenol and p-cresol urinary excretion as shows in Fig. 1.
These seasonal variations can be attributed to the improvement of
environmental conditions as observed with mandelic acid.
     In a previous investigation we have observed an unexpected
data in contrast with data previously stated[3]: the urinary
excretion of phenol and p-cresol shows a daily variation and the
amount found in the morning samples are higher than those found in
the end-shift samples.

     Analogous variations were already observed in the urinary
excretion of coproporphyrins[5]. Our data are reported in Table 1 in
comparison with data concerning the urinary excretion of hippuric
acid: this metabolite also shows the highest concentration in
morning samples. This observed difference between morning and
afternoon samples involves more than 60% and 80% of examined
subjects respectively for volatile phenols and hippuric acid
excretion.

     In four non-exposed volunteers we were able to control the
phenol urinary excretion during the whole day, obtaining a
confirmatory result as shown in Fig. 2. This phenomenon can be
attributed to dietary influence as observed in hippuric acid
excretion[3].

     These results seem to be of some importance. Many problems
can rise from these observed daily and seasonal variations because
reference interval values of biological indicators are necessary
in the evaluation of results of biological monitoring data. It is
possible to avoid to these difficulties adopting a double samples
collection, at the start and at the end of the weekly work-period.

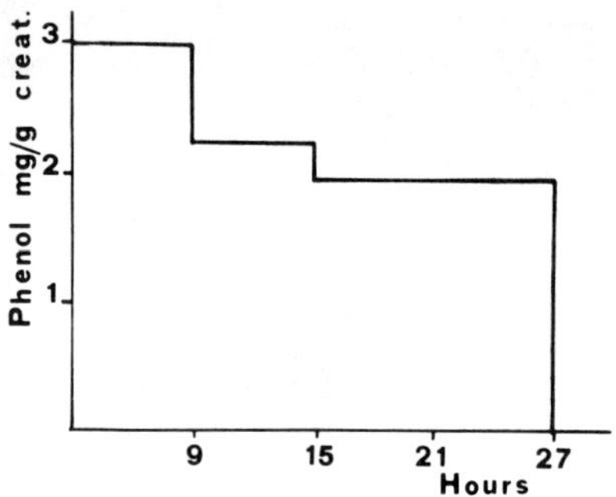

Fig. 2. Daily variation of phenol urinary excretion. The
concentrations of metabolites have been corrected by
creatinine levels in urines.

REFERENCES

1. E. Bone, A. Tamm and M. Hill, The production of urinary phenols
   by gut bacteria and their possible role in the causation of
   large bowel cancer, Amer. J. Clin. Nutr. 29:1448 (1979).
2. H.A. Backer, Amino acid degradation by anaerobic bacteria, Ann.
   Rev. Biochem. 50:23 (1981).
3. R.R. Lauwerys, Industrial chemical exposure: guidelines for
   biological monitoring, Biomedical Publ., Davis (1983).
4. F. Sanguinetti, S. Mantovani and M.C. Sanguinetti, The
   biological monitoring of low exposure to benzene and toluene:
   some interferences impairing its validity, Int.Symposium:
   Biochemical and cellular indices of human toxicity in
   occupational and environmental medicine. Milan, 19-22 May 1986.
5. F. Sanguinetti and S. Mantovani, Ritmi circadiani e controllo
   biochimico del danno saturnino, Atti del Seminario Int.:
   Problemi da inquinamento per piombo e fluoro di origine
   industriale con particolare riferimento all'industria ceramica.
   Bologna, 26-27 Jan. 1978.

BIOTRANSFORMATIONS OF A K REGION METABOLITE OF BENZO(A)PYRENE

G. Lhoest, M. Lesne, J. De Graeve*,
and J. Van Cantfort*

Mass Spectrometry Section, School of Pharmacy - U.C.L.
Brussels
* Laboratory of Medical Chemistry and Toxicology
Institute of Pathology - ULg
Liege, Belgium

INTRODUCTION

Polycyclic aromatic hydrocarbons are largely distributed in
the atmosphere soil, water[1,2], in the food chain[3] and epidemiological studies indicate that the environment is a significant
determinant in the incidence of human cancer.

The initial biological receptors for the polycyclic aromatic
hydrocarbons have been identified as the microsomal polysubstrate
mixed-function oxidases consisting of several forms of cytochrome
P450 and the associated NADPH-cytochrome-P450 reductase[4,5].

Benzo(a)pyrene is a much-studied polycyclic aromatic
hydrocarbon and the metabolism of this xenobiotic is mediated by
enzymes such as the mixed-function oxygenases system, the epoxide
hydratase, the glutathion S-epoxide-transferase, the UDP-
glucuronic acid transferase and sulfotransferase.

The metabolites commonly reported[6-10] and resulting from cell
or tissue preparations are the 1,3,7,9 phenols derivatives, the
4,5; 7,8, and 9,10-dihydrodiol, the 1,6; 3,6 and 6,12 quinones,
the 4,5 epoxide and 7,8-dihydrodiol-9,10-epoxide of
benzo(a)pyrene.

It is generally accepted that epoxides mediate most of the
mutagenic and carcinogenic effects exerted by polycyclic aromatic
hydrocarbons and ascribed to their capacity to bind covalently to
critical nuclear macromolecules, particularly DNA.

Benzo(a)pyrene is reported[10], to form an ultimate carcinogenic
and mutagenic metabolite, benzo(a)pyrene-diol-epoxide (BaPDE).

Because the 7,8-diol formed enzymically from benzo(a)pyrene
is the trans isomer[11], there are four possible stereoisomeric
9,10-epoxides derived from the trans-7,8-diol existing as a pair
of diastereoisomers, anti-BaPDE or diol epoxide I (benzylic
hydroxyl and epoxide on opposite faces of the molecular plane) and

syn-BaPDE or diol epoxide II (benzylic hydroxyl and epoxide on the same face of the molecular plane).

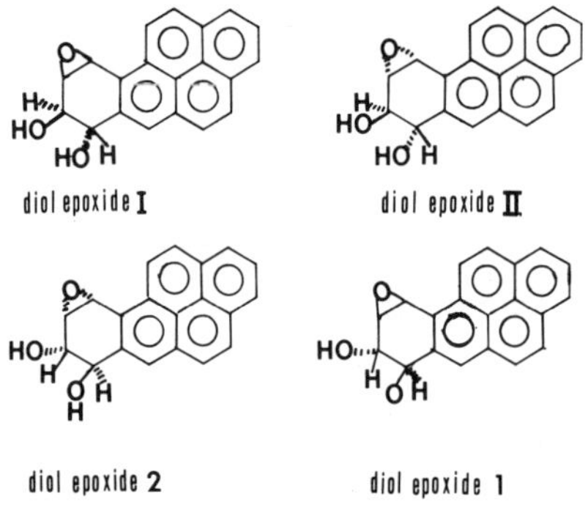

Fig. 1. Stereoisomeric epoxides derived from the 7,8-diol.

Each diastereoisomer consists of two enantiomers (+) and (−) as shown in Fig. 1.

Recently, D.P. Schwartz and W.M. Baird[12] mentioned that a BaP-deoxy-ribonucleoside adduct formed in Wistar rat embryo cell cultures frequently identified as the (−) anti-BaPDE:dGuo adduct based on chromatographic properties results from an entirely different unidentified metabolite of BaPDE as shown in Fig. 2 by acid hydrolysis observations proving the multiple pathways of activation of benzo(a)pyrene to DNA-binding metabolites in rat embryo cell cultures. Since a new K region metabolite of benzo(a)pyrene, 4-hydroxy-5-oxo-4,5-dihydro-benzo(a)pyrene of mass m/z=284 was isolated in our laboratory from rat liver microsomes incubation media and since the same compound of mass m/z=284, directly active in the Ame's test but of unidentified structure, was also detected in the atmosphere with different quinones by J.N. Pitts[13], we decided to compare in vitro the further metabolism of 4-hydroxy-5-oxo-4,5-dihydro-benzo(a)pyrene using rat and human liver microsomes in order to see if non K region metabolites were produced. The final goal in the future being the isolation and identification of new DNA-binding metabolites of benzo(a)-pyrene.

MATERIALS AND METHODS

<u>Chemicals</u>

Benzo(a)pyrene was purchased from Aldrich (Milwaukee, Wi, U.S.A.). Sodium hydrogen sulfite, osmium tetroxide and potassium cyanide were purchased from Merck (Darmstadt, FRG). NADH, NADP, glucose-6-phosphate dehydrogenase (350 U/ml) were purchased from Boehringer (Mannheim, FRG). 4-hydroxy-5-oxo-4,5-dihydro-benzo(a)pyrene was synthesized following a technique described elswhere[14], and labelled compound (generally tritiated, 1.6 Ci/mmol) was obtained from IRE (Fleurus, Belgium). Purity of both product was found greater than 99%. Picofluor 30™ was purchased from Packard. BSTFA and TMS were purchased from Macherey-Nagel (Duren, FRG). All other chemicals and solvents were of analytical grade and obtained from Merck.

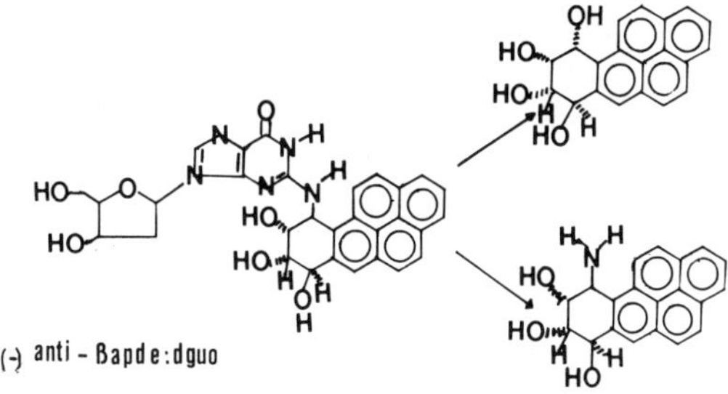

Fig. 2. DNA-binding metabolite.

Animals

Male Wistar rats, R strain (250 g), were injected intra-
peritonealy with 40 mg/Kg 3-methylcholanthrene on 48 and 24 hours
before sacrifice. Human liver was obtained from a "kidney donor"
after clinical death.

Enzymic Preparations

Rat livers. After exsanguination by decapitation the livers
were removed, weighed, washed and fractionated according to the
method described by Amar-Costesec[15] to give a microsomal fraction
containing between 0.4 and 1.9 nmol of cytochrome P450/mg of
protein.

Human livers. The liver was perfused with a cold physiologic
saline solution, cut into several pieces (± 20 g) cooled in liquid
nitrogen and stored at -80°C for subsequent microsomal preparations. For this particular procedure, the pieces of liver were
left over night at 4°C, extruded from the holes (1.5 mm I.D.) of a
metal disc in order to eliminate conjonctive tissues and vessels.

Table 1. Characterization of the Human Microsomal Preparation

Assays	Unit	Rat liver microsomes	Human liver microsomes
Protein concentration	mg/ml	41	43
Cyt.P450	nmol/mg prot.	0.48	0.49
Aryl hydrocarbon hydroxylase	nmol/min mg prot.	0.40	0.03
Ethoxycoumarin deethylase	nmol/min mg prot.	1.55	0.18
Aldrin epoxidase	nmol/min mg prot.	2.35	0.29

The resulting homogenate was diluted four times with a Tris (0.1 mol/l, pH 7.6)/Sucrose (0.25 mol/l) solution and homogenized in a Potter-Elvekjem tube with a teflon pestle. The homogenate was centrifuged for 20 min at 12,000 g in a refrigerated Sorvall and the supernatant was then centrifuged for 1 h at 105,000 g in a Spinco RC2B ultracentrifuge. The pellet was finally resuspended in 0.1 mol/l Tris buffer (pH 7.6) in such a way that 1 ml of the final microsomal suspension corresponds to 2 g of liver homogenate. The suspension was stored in small volumes at -80°C under argon gas.

Optimal conditions for the monooxygenase assays. The same incubation medium was prepared for all the momooxigenase assays. One ml of the final incubation mixture contains an enzyme preparation equivalent to 200 mg of human liver or 20 mg of rat liver microsomes, a NADPH-generating system (glucose-6-phosphate 2.5 mmol/l; NADP 0.5 mmol/l; glucose-6-phosphate dehydrogenase 2 IU; NAD 0.5 mg; bovine serum albumine 1 mg, $MgCl_2$ (5 mmol/l), Tris buffer (0.1 mol/l, pH 7.6) and from 20 to 90 µg of the substrate.

Biochemical assays. Protein and cytochrome P450 concentration[16,17], arylhydrocarbon-hydroxylase[18], ethoxycoumarin deethylase[19] and aldrin epoxidase[20] were determined according to published standard procedures.

Characterization of the human microsomal preparation. The study of the biotransformation of the K region metabolite of benzo(a)pyrene, 4-hydroxy-5-oxo-4,5-dihydro-benzo(a)pyrene, was only performed after characterization of the human microsomal preparation. Table 1 shows that the human liver monooxygenase system is less active than the same rat liver monooxygenase system. Nevertheless, detectable enzymic values were obtained for each assay.

<u>Product Incubation And Isolation</u>

Using rat liver microsomes. To the NADPH-generating medium (100 ml) containing 0.25 mmol/l NADH, 0.16 mmol/l NADP, 0.025 mmol/l $MnCl_2$, 5.3 mmol/l glucose-6-phosphate in histidine buffer (0.1 mol/l, pH 7.6) were added 300 µl of glucose-6-phosphate dehydrogenase. The solutions were preincubated in a Gallenkamp shaking incubator for 15 min at 37°C in glass-stoppered tubes, then 5 ml of the rat liver microsomal fraction were added together with 250 µg of an acetone solution of 4-hydroxy-5-oxo-4,5-dihydro-benzo(a)pyrene. The resulting solution was pre-incubated for 1 h at 37°C, then extracted three times with 100 ml of chloroform. The chloroform extracts were evaporated to dryness under vacuum and the residue, dried in vacuo, was chromatographed on Kieselgel 60 plates using chloroform as the solvent. Fluorescent spots at Rf 0.028 and 0.45 (unchanged metabolite) appeared under U.V. light. The more polar fraction of Rf 0.028 was scraped off and the powder obtained ultrasonicated in methanol for 5 min. After filtration of the suspension, the residue was dried under vacuum and submitted to HPLC analysis.

Using human liver microsomes. 4-hydroxy-5-oxo-4,5-dihydro-benzo(a)pyrene was incubated in the predefined standard conditions in presence of human liver microsomes. Microsomal incubations were performed using different substrate concentrations (10 µg/ml and 40 µg/ml) and different incubation times with the same microsomal

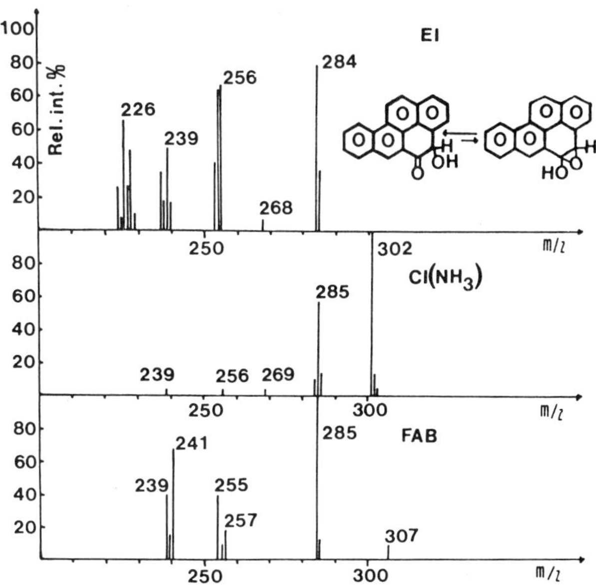

Fig. 3. Mass spectrum (EI,CI and FAB) of 4-hydroxy-4,5-dihydrobenzo(a)pyrene.

concentration. The incubation medium was extracted twice with three volumes of chloroform. The organic phase was evaporated under nitrogen and the residue was dissolved in methanol for HPLC analysis.

<u>Instrumentation and Operating Conditions</u>

High performance liquid chromatography was performed on a Waters liquid chromatograph equipped with a Pye Unicam LC3, UV detector and with a radioactivity monitor (Berthold, Gosheim, FRG) coupled to the HPLC system. The residues obtained from the rat liver microsomes assays were injected on a Radial-Pak μBondapak precolumn module and C18 inserts. The eluent was methanol:water (75:25) and the flow rate 1.5 ml/min. The residues obtained from the human liver microsomes assays were injected on a 7 μ C18 Macherey Nagel column. The eluent was a mixture of phosphate buffer pH 2.5 (10 mmol/l) and acetonitrile. The acetonitrile gradient was changed from 20 to 100% over a total time of 30 min.

Mass spectrometric analysis was carried out with a LKB 9000S (Bromma, Sweden), a VG 7070 EQ (Manchester,UK) or a Kratos MS80 RFA (Ramsey, NY, USA) instrument under EI, CI or FAB mode.

GC/MS analysis was performed on a Chrompack Sil 5 OV1 capillary column (Middelburg, NL). The oven temperature was 200°C.

RESULTS AND DISCUSSION

As reported earlier[14], a K region metabolite of benzo(a)pyrene, 4-hydroxy-5-oxo-4,5-dihydro-benzo(a)pyrene was detected from phenobarbital-induced rat liver microsomes incubation media and was also found in the atmosphere[13].

173

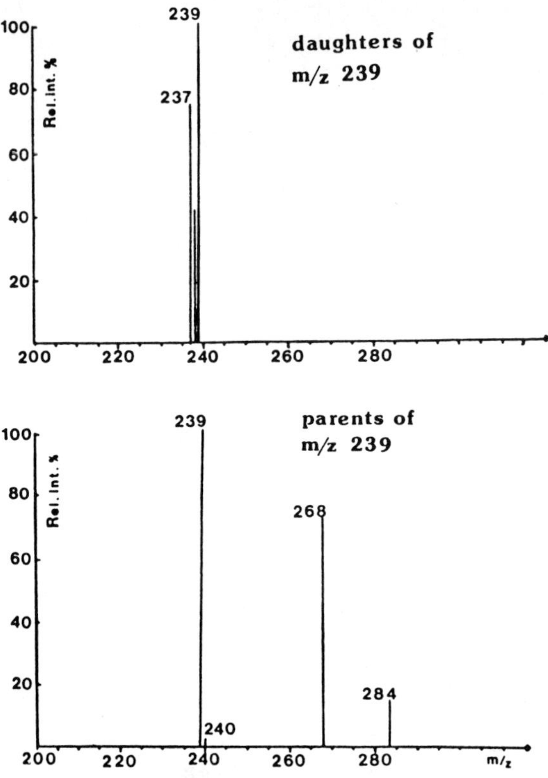

Fig. 4. Daughters and parents of m/z = 239.

The EI mass spectrum (Fig. 3) of this metabolite shows a molecular ion of mass m/z=284 and a series of relevant fragmentation ions of mass m/z=268 (M-O), 256 (M-CO), 255 (M-CO-H), 254 (M-CO-H2), 239 (M-CO-OH), 228 (M-2CO), 227, 226. The fact that a loss of sixteen mass units is observed from the molecular ion prompted us to look at certain ionic reaction mechanisms from tandem mass spectrometry (MS/MS) making possible the study of collisionally induced decompositions (CID).

We demonstrated, focusing our attention on fragment ion m/z=239, that the daughters were 238, 237 and the parents of the same ion 240, 268 and 284 (Fig. 4). Taking into account that a fragment ion of mass m/z=256 (M-CO) is also observed in the mass spectrum of this compound, the overall fragmentation pathway (Fig. 5) is easily explained from two distinct tautomers of 4-hydroxy-5-oxo-4,5-dihydro-benzo(a)pyrene, the ketol and the hemiketal form which may be considered also as an epoxide-ol function.

So we briefly demonstrated that starting from the hemiketal tautomer, successive losses of oxigen, CO and CO+H give rise the fragment ions of mass m/z=268, 240 and 239. Since this compound has been detected in the atmosphere[13] and is also considered as a direct mutagen, it would be of greater interest on the one hand to study the DNA covalent binding capacities as such and in the other hand to demonstrate if this compound is metabolized in non-K region metabolites such as the 7,8-diol-epoxide of benzo(a)pyrene

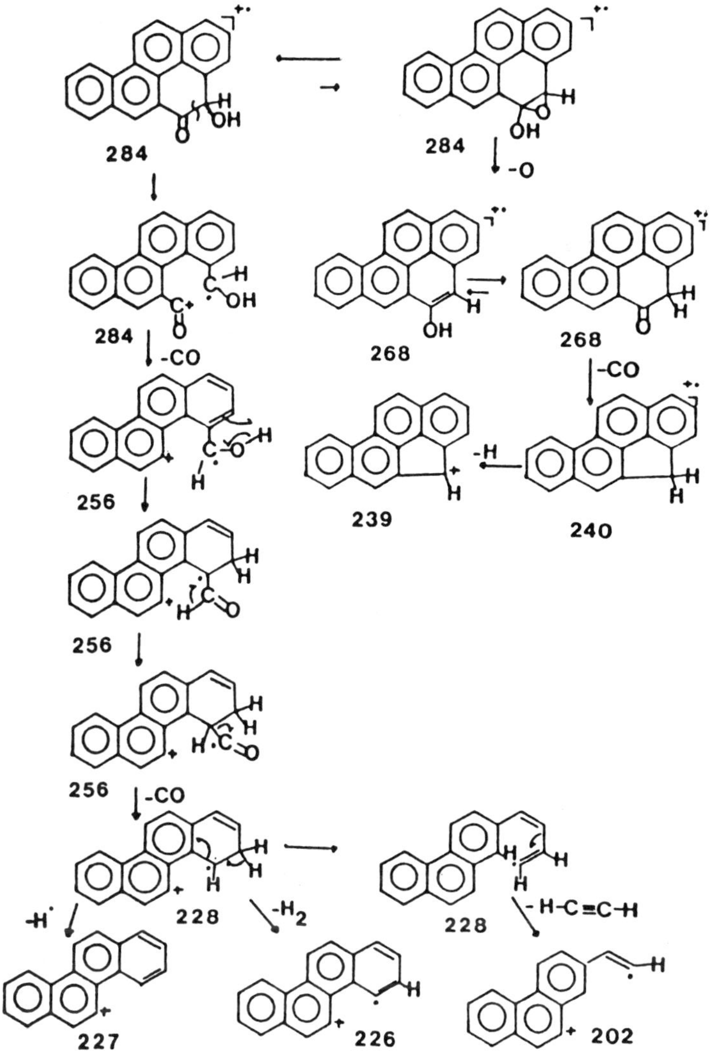

Fig. 5. Fragmentation pathway of the K region metabolite.

by rat and human liver microsomes. This last approach will be
presently discussed.

4-Hydroxy-5-oxo-4,5-dihydro-benzo(a)pyrene was synthesized in
a three step synthesis and the synthesized compound was compared
to the metabolite isolated from rat liver microsomes incubation
media. Mass spectra (Fig. 6) reveals the identity of the metabolite and the synthetic compound.

When the residues obtained after incubation of synthetic
4-hydroxy-5oxo-4,5-dihydro-benzo(a)pyrene with rat liver
microsomes in presence of the usual cofactors were submitted to
HPLC analysis, essentially three main peaks with retention times
3 min 20 s (M1), 4 min 10 s (M2) and 10 min (M3, unchanged
compound), were observed and collected systematically. The EI mass
spectrum of metabolite M1 shows a molecular ion of mass m/z=318

175

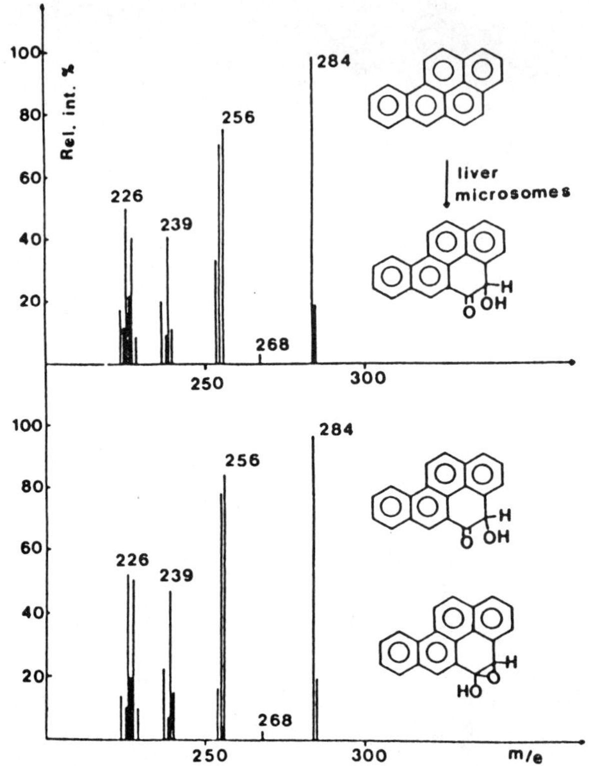

Fig. 6. Comparison of mass spectra of synthesized and isolated 4-hydroxy-5-oxo-4,5-dihydro-benzo(a)pyrene from biological media.

and fragmentation ions of mass m/z=300, 287, 284, 273, 272, 268, 256 and 244 demonstrating that vicinal hydroxy groups are most probably located in the 7,8 position of the benzo(a)pyrene ring. Metabolite M_1 is then consequently 4-hydroxy-5-oxo-4,5-dihydro-trans-7,8-dihydroxy-7,8-dihydro-benzo(a)pyrene. Moreover, the CI (NH3) mass spectrum of the same compound reveals quasi-molecular ions of mass m/z=319 (M+H)$^+$, and 336 (M+NH$_4$)$^+$ (Fig. 7).

The EI mass spectrum of metabolite M_2 (Fig. 8) exhibits a molecular ion of mass m/z=300 and relevant fragmentation ions of mass m/z=284 (M-O), 268, 256, 255, 239, 228.

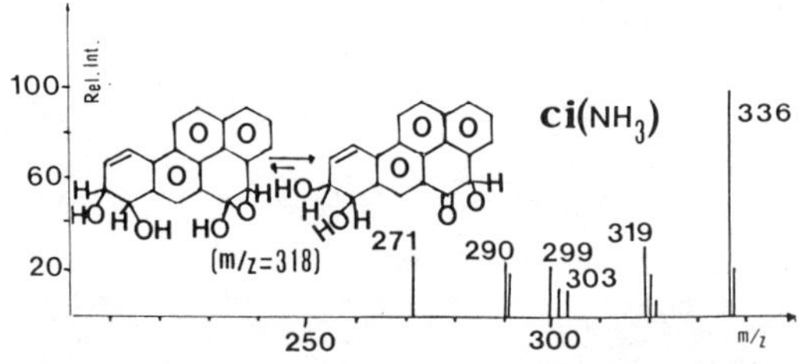

Fig. 7. CI(NH$_3$) mass spectrum of metabolite M_1.

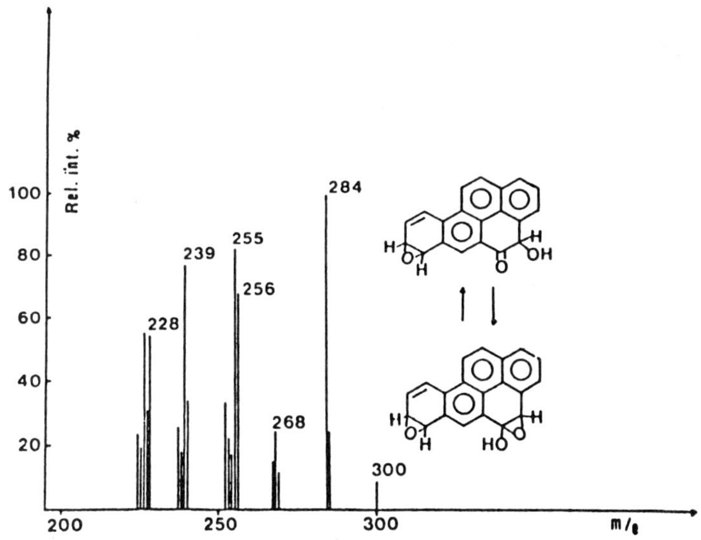

Fig. 8. Mass spectrum (EI) of metabolite M_2.

When the residues, resulting from the incubation of a mixture of cold and tritiated 4-hydroxy-5-oxo-4,5-dihydro-benzo(a)pyrene with human liver microsomes were submitted to HPLC analysis essentially 4 fractions were collected. Fraction 4 was the unchanged K region metabolite eluting after 26 min and the remaining fractions were collected between 11 and 21 min. The relative importance of each fraction I to III (Table 2) was determined on the base of the radioactivity recorded by the radioactivity monitor coupled to the HPLC system. Fractions III, II and I were repeatedly collected in order to be able to perform mass spectrometric analysis. Each fraction was trimethylsilylated and fraction III reveals a molecular ion of mass m/z=372 and fragmentation ions of m/z=344, 327, 283, 255, 239 (Fig. 9).

Looking at the fragmentation pattern (Fig. 10) the molecular ion indicates that a phenolic function fixed most probably in the 7,8 or 9,10 position of the benzo-ring was silylated and this observation is also well corroborated by the presence of fragments corresponding to respective losses of $(M-CO)^+$, $(M-CO-OH\cdot)^+$, $(M-OSI(CH_3)_3-CO)^+$, $(M-OSI(CH_3)_3)^+$, $(M-OSI(CH_3)_3-O-CO)^+$. Moreover in

Table 2.   % of 4-hydroxy-5-oxo-4,5-dihydro-Benzo(a)pyrene metabolized into Fraction I, II and III

Substrate conc.	Incubation time	% of K region metabolite metabolized into:		
		Fraction I	Fraction II	Fraction III
40 µg/ml	30 min	10%	15%	–
40 µg/ml	120 min	15%	15%	5%
10 µg/ml	30 min	25%	25%	–
10 µg/ml	120 min	35%	25%	–

177

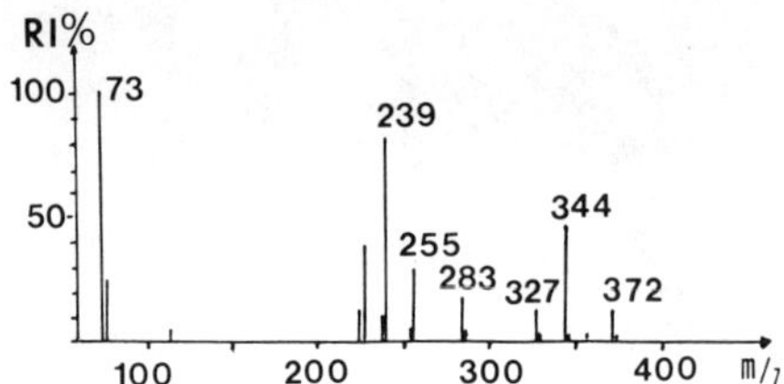

Fig. 9. Mass spectrum (EI) of the trimethylsilylated derivative of fraction III.

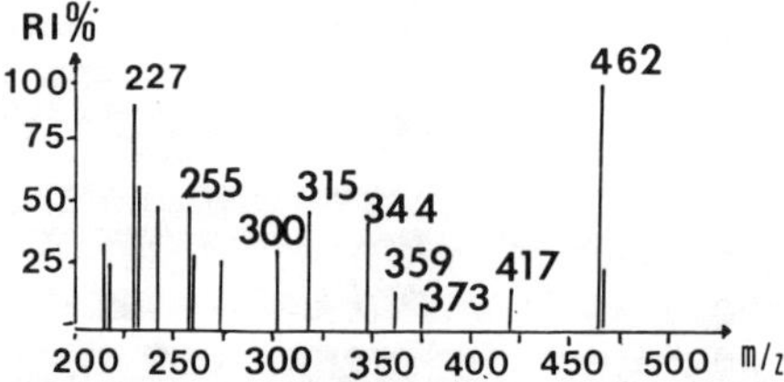

Fig. 10. Fragmentation pattern of the silylated phenol of fraction III.

Fig. 11. Mass spectrum of a silylated dihydrodiol of 4-hydroxy-5-oxo-4,5-dihydro-benzo(a)pyrene.

178

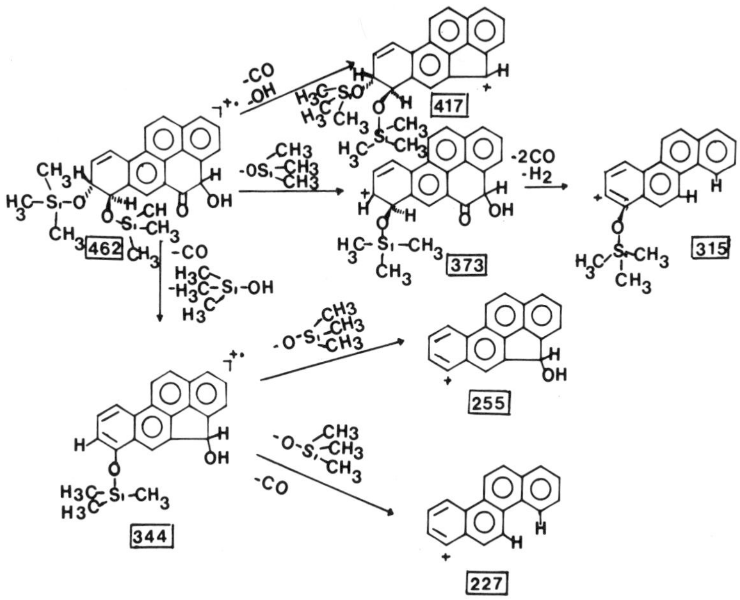

Fig. 12. Fragmentation pattern of a silylated dihydrodiol of
4-hydroxy-5-oxo-4,5-dihydro-benzo(a)pyrene.

fraction I and II two types of derivatives were found, phenol
derivatives and dihydrodiol derivatives. Mass spectrum (Fig. 11)
of a silylated derivative of fraction I reveals mainly the
presence of a molecular ion of mass m/z=462 and fragment ions of
mass m/z=344, 315, 300, 285, 271, 255, 227 and 213 proving the
existence of a 7,8 or 9,10 trans-dihydrodiol of 4-hydroxy-5-oxo-
4,5-dihydro-benzo(a)pyrene.

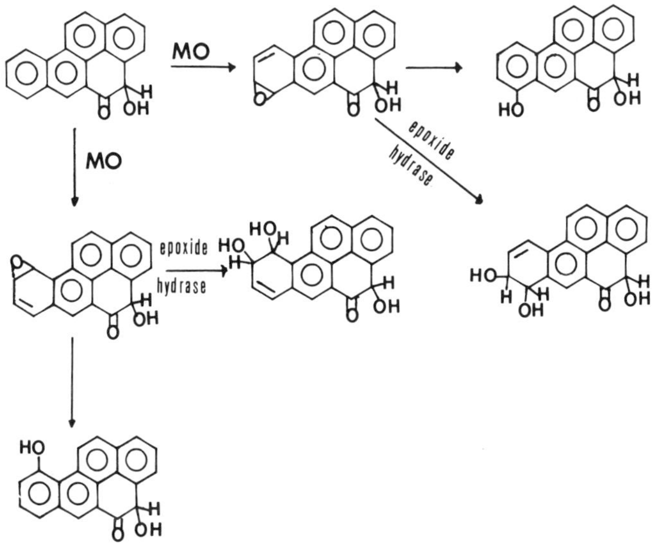

Fig. 13. Metabolic pathway of 4-hydroxy-5-oxo-4,5-dihydro-
benzo(a)pyrene.

As shown in Fig. 12, the observed fragmentation pattern is in
good agreement with the presence of a dihydrodiol since a simple
fragmentation pathway corresponding to successive fragment losses
of CO and $(CH_3)_3$-SiOH giving rise to the ionic species m/z=344 and
again of CO and O-Si$(CH_3)_3$ leading to the fragment ion m/z=227,
proves the presence of two hydroxy groups in the molecule.

Fraction II contains also a phenol and a dihydrodiol of
4-hydroxy-5-oxo-4,5-dihydro-benzo(a)pyrene which are isomeric
compounds of the metabolites isolated in fraction I. Since
4-hydroxy-5-oxo-4,5-dihydro-benzo(a)pyrene is metabolized by human
liver microsomes in different phenols and dihydrodiols of the
benzo-ring, some intermediates are most likely the 7,8 and/or the
9,10 epoxide. Also it may be anticipated that most probably the
7,8-diol-9,10-epoxide does exist by reason of the fact that
fraction I was inhomogeneous and may contain some tetrol
derivatives of this K region metabolite. Metabolism of
4-hydroxy-5-oxo-4,5-dihydro-benzo(a)pyrene by rat and human liver
microsomes leads to the conclusion that this K region polycyclic
aromatic hydrocarbon may be metabolized into different phenols
epoxides and dihydrodiols of the benzo-ring as illustrated in
Fig. 13. Since diol-epoxides are reported[10] as ultimate
carcinogens, it would be of great interest to evaluate the
carcinogenic potency of these derivatives and their DNA binding
capacities in order to be able to prove multiple pathways of
activation of BaP to DNA-binding metabolites as was suggested by
D.P. Scwhartz and W.M. Baird[12].

REFERENCES

1. J.B. Andelman and M.J. Sues, Bull. W.H.O. 43:479 (1970).
2. L.M. Shabad, Y.L. Cohan, A.P. Illnitshyk, A.Y. Khesina, N.P.
   Sherbak and G.A. Smirnov, J. Natl. Cancer Inst. 47:1179
   (1971).
3. "Particulate Polycyclic Organic Matter Committee", National
   Academy of Science Reports (1972).
4. B.S.S. Masters and R.T. Okita, in: "Enzymatic Basis of
   Detoxification", 1:183, Acad. Press, N.Y. (1979).
5. J.C. Kawalek, W. Lewis, D. Ryan, P.E. Thomas and A.Y.H. Lu,
   Mol.Pharmacol. 11:874 (1975).
6. R.I. Freudenthal, A.P. Leber, D. Emmerling and P. Clarke,
   Chem. Biol. Interact. 11:449 (1975).
7. G. Holder, H. Yagi, P. Dansette, D.M. Jerina, W. Levin,
   A.Y.H. Lu and A. Conney, Proc. Natl. Acad. Sci. USA 71:4356
   (1975).
8. M. Rojas and K. Alexandrov, Carcinogenesis 7:235 (1986).
9. J.K. Selkirk, R.G. Cray and H.V. Gelboin, Arch. Biochem.
   Biophys. 168:322 (1975).
10. O. Pelkonen and D.W. Nebert, Pharmacol. Rev. 34:1989 (1982).
11. S.K. Young, Mc. Court, J.C. Leutz and H.V. Gelboin, Science
    196:1199 (1977).
12. D.P. Schwartz and W.M. Baird, Cancer Research 46:545
    (1986).
13. J.N. Pitts, J.R. Karel, A. Van Cauwenberghe, D. Grosjean,
    J.P. Schmid, D.R. Fitz, W.L. Belser, J.R. Gregory, B. Knudson
    and P.M. Hynds, Science 202:518 (1978).
14. G. Lhoest, Recent Developments in Mass Spectrometry in
    Biochemistry and Medicine 6:523 (1980).
15. A. Amar-Costsec, H. Beaufay, M. Wibo, D. Thinès-Semfoux, E.
    Feytmars, M. Robbi and J. Berthet, J. Cell. Biol. 61:201

(1974).

16. O. Lowry, N. Rosebrough, A. Farr and R. Randall, J. Biol. Chem. 193:265 (1951).

17. R.A. Omura and R. Sato, J. Biol. Chem. 239:2370 (1964).

18. J. Van Cantfort, J. De Greave and J.E. Gielen, Biochem. Biophys. Res. Comm. 79:505 (1977).

19. W.F. Greenlee and A. Poland, J. Pharmacol. Exp. Ther. 205:596 (1978).

20. J. Van Cantfort, M. Lèonard-Poma, J. Sèle-Doyen and J.E. Gielen, Biochem. Pharmacol. 32:2697 (1983).

AN INTEGRATED GAS CHROMATOGRAPHIC-MASS SPECTROMETRIC SCREENING

METHOD FOR ANABOLIC STEROID URINARY METABOLITES IN MAN

R. Massé, C. Ayotte, and R. Dugal

National Institute of Scientific Research
University of Quebec
Quebec, Canada

INTRODUCTION

For at least 3000 years, it has been a common practice to use various substances in attempts to increase physical prowess[1]. Since the mid-1970's, an upsurge in the use of anabolic steroids has occured, due in part to a world-wide growth in sports interest and participation, and also because of social and financial implications of athletic achievements. The continued climb of records and anecdotal evidence on the beneficial effects of anabolic steroids have pushed their use to epidemic proportions[2]. For ethical reasons and due to their toxic and hormone-like side effects, the use of anabolic steroids in sports has been prohibited by the International Olympic Committee[3] and all International Sports Governing Bodies.

Various techniques have been developed for the detection of anabolic steroids and their metabolites in urine. Thus, the detection of these substances was originally performed by sensitive radioimmunoassays, but mainly for screening purposes[4-6]. Concurrently, gas chromatographic-mass spectrometric (GC/MS) methods were developed and applied to the characterization of one or several anabolic steroids using packed column gas chromatography and selected-ion monitoring mass spectrometry[7-14]. More recently, several GC/MS methods using fused silica capillary columns were reported for the metabolic profiling and characterization of anabolic steroid urinary metabolites in man[15-20].

A general method for the analysis of anabolic steroids by GC/MS may be subdivided into four main steps: (a) extraction, group separation or purification; (b) hydrolysis of conjugates, group separation or purification; (c) derivatization; (d) GC/MS analysis. The analysis of urine samples for anabolic steroids in doping control situations is normally subjected to various logistic imperatives and time constraints. Thus, different steps in this procedure are omitted or optimized in order to cope with doping control specific analytical challenges.

Although many anabolic steroids are commercially available at present and most are currently used in sports, no extensive and comprehensive GC/MS method describing the specific and simultaneous detection of more than five anabolic steroids has never been reported. The purpose of this paper is to report the first comprehensive method, developed in our laboratoty, for the specific screening of more than twenty anabolic steroid urinary metabolites in man using high resolution gas chromatography combined with mass spectrometry. The general method comprises an efficient extraction of the steroids of interest on Sep-Pak C-18 solid support followed by enzymatic hydrolysis of the crude steroidal extract. The free and unconjugated steroids are then transformed into TMS-enol-TMS-ether derivatives and analyzed by GC/MS in the selected-ion monitoring mode.

EXPERIMENTAL

Chemicals and Solvents

Inorganic salts and solvents were of analytical grade. Solvents were distilled before use. Trimethyliodosilane was kept in a dry dichloromethane solution under nitrogen and stored in the dark at 4°C. Helix Pomatia digestive juice (Sigma Chemical Co., St. Louis, Missouri, USA) and MSTFA (Pierce Chemical Co., Rockford, Illinois, USA) were used as received.

Steroids

Most of the anabolic steroid pharmaceutical preparations and some of their urinary metabolites were kindly supplied by pharmaceutical companies. The other steroids were obtained from Sigma Chemical Co., and Steraloids Inc. (Wilton, New Hampshire, USA).

Human Studies

In order to identify the urinary metabolites of anabolic steroids, urinary excretion studies were performed after the administration of single or multiple doses in normal and healthy male volunteers, using minimum recommended therapeutic doses of the pharmaceutical forms of commonly used and commercially available anabolic steroids. Volunteers were requested to collect daily random urine in prelabelled bottles for seven days. All samples were kept frozen at −20°C until analysis in order to inhibit potential bacterial degradation and artefact formation.

Extraction of Urinary Anabolic Steroids

Urine (5 ml) was applied over a Sep-Pak C-18 cartridge (previously washed with 5 ml of methanol and 5 ml of water) followed by a 5 ml wash with water. The free and conjugated steroids were then eluted with 5 ml of methanol. The solvent was evaporated to dryness at 40°C under a nitrogen stream. The residue was dissolved in 1 ml of 0.2 mol/l sodium acetate buffer (pH 4.5) and 150 µl of Helix Pomatia juice (ß-glucoronidase 5 U/ml, aryl sulfatase 2.5 U/ml) was added. The resulting mixture was incubated at 55°C for 3 hours or at 37°C for 16 hours. The hydrolysate was cooled at room temperature and 100 mg of potassium carbonate and 5 ml of diethylether were added. The mixture was shaken in an Eberback shaker for 10 min and centrifuged at 2500 rpm for 5 min. The etheral layer was decanted and dried over anhydrous sodium sulfate. This extract, containing the free and the hydrolyzed

steroids was concentrated under nitrogen stream to a final volume
of about 100 μl, transferred to a 1 ml reactivial and evaporated
to dryness at 40°C. This vial was then stoppered with a teflon
lined septum under nitrogen. The extraction procedure was modified
as follows when the free steroids fraction was specifically
analyzed. The aqueous sodium acetate buffered solution was in this
case extracted with 5 ml of diethylether before enzymatic
hydrolysis. The etheral layer and the resulting residues were
processed as described above.

<u>Derivatization</u>

To the dry urinary extract were successively added 0.5 to
1 mg of dithioerythritol, 100 μl of MSTFA and 1 μl of a 0.1 mol/l
solution of trimethylsilyliodositane (TMSI) in dichloromethane.
The vial was then stoppered under nitrogen and the mixture heated
at 70°C for 15 min. One μl is injected in the splitless mode for
GC/MS analysis. Potential stanazolol positive sample which are
flagged out by the specific detection of its hydroxylated
metabolites as N,0-TMS derivatives are confirmed as their N-TFA,
0-TMS derivatives. The conversion of the N-TMS to the N-TFA
derivative is done by the addition of 20 μl of N-methylbis-
trifluoroacetamide (MBTFA) to the MSTFA silylation mixture and
heating at 70°C for 10 min.

<u>GC/MS Analysis</u>

A HP-5970 mass selective detector (MSD) linked to a HP-5890
gas chromatograph equipped with a 12 m (0.2 mm I.D.) cross-linked
methyl silicone fused silica capillary column (film thickness
0.33 μm) was used. The injector and transfer line temperatures
were held at 270° and 290°C respectively. The capillary column
temperature is programmed from 100°C (1 min hold) up to 200°C at
16°C/min and at 5°C/min up to 290°C and held for 4.5 min.
Potential testosterone positive samples are confirmed on a 25 m
fused silica capillary column (identical in nature to the 12 m
column used for screening purposes) under the chromatographic
conditions described above.

RESULTS AND DISCUSSION

A thorough knowledge of the urinary excretion profiles of
anabolic steroids in man is an essential prerequisite for the
development of a comprehensive and specific GC/MS screening method
of their urinary metabolites. In this respect, we have designed a
series of analytical and metabolic studies in order to obtain
specific information about the identity of anabolic steroids
urinary metabolites and study their mass spectral properties. Only
few such studies were reported in the literature[11,19,21].
We present in Table 1 a partial listing of the most common
anabolic steroids misused in sports and their corresponding
urinary metabolites which are detected by our GC/MS screening
method. Although only one and occasionally two metabolites are
selected for the GC/MS screening of each anabolic steroid, their
urinary excretion profiles are in most cases relatively complex.

For example, methanedienone, methenolone, oxymesterone and
oxymetholone are extensively metabolized in man yielding seven,
eight, four and seven urinary metabolites respectively. In

Table 1. Partial List of the Anabolic Steroids and some of their
          Urinary Metabolites detected by SIM GC-MS

Steroids	Screening Group	Urinary[a] Metabolites	RRT[b]	M+[c]
19-nortestosterone	1	19-norandrosterone	0.82	420,405,315
	1	19-noretiocholanolone	0.86	420,405,315
	1	19-norepiandrosterone	0.87	420,405,315
Methyltestosterone	2	17α-methyl-3α,17ß-dihydroxy androstane	0.96	450,435,143
Mesterolone	2	mesterolone	0.98	448,433
	2	1-methyl-3α-hydroxy-androstan-17-one	0.95	448,433
Methenolone	2	methenolone	1.02	446,431
	2	1-methylen-3α-hydroxy-androstan-17-one	0.94	446,431
Epitestosterone	2	epitestosterone	0.96	432,417
Testosterone	2	testosterone	1.00	432,417
4-chlorotestosterone	3	4-chlorotestosterone	1.02	468,453
Oxandrolone	3	oxandrolone	1.09	378,363,143
Fluoxymesterone	4	6ß-hydroxyfluoxymesterone	1.13	642,552,143
Norethandrolone	4	19-nor-5ξ,17α-pregnane-3α 17ß,21-triol	1.13	538,421,331
Ethylestrenol	4	19-nor-5ξ,17-pregnane-3α,17ß,21-triol	1.13	538,421,331
Methanedienone	5	6ß-hydroxymethanedienone	1.16	532,517,143
Oxymesterone	6	3ξ,4ξ,17ß-trihydroxy-17α-methyl-5-ξandrostane (2 isomers)	1.20	538,523,143
			1.24	538,523,143
Oxymetholone	6	2-hydroxymethyl-3α,6ß,17ß trihydroxy-17α-methyl-androstane	1.25	640,550,143
	6	2-hydroxymethyl-17ß-hydroxy-17α-methyl-androstan-3-one	1.22	550,460,143
4-chloromethanedienone	6	6ß-hydroxyturinabol	1.25	494,404,143
Stanozolol	7	4ξ-hydroxystanazolol	1.49	560,545,143

a. urinary metabolites selected for GC/MS screening and positive identification of parent drug administration.
b. retention times relative to that of testosterone TMS-enol-TMS-ether derivative.
c. Molecular ions and selected ions of the TMS-enol-TMS-ether derivatives.

general, only the most prominent ones are selected for screening purposes, but one has also to take into account their mass spectral properties and relative gas chromatographic retention when selecting the ion groups and corresponding scanning intervals. In order to maximize sensitivity and specificity of the GC-MS screening method, the steroids were divided into seven[7] groups and sets of 2 to 4 ions were selected for each compound. Each group of ions was consecutively scanned in the SIM mode in accordance with the retention times of the corresponding steroid TMS derivatives (Table 1). The steroids are analyzed as the TMS-enol-TMS-ether derivatives since, in most cases, their mass spectral fragmentation is relatively light and characteristic ion at M+· and [M-15]+ are carrying a very large proportion of the total ionic current. This feature allows for lower detection limits and increased specificity in detection of the steroids of interest.

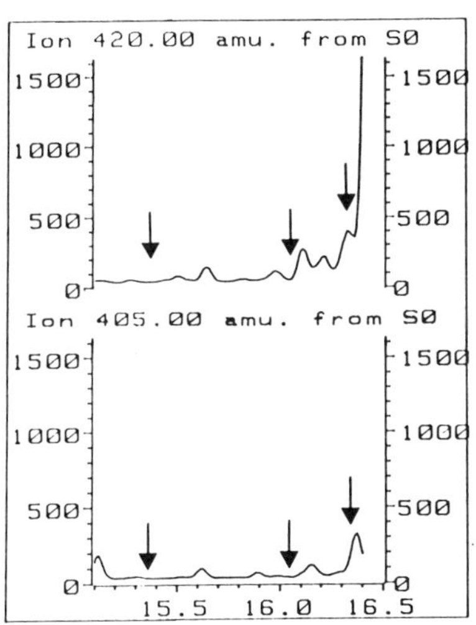

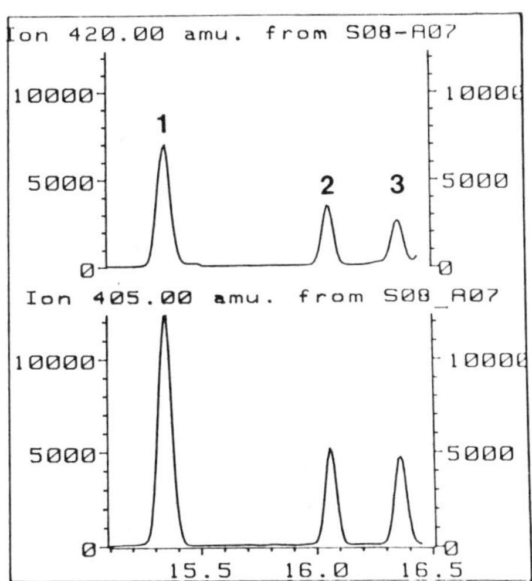

Fig. 1. Computer reconstructed ion-fragmentograms from analysis of TMS-enol-TMS-ether derivatives of the combined conjugated and unconjugated neutral steroid in: a) normal blank urine and b) urine sample from a young male athlete; m/z values typical for the derivatives of 19-nortestosterone urinary metabolites (1) 19-norandrosterone (2) 19-noretiocholanolone and (3) 19-norepiandrosterone.

The programming of the capillary column up to 290°C permits the elution of the long retention time compounds. No sample to sample contamination has been observed after the injection of several thousand urinary extracts over a period of 30 months. Furthermore, these data have shown that consecutive sample injection (about 20 samples per day) have no measurable effect on the reproducibility of retention times and the sensitivity of the assay over a period of time as long as 3 to 4 months.

The specific detection of 19-nortestosterone urinary metabolites using selected-ion monitoring GC/MS is illustrated in Fig. 1. The ions fragmentograms at m/z 420 and 405 of an urinary blank (Fig. 1a) and positive urine (Fig. 1b) samples cleary show that no peak interferes with those corresponding to the metabolites of interest. However, in cases where the relative concentration of these steroids is much lower and where interfering peaks are emerging from the background noise at the retention time of one, two or all three metabolites of 19-nortestosterone, the positive assessment of these exogeneous steroids must be based on an extensive mass spectrometric analysis of the sample. This includes:reextraction of a larger volume of urine, preparation of other type of derivatives and comparison of the gas chromatographic properties of the steroid derivatives with those of authentic compounds, selected ion monitoring analysis of the steroid derivatives with the determination of the relative intensities of the selected ions and comparison with those of authentic compounds and with those of urinary metabolites obtained from human studies and finally repetitive scanning analysis with background substraction in order to obtain the characteristic mass spectrum of at least 19-norandrosterone, the major metabolite of 19-nortestosterone in man.

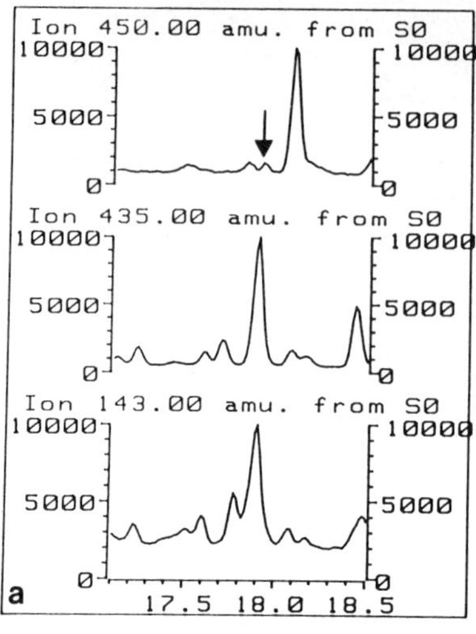
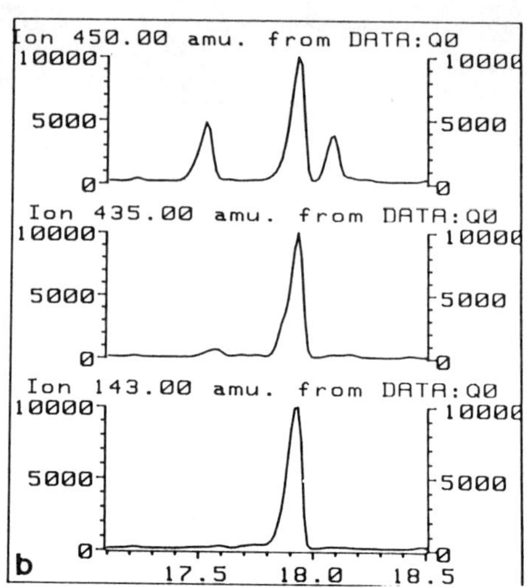

Fig. 2. Computer reconstructed ion-fragmentograms from analysis of TMS-ether derivatives of the combined conjugated and unconjugated neutral steroids in: a) normal blank urine and b) urine sample from young male athlete; m/z value typical for the TMS derivative of methyltestosterone urinary metabolite.

The detection of some anabolic steroids requires from the analyst the specific observation of the corresponding ion-fragmentogram profiles which are often very similar for both normal (negative) and potential positive urine sample. This is the case for methyltestosterone urinary metabolite where the characteristic ion profiles at m/z 435 and 143 in the blank urine sample (Fig. 2a) show prominent peaks at the retention time of tetrahydromethyltestosterone TMS derivative. The presence of an intense peak at m/z 450, the elution pattern of which coincides with that of its daughter ions at m/z 435 and 143 (Fig. 2b) is indicative of the potential presence of this anabolic steroid.

The same rationale is used for the detection and positive identification of the other steroids listed in Table 1. Some of the characteristic ions of their TMS derivatives are monitored within discreet and specific time intervals. The resulting ion-fragmentogram profiles are then compared with those from normal urine samples.

Testosterone (T) abuse is detected by measuring selectively and specifically its relative concentration to that of epitestosterone (E). In normal and healthy individuals, the average T/E ratio is about 1.0. A urine sample is declared positive to testosterone when this ratio is greater than 6. The ratio is determined using the corresponding peak areas obtained from ion fragmentograms of m/z 432, and 417 ($M^{+\cdot}$ and $[M-15]^+$ ions of T and E). The molecular ion of 11ß-hydroxyandrosterone at m/z 522 is also monitored to assess the chromatographic homogeneity of the testosterone peak. Fig. 3 shows the characteristic profile of the fragment-ion current chromatogram for the ion at m/z 432 obtained by SIM GC/MS analysis. One can observe that the steroids of interest are chromatographically separated from the other endogeneous steroids. This enables the accurate determination of

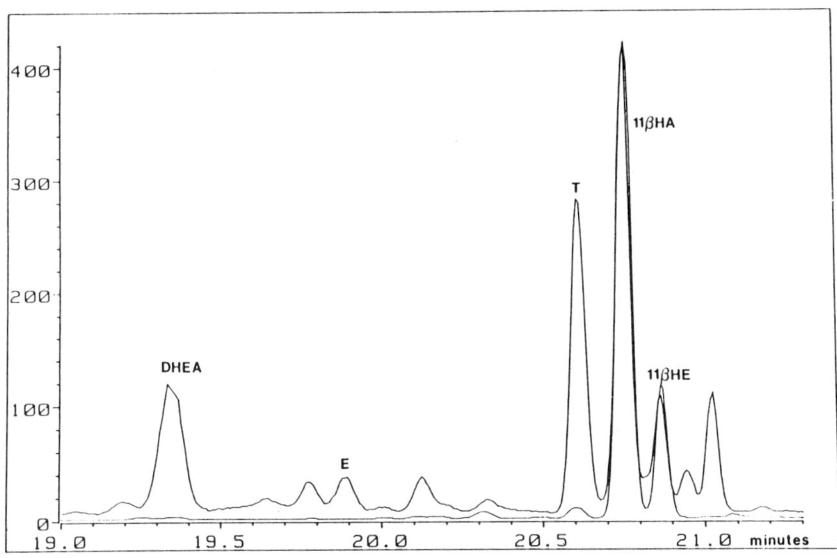

Fig. 3. Determination of T/E ratio by recording the molecular
         ion of testosterone (T) and epitestosterone (E) TMS-
         enol-TMS-ether derivatives using SIM GC-MS. In this case
         T/E = 7.2. DHEA: dehydroepiandrosterone, 11 HA = 11ß-
         hydroxyandrosterone, 11 HE = 11ß-hydroxyetiocholanolone.

the T/E ratio which is further assessed by triplicate SIM
analysis. In this case the ratio (7.2) indicates that the urinary
sample was positive to testosterone, owing to the fact that
epistestosterone level was within normal concentration range.

In conclusion, the present method provides a sensitive and
reliable technique for the specific detection of anabolic steroid
urinary metabolites in man. Although routinely used in our
laboratory for the analysis of several thousand samples in the
last three years, the method is periodically revised and updated
in order to increase the number of anabolic steroids screened for
and improve its proficiency level and sensitivity.

Acknowledgments

This work was supported by the Sport Medicine Council of
Canada and in part by the Natural Sciences and Engineering
Research Council of Canada. The skilful assistance of Mrs. Pauline
Fournier is gratefully acknowledged.

REFERENCES

1. E.I.Emmanouel, "History of Pharmacy" (Istoria Pharmakeutikis),
   Pryssos, Athens (1947).
2. D.R. Lamb, Am. J. Sports Med. 12:31 (1984).
3. Report of the Medical Commission of the IOC, Innsbruck,
   Austria (1974).
4. N.A. Sumner, J. Steroid Biochem. 5:307 (1974).

5. R.V. Brooks, R.G. Firth and N.A. Sumner, Br. J. Sports Med.
    9:89 (1975).
6. R.V. Brooks, G. Jeremiah, W.A. Webb and M. Wheeler, J.
    Steroid Biochem. 11:913 (1979).
7. R.J. Ward, C.H.L. Shackleton and A.M.Lawson, Br. J. Sports
    Med. 9:93 (1975).
8. R.J. Ward, A.M. Lawson and C.H.L. Shackleton, J. Steroid
    Biochem. 8:1057 (1977).
9. C.J.W. Brooks, A.R. Thawley, P. Rocher, B.S. Middleditch,
    G.M. Anthony and W.G. Stillwell, J. Chromatogr. Sci. 9:35
    (1977).
10. H.W. Durbeck, I. Buker, B. Scheulen and B. Telin, J.
    Chromatogr. 167:117 (1978).
11. M. Bertrand, R. Massé and R. Dugal, Belgisch. Farm.
    Tidschrift 55:55 (1978).
12. I. Bjorkhem, O. Lantto and A. Lof, J. Steroid Biochem.
    13:169 (1980).
13. O. Lantto, I. Bjorkhem, H. Ek and D. Johnston, J. Steroid
    Biochem. 14:721 (1981).
14. I. Bjorkhem and H. Ek, J. Steroid Biochem. 18:481 (1983).
15. H.W. Durbeck, I. Buker, B. Scheulen and B. Telin,
    J. Chromatogr. Sci. 21:405 (1983).
16. G.P. Cartoni, M. Ciardi, A. Giarrusso and F. Rosati,
    J. Chromatogr. 279:515 (1983).
17. V.P. Uralets, V.A. Semenova, M.S. Yakushin and V.A. Semenov,
    J. Chromatogr. 279:695 (1983).
18. G.P. Cartoni, A. Giarrusso, M. Ciardi and F. Rosati, J. High.
    Res. Chromatogr. Commun. 8:539 (1985).
19. R. Massé, C. Laliberté, L. Tremblay and R. Dugal, Biomed.
    Mass Spectrom. 12:115 (1985).
20. R. Massé, C. Ayotte, C. Laliberté and R. Dugal, J. Pharm.
    Biomed. Anal., in press.
21. R. Massé, C. Laliberté and L. Tremblay, J. Chromatogr.
    339:11 (1985).

METHOXYIMINO-<u>tert</u> -BUTYLDIMETHYLSILYL ETHERS AS DERIVATIVES FOR THE GAS CHROMATOGRAPHY-MASS SPECTROMETRY OF 3 (or 17)-HYDROXY-C_{19} - AND 20-HYDROXY-C_{21}-KETOSTEROIDS

B.P. Lisboa

Clinic of Obstetrics and Gynecology
University of Hamburg
Hamburg, FRG

INTRODUCTION

Since the application of <u>t</u>.-butyldimethylsilyl (<u>t</u>.BDMSi) ether derivatives to the GC-MS analysis of steroids by Kelly and Taylor[1] in 1974, this silylation has been widely used in the selective detection of hydroxylated steroids[2-22].

During the silylation of ketosteroids an enolization occurs as a side-reaction. Unsaturated Δ^4 -3-keto-[3,8,13,14] and saturated 5α (H)-3-ketosteroids[8,12,14] are transformed to some extention to 2-dehydro-3-hydroxy-compounds, whereas reversible enolization of the 20-keto group leads to an epimerization of the side-chain in steroids of the pregnane series[13]. This enolization also observed during the formation of trimethylsilyl ethers, can be avoided by protecting the keto group through the formation of methoxyimino (methyloxime, MO) derivatives prior to silylation.

Methoxyimino-<u>t</u>.BDMSi ethers have previously been employed during the investigation by GC-MS of ketonic sterols at 22.5 eV[7] and $C_{19}O_2$-steroids at 70 eV[3,15]. However, as indicated previously by Gaskell and Brooks[3] and reported later by Ballhorn and collaborators[16], the spectra obtained at 70 eV (11.2 aJ) for such derivatives present a great number of fragment ions and base peaks of lower intensity, when compared to those obtained for the corresponding <u>t</u>.BDMSi ethers.

The purpose of this study was to investigate the GC-MS behaviour of a number of 3- and 17-keto-$C_{19}O_2$-, 3-keto-Δ^4 -$C_{19}O_2$- as well as 3- and 20-keto-$C_{21}O_2$-steroids of biological importance when subjected to low-resolution electron impact mass spectrometry using lower ionization energy (22.5 eV).

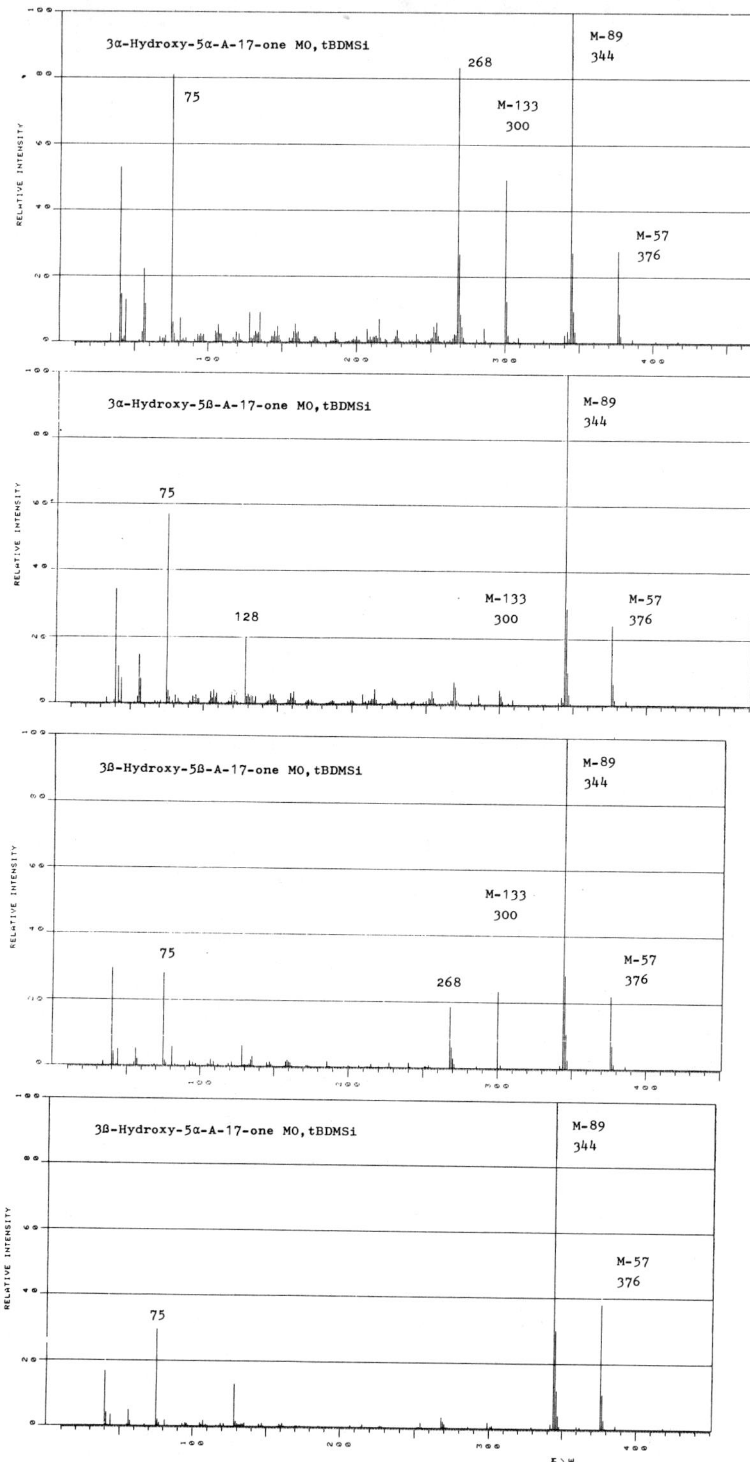

Fig. 1. Mass spectra of the methoxyimino-t.BDMSi ether derivatives of isomeric 3-hydroxy-androstan-17-ones.

MATERIALS AND METHODS

Chemicals and solvents of analytical grade were obtained from Merck AG (Darmstadt, FRG) unless otherwise stated. 3α-Hydroxy-5α-pregnan-20-one and 20α-hydroxy-4-pregnen-3-one were from Sigma Chemical Co. (St. Louis, Mo, USA); all other steroids were purchased from Steraloids Inc. (Wilton, NY, USA). Sephadex LH 20 was from Pharmacia (Uppsala, Sweden).

The reagent mixture used for silylation (t.BDMSi-reagent) containing tert.-butyldimethylsilyl chloride (1 mmole), imidazole (2.5 mmoles) and dimethylformamide (to 1.5 ml) was purchased from Applied Science Laboratories (State College, Pa, USA).

The methoxymino (methyloxime)-tert.-butyldimethylsilyl ether (t.BDMSi ether) derivative was prepared by adding 0.5 ml of a 3% solution of methoxyamine hydrochloride (Eastman Kodak Inc., Rochester, USA) in dry pyridine (17 mg/ml) and leaving for 24 h at room temperature. After evaporation of pyridine under nitrogen, the steroid sample was treated with 20 µl of t.BDMSi-reagent for 30 min at room temperature. The mixture was filtered through a small column (0.5 x 2 cm) of Sephadex LH 20 swollen in chloroform/n-hexane 1:1[2]; the first 2 ml eluted with the solvent were dried down under nitrogen and the residue dissolved in 100 µl of ethyl acetate for analysis by GC-MS.

An LKB 9000S gas chromatograph-mass spectrometer instrument was used, equipped with a 25 m open tubular capillary column coated with OV-101 (0.32 I.D.). Operating conditions: column temperature 240° or 260°, separator temperature 270°, ion source temperature 290°C, ionization voltage 22.5 eV (3.6 aJ), ionizing current 120 µA. A solventless solid injection system connected to the capillary column with shrinkable PTFE tubing, was used.

Successively scanned mass spectra (m/z 40-800) were recorded on magnetic disk every 16s via an LKB 2130 data system connected on line to the mass spectrometer.

RESULTS

The general feature of the mass spectra obtained for methoxyimino-t.BDMSi ethers indicate in most cases smaller or no molecular ions and four groups of primary fragment ions resulting from:
a) loss of CH_3 group (M-15) either from the steroid nucleus or the silyl derivative;
b) the methoxyimino-derivative by loss of 30 (CH_2O), 31 (CH_3O) or 32 (CH_3OH) daltons;
c) the silyl radical, by loss of 57 (tert.-butyl moiety, C_4H_9-) or 132 [tert.-butyldimethylsilanol, $C_4H_9-Si(CH_3)_2OH$] daltons;
d) for C_{21}-steroids, by cleavage of the side-chain, M-73 (20-methoxyimino, 21-desoxy-) or M-159 (20-t.BDMSi ether, 21-desoxy-steroids).

Many fragments contribute to the formation of secondary ions, as for instance M-87 (increments 57 and 30), M-88 (57 and 31), M-89 (57 and 32), M-133 (57 and 76, this latter corresponding to dimethylsilanol), M-147 (15 and 132), M-148 (15,57 and 76), M-163 (31 and 132), M-164 (31 and 133), M-178 (15,31 and 132). In some cases two groups of increments can be hypothetically responsible

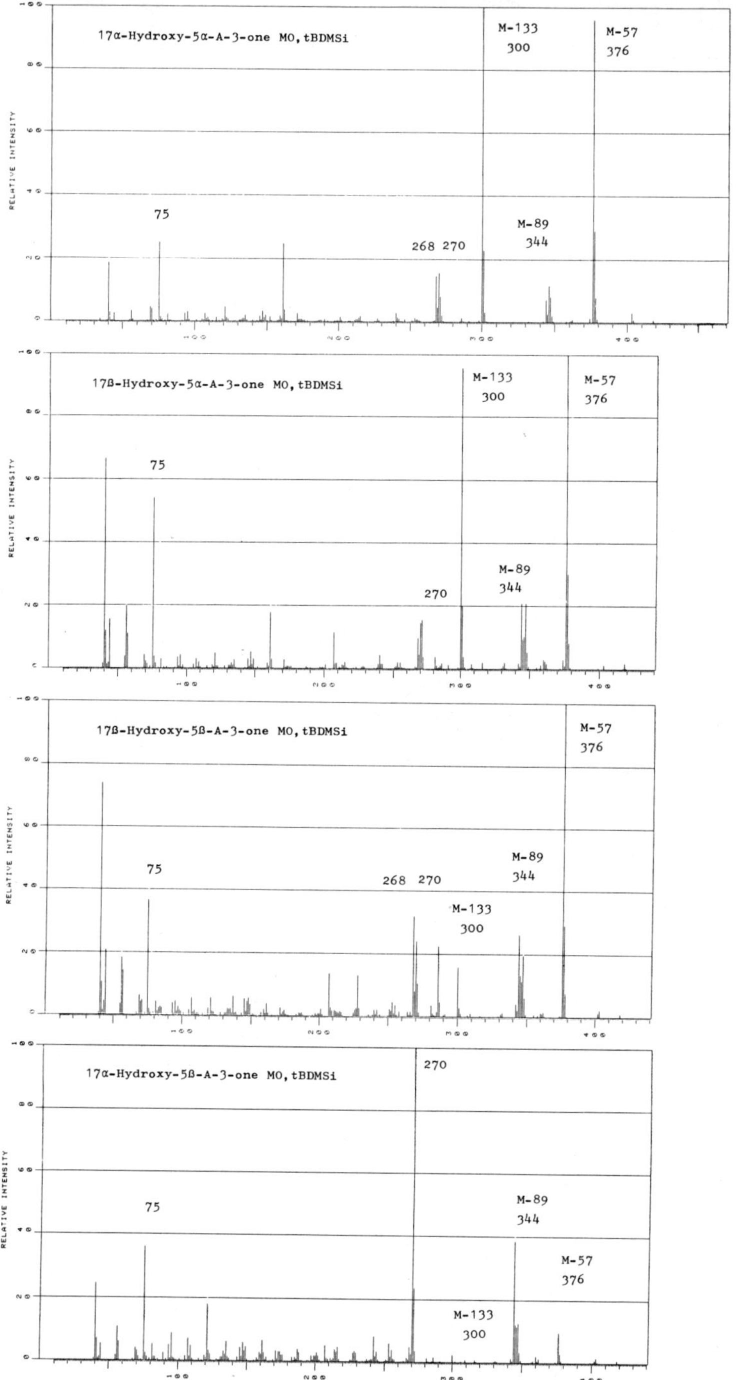

Fig. 2. Mass spectra of the methoxyimino-t.BDMSi ether derivatives of isomeric 17-hydroxy-androstan-3-ones.

for the formation of the daughter ion. The elemental composition
of most of the fragment ions has been studied by several
investigators as referred to in the Introduction[5,12,15].

I. $C_{19}O_2$-Steroids

Electron-impact mass spectra obtained at 22.5 eV for the
methoxyimino-$\underline{t}$.BDMSi ether derivatives of isomeric 3-hydroxy-17-
keto-C_{19}-steroids as well as 17-hydroxy-3-keto-C_{19}-steroids are
reproduced in Fig. 1 and 2 respectively. The base peak in all the
four isomeric 17-ketosteroids is the daughter ion m/z 344
originating from the loss of fragments m/z 57 and 32. In the
spectrum of 3α-hydroxy-5α-androstan-17-one (androsterone) this
fragment accounts for 17% of the total ion current for values
above m/z 50 (%Σ_{50}). The two other important ions are m/z 376
(M-57) and m/z 75. The spectra of androsterone and
epietiocholanolone (3ß-hydroxy-5ß-androstan-17-one) present
significant daughter fragment ions at m/z 300 (M-133) and 268
[M-(133+32)] formed by losses of 57, 76 and (for m/z 268) 32
daltons, as described before by Gaskell and Pike[15]. The spectra of
etiocholanolone (3α-hydroxy-5ß-androstan-17-one) and
epiandrosterone (3ß-hydroxy-5α-androstan-17-one) have high
abundance of the fragment m/z 128. A summary of the relative

Table 1. Abundance of Significant Fragment Ions, mean and (S.D.),
found in the Electron-Impact Mass Spectra (22.5 eV) of
the Isomeric 3-Hydroxy-Androstan-17-ones after
Methoxyimino-<u>tert</u>.Butyldimethylsilyl Ether
Derivatization. n = Number of Spectra

	Steroids			
	3α/5α (n=8)	3ß/5ß (n=8)	3α/5ß (n=7)	3ß/5ß (n=7)
m/z	t_R 0.90	t_R 1.21	t_R 0.93	t_R 0.97
402 (a)	0.5 (0.3)	0.1 (0.1)	0.2 (0.1)	0.1 (0.1)
376 (b)	28.1 (3.7)	28.6 (5.8)	27.0 (7.1)	23.5 (3.2)
344 (c)	(bp)	(bp)	(bp)	(bp)
300 (d)	50.9 (6.5)	1.2 (0.9)	4.3 (1.5)	26.6 (3.1)
270 (e)	7.8 (0.9)	2.1 (0.7)	5.5 (1.1)	4.1 (0.6)
269 (f)	24.6 (3.0)	3.8 (2.0)	4.4 (2.3)	9.3 (2.3)
268 (g)	84.8 (4.6)	3.6 (0.7)	1.8 (0.6)	19.1 (1.2)
254	4.8 (1.5)	2.2 (1.2)	4.0 (1.2)	2.8 (1.4)
253	2.1 (0.8)	0.4 (0.2)	2.0 (0.1)	1.0 (0.5)
135	8.7 (1.7)	1.6 (0.6)	2.0 (0.9)	3.4 (1.2)
128	9.4 (2.1)	14.6 (2.8)	22.0 (8.3)	6.1 (1.4)
75	73.5 (8.6)	37.6 (6.1)	55.1 (7.4)	43.9 (6.6)

t_R = relative retention time to 5α-cholestane (658s, 260°C) of
the principal geometric isomer (3α/5α, <u>a</u>; 3ß/5α, <u>c</u>; 3α/5ß, <u>d</u>; 3ß/5ß,
<u>f</u>); see Fig. 3. Fragment ions: (a) M-31 (b) M-57 (c) M-89 (d) M-
133 (e) M-(132+31) (f) M-(133+31) (g) M-(133+32).
(bp)= base peak.

intensities of the twelve principal fragments found for the four
isomers and their t$_R$ to 5α-cholestane on an OV-101 capillary column
is presented in Table 1.

The spectra of 5α-dihydro-testosterone (17ß-hydroxy-5α-
androstan-3-one) and 5α-dihydro-epitestosterone (17α-hydroxy-5α-
androstan-3-one), as shown in Fig. 2 and Table 2 are characterized
by very intense fragment ions m/z 376 (M-57) and m/z 300 (M-133).
Ions with lower abundance at m/z 270 [M-(31+132)], m/z 268
[M-(31+133)], m/z 344 (M-89) and m/z 347 (M-87) all originating in
part from the methoxyimino function of the steroid. The spectra of
5ß-dihydro-testosterone (17ß-hydroxy-5ß-androstan-3-one) and 5ß-
dihydro-epitestosterone (17α-hydroxy-5ß-androstan-3-one) are
characterized by base peaks at m/z 376 (M-57) and m/z 270
[M-(31+132)] respectively and a lower relative intensity for the
fragment ion m/z 300 (M-133). The base peak (m/z 270) in the
spectrum of 5ß-dihydro-epitestosterone accounts for intensity
values between 15 and 18.8 per cent of Σ$_{50}$. The fragment ion m/z

Table 2. Abundance of Significant Fragment Ions, mean and (S.D.),
found in the Electron-Impact Mass Spectra (22.5 eV) of
the Isomeric 17-Hydroxy-Androstan-3-ones after
Methoxyimino-<u>tert</u>.Butyldimethylsilyl Ether
Derivatization. n = Number of Spectra

	Steroids			
	17α/5α (n=11)	17ß/5α (n=12)	17α/5ß (n=13)	17ß/5ß (n=9)
m/z	t$_R$ 1.01	t$_R$ 1.33	t$_R$ 0.86	t$_R$ 1.16
403	1.6 (1.2)	1.8 (1.0)	1.9 (0.7)	2.6 (1.5)
402	*	*	0.9 (0.3)	1.0 (0.4)
376 (a)	98.6 (2.1) (bp)	99.8 (0.6) (bp)	8.5 (2.6)	(bp)
347	14.5 (6.6)	21.3 (5.1)	11.8 (4.9)	19.9 (6.3)
346 (b)	8.8 (1.5)	10.0 (5.5)	13.2 (3.9)	16.1 (5.0)
345 (c)	3.9 (2.0)	9.9 (5.4)	12.2 (2.0)	11.0 (3.9)
344 (d)	12.8 (5.1)	18.8 (5.1)	37.6 (4.4)	23.6 (5.6)
300 (e)	98.0 (2.5)**	95.4 (4.2)**	2.0 (0.6)	17.6 (5.5)
271	12.4 (6.4)	16.9 (3.1)	23.9 (1.9)	10.1 (2.1)
270 (f)	11.5 (2.6)	16.9 (2.7)	(bp)	25.0 (2.7)
268	15.4 (3.2)	12.0 (2.5)	4.2 (1.8)	31.3 (2.8)
253	1.6 (1.2)	1.8 (0.7)	5.6 (1.8)	4.8 (1.9)
161	22.1 (6.1)	19.4 (2.8)	7.3 (1.2)	3.5 (0.7)
147	3.4 (1.0)	5.9 (1.0)	6.2 (1.2)	4.7 (1.2)
121	3.9 (0.7)	5.5 (0.9)	19.7 (3.6)	6.4 (1.4)
107	2.3 (0.6)	3.9 (1.1)	8.5 (1.8)	6.4 (1.2)
75	34.5 (5.9)	54.6 (7.4)	33.4 (5.8)	30.0 (5.5)

* less than 1%
** base peak in some spectra.
t$_R$ = relative retention time to 5α-cholestane (658s, 260°C).
Fragment ions: (a) M-57 (b) M-87 (c) M-88 (d) M-89 (e) M-133
(f) M-163.

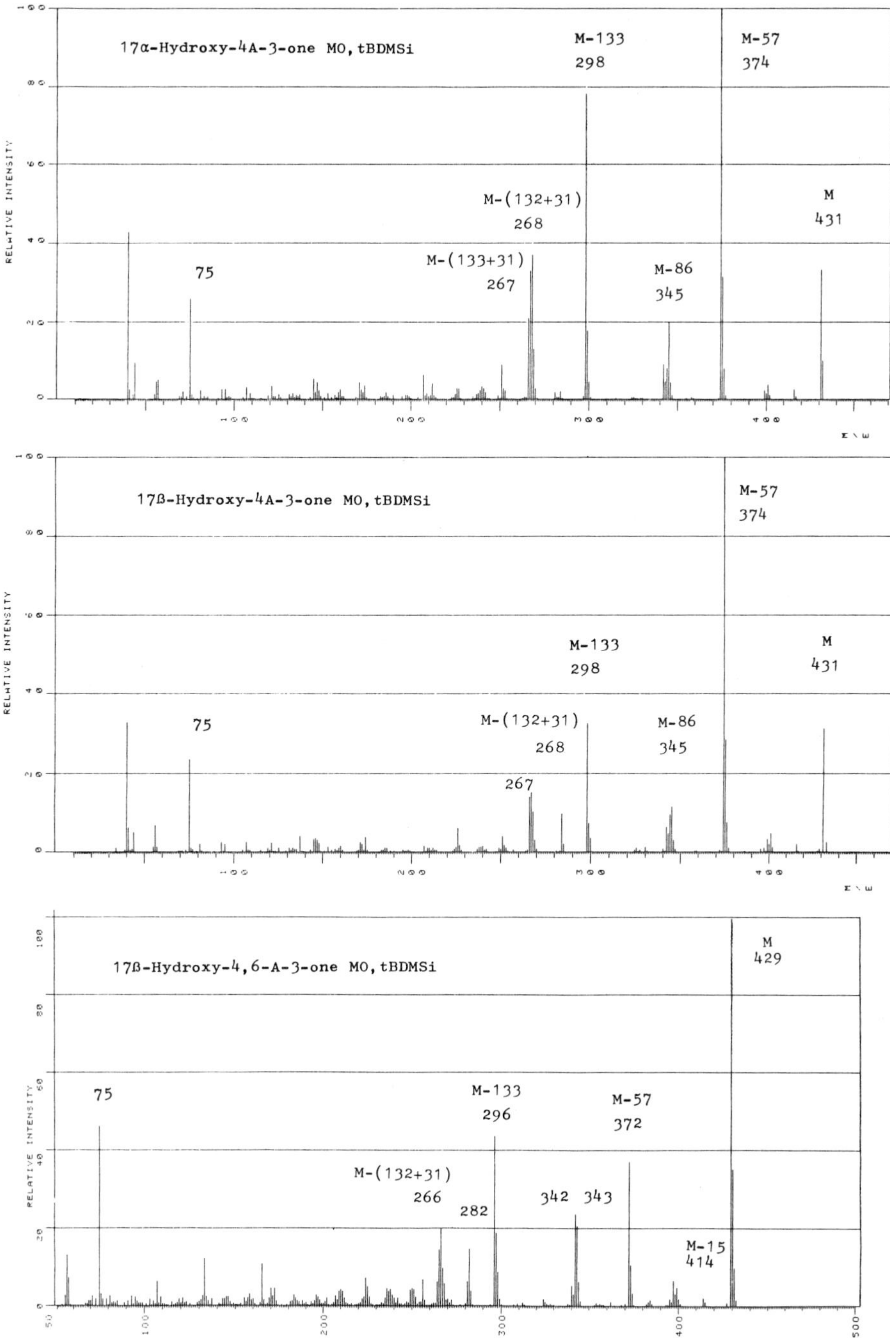

Fig. 3. Mass spectra of the methoxyimino-t.BDMSi ether derivatives of epitestosterone, testosterone and 6-dehydrotestosterone.

161 reported by Quilliam and Westmore[12] for several sterically crowded silyl ethers of 5α-dihydro-testosterone was found in the spectra of the four isomers investigated as methoxyimino-$\underline{t}$.BDMSi ether derivatives. The ions m/z 145 and 147 observed by Quilliam and Westmore[12] in the spectrum of 5α-dihydro-testosterone and Lisboa and Ganschow[11] in all the four epimers as $\underline{t}$.BDMSi ethers could also be observed after methoxyimino-$\underline{t}$.BDMSi ether derivatization (Fig.2). The abundances of significant fragment ions (mean and S.D.) found at 22.5 eV for the isomeric 17-hydroxy-3-ketosteroids, as well as their t_R-values to 5α-cholestane are summarized in Table 2.

Fig. 3 and Table 3 give the data obtained for the mass spectra of testosterone (17ß-hydroxy-4-androsten-3-one), epitestosterone (17α-hydroxy-4-androsten-3-one) and 6-dehydro-testosterone (17ß-hydroxy-4,6-androstadien-3-one) acquired at 22.5 eV as methoxyimino-$\underline{t}$.BDMSi ethers. Significant molecular ions were observed for all three steroids, being base peak for 6-dehydro-testosterone. The high abundances of fragment ions at M-57 (base peak for testosterone and epitestosterone) and the daughter ion M-133 (m/z 298 for testosterone and epitestosterone, m/z 296 for

Table 3. Abundance of Molecular and Significant Fragment Ions, mean and (S.D.), found in the Electron-Impact Mass Spectra (22.5 eV) of Testosterone (T, n=17), Epitestosterone (epi-T, n=11) and 6-Dehydro-Testosterone (dehydro-T, n=8) after Methoxyimino-$\underline{tert}$.Butyldimethylsilyl Ether Derivatization

m/z		T	epi-T	m/z		dehydro-T
431	(a)	32.2 (3.8)	33.2 (1.6)	429	(a)	(bp)
416		2.7 (0.6)	2.3 (0.8)	414		2.6 (0.3)
401		5.1 (2.3)	2.6 (1.6)	399		2.9 (1.4)
400		1.9 (0.8)	1.3 (0.2)	398		2.4 (0.6)
375		30.8 (3.2)	29.9 (1.2)	397		3.8 (2.0)
374	(b)	(bp)	(bp).	373		11.2 (0.3)
346		7.3 (3.7)	4.3 (3.2)	372	(b)	37.3 (2.1)
345	(c)	19.3 (6.1)	16.0 (6.3)	344		5.0 (1.0)
344		10.5 (3.8)	8.6 (2.9)	343	(c)	14.8 (3.8)
343		7.7 (2.6)	4.0 (2.0)	342		12.7 (6.8)
342		6.8 (2.1)	8.2 (1.1)	296	(e)	41.9 (2.2)
299	(d)	6.6 (2.6)	17.5 (1.1)	297	(d)	19.3 (1.4)
298	(e)	34.1 (4.1)	73.8 (5.4)	296	(e)	41.9 (2.2)
284	(f)	11.6 (1.9)	1.5 (0.6)	282	(f)	14.8 (1.0)
269		4.3 (1.3)	13.1 (2.5)	266	(g)	16.6 (3.0)
268	(g)	11.7 (2.4)	36.5 (4.9)	265	(h)	14.9 (1.5)
267	(h)	17.9 (3.1)	31.2 (4.6)	165		11.5 (1.4)
266		14.2 (3.9)	20.9 (2.1)	133		11.4 (1.7)
251		4.6 (1.2)	7.9 (3.0)	107		5.9 (2.6)
75		27.0 (5.5)	28.8 (6.9)	75		38.5 (6.8)

Molecular and fragment ions: (a) M (b) M-57 (c) M-86 (d) M-132 (e) M-133 (f) M-(132+15) (g) M-(132+31) (h) M-(133+31). n = number of spectra. Relative retention time to 5α-cholestane (646s, 240°C): T (1.49), epi-T (1.15).

6-dehydrotestosterone) were noteworthy. Primary (M-30, M-32) and
secondary (M-87, M-89) ions, the latter including part of
methoxyimino derivatization, were found in the spectra of epimeric
testosterones (m/z 401, 400, 344 and 342) and 6-dehydro-
testosterone (m/z 399, 398, 342 and 340), as indicated in Table 3.
Intensities in per cent of Σ_{50} for the fragment ions 374 (base peak
for epimeric testosterones) and molecular ion 429 (base peak for
6-dehydro-testosterone) were calculated between 13.3 and 16.6 for
the first and 11-14.4 for the latter.

Fig. 4 reproduces two computer reconstructed fragment ion
current chromatograms from GC-MS obtained after repetitive
scanning from a mixture of isomeric 3-hydroxy-androstan-17-ones
carried out on an OV-101 25 m capillary column at 240°C, after
methoxyimino-t.BDMSi ether derivatization. The chromatogram on the
left was reconstructed from the ions at m/z 300 and 260 and that
to the right for the ions at m/z 300, 344 and 376. Androsterone
(two isomers a and b) and epietiocholanone (two isomers d and e)
were detected by recording the ion at m/z 300 whereas the ion m/z
268 was chosen for epiandrosterone (two isomers f and g).
Etiocholanolone (scan number 186, right hand picture) was detected
by the fragments at m/z 376 and 344.

II. 3-Keto-Δ_4-C$_{19}$O$_3$-Steroids

The mass spectra of four C$_{19}$O$_3$-steroids as methoxyimino-bis-
t.BDMSi ethers, 19-hydroxy-androstenedione (19-hydroxy-4-
androstene-3, 17-dione), 16ß-hydroxy-testosterone (16ß, 17ß-
dihydroxy-4-androsten-3-one), 6ß-hydroxy-testosterone (6ß, 17ß-

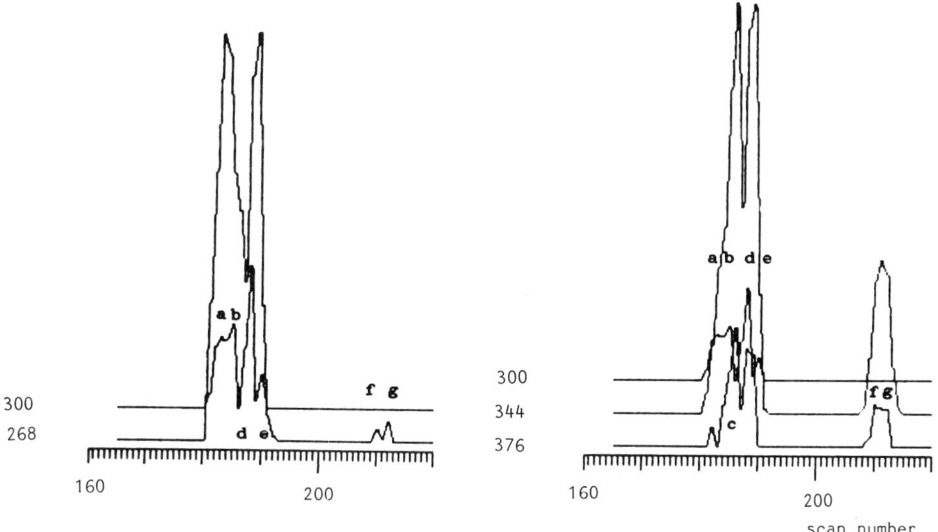

Fig. 4. GC-MS analysis of a mixture of isomeric 3-hydroxy-
androstan-17-ones as methoxyimino-t.BDMSi ether derivatives. Computer reconstructed fragment ion current
chromatograms for m/z values 300 and 268 (left) as well
as 300, 344 and 376 (right). Syn- and anti- geometric
isomers of androsterone (a, b) and epietio-cholanolone
(d, e) are detected by recording the m/z value 300, those
of epiandrosterone (f,g) using m/z 268. Peaks at scan
number 186 (right) recorded for the fragments at m/z 376
and 344 correspond to etiocholanolone (c).

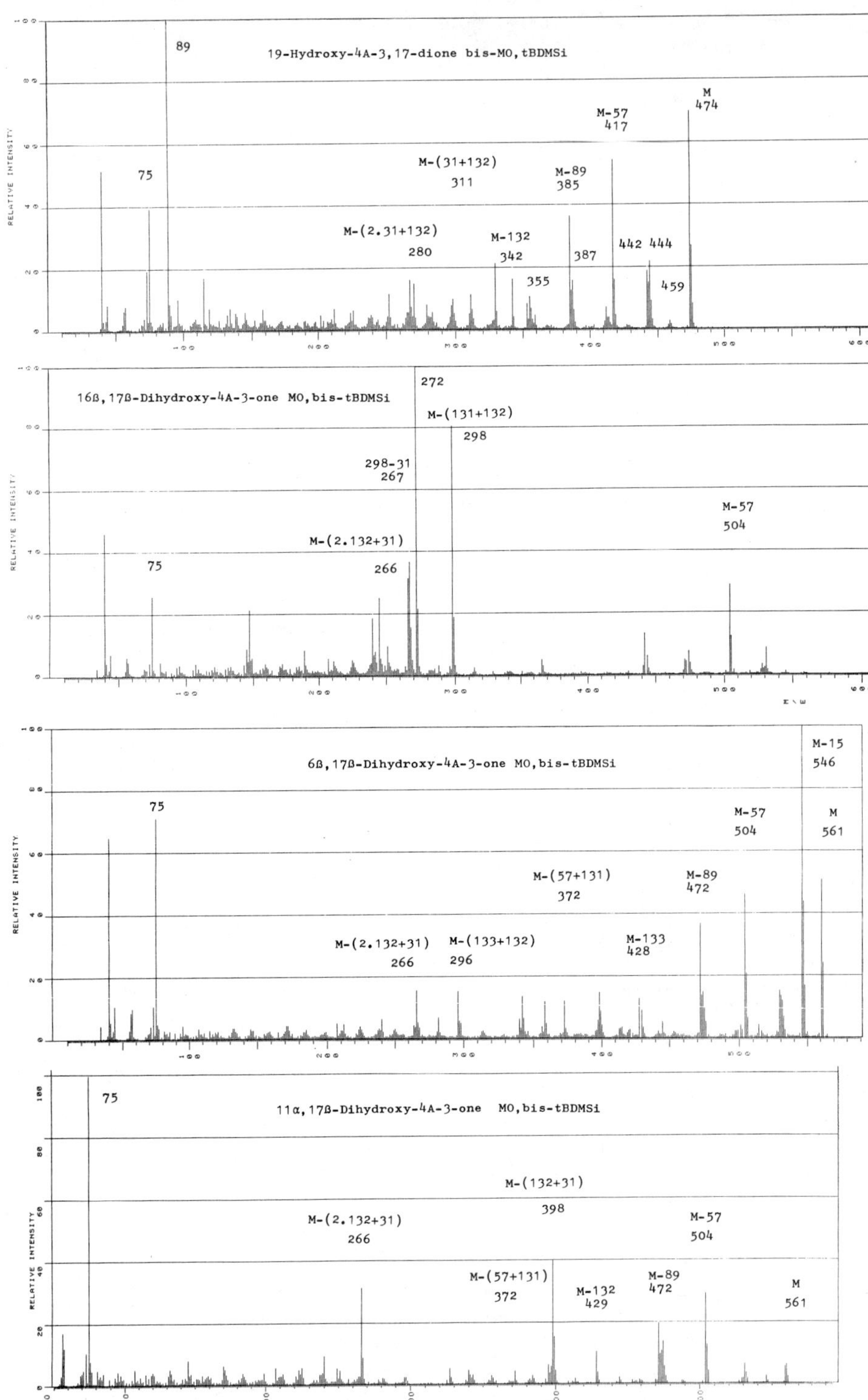

Fig. 5. Mass spectra of the methoxyimino-bis-t.BDMSi ether derivatives of 19-hydroxy-4-androstene-3,17-dione, 16ß-hydroxy-, 6ß-hydroxy- and 11α-hydroxy-testosterones.

dihydroxy-4-androsten-3-one) and 11α-hydroxy-testosterone (11α,
17ß-dihydroxy-4-androsten-3-one) are reproduced in Fig. 5.

All the spectra show a great number of fragment ions of
diagnostic interest. The spectrum of 19-hydroxy-androstenedione
presents a series of interesting fragments originating from both
derivatizations: m/z 444 (M-30), 442 (M-32), 417 (M-57), 387
(M-87), 385 (M-89), 355 [M-(89+30)], 353 [M-(89+32)], 342 (M-132),
311 [M-(31+132)], 280 [M-(2 x 31 + 132)] and 277 [M-(2 x 32 +
133)]. The ion m/z 89 (base peak) originates from breakdown of the
silylated primary hydroxyl group after loss of its tert.-butyl
moiety; a similar fragmentation in the molecules containing a
primary trimethylsilyloxy-group is indicated by the formation of
M-103 ions on 19-hydroxy-steroids[17,18] and 18-hydroxy steroids[19,20]
as well as m/z 103 and M-103 ions in the spectra of 18-hydroxy-
C_{19}-[19,21] and 18-hydroxy-C_{21}-steroids[22].

The spectra of 16ß-, 6ß and 11α-hydroxy-testosterones each
present particular fragmentation patterns. The base peak in the
spectrum of 16ß-hydroxy-testosterone was the ion m/z 272 the
formation of which require steps including losses of both silyl
radicals and part of the ring D (carbons 16 and 17). Other
relevant fragment ions are m/z 298 [M-(131+132)], 267
[M-(131+132+31)], 266 [M-(131+132+32) or M-(2 x 132 + 31)] and 240
(base peak -32).

The spectrum of 6ß-hydroxy-testosterone presents as base peak
the fragment ion at m/z 546 (M-15) and significant molecular
(M^+,561) and fragment ions of m/z 504 (M-57) and 472 (M-89). Other
interesting secondary ions are m/z 428 (M-133), 372 [M-(57+131)],
296 [M-(57+76+132)] and 266 [M-(2 x 132+31)].

Besides the fragment ion m/z 75 as the most abundant (per
cent of Σ_{50}: between 9 and 10.1) the spectrum of 11α-hydroxy-
testosterone, methoxyimino-bis-t.BDMSi ether shows interesting
fragment ions with m/z values 504 (M-57), 472 (M-133), 429
(M-132), 398 [M-(132+31)], 372 [M-(57+131)] and 266 [M-(2 x 132+
31)]. Partial derivatization of 11α-hydroxy-testosterone gives a
mono-ether with the following spectrum: molecular ion m/z 447 (M+,
22.2%), fragment ions at m/z 429 (M-18, 72.4%), 414 [M-(18+15),
27.2%], 399 [M-(18+30), 23%], 340 (M-89, 26.8%), 298 [M-(131+18),
15.8%], 296 [M-(133+18), 26.7%), 282 (24.1%), 266 [M-(132+31+18),
34.6%], 264 [M-(133+32+18), 12.7%], 224 (16.4%), 207 (12.5%), 171
(10.4%), 145 (19.9%) and 75 (76%). This abundant fragmentation
resulting from the free hydroxyl group lead to lower per cent of
Σ_{50} values for the base peak of only 5.8 to 6.1.

III. $C_{21}O_2$-Steroids

Fig. 6 reproduces the mass spectra obtained at 22.5 eV for 3ß-
and 3α-hydroxy-5α-pregnan-20-ones as methoxyimino-t.BDMSi ethers.
The abundance of the molecular and significant fragment ions and
the relative retention time to cholestane recorded at 260°C are
summarized in Table 4. Fragments of low intensity were recorded
for the molecular ion and ions resulting from losses of a methyl
group (m/z 445, M-15) or originating from the methoxyimino
substituent (M-31, m/z 430; M-30 m/z 431 and the daughter ion
M-(132+31), m/z 298]. The base peak is represented by the fragment
ion at m/z 404, M-57 (per cent Σ_{50} : 10.6-15 measured for the 3α-
hydroxy-5α-pregnan-20-one and 12.3-24.5 for its 3ß-isomer).

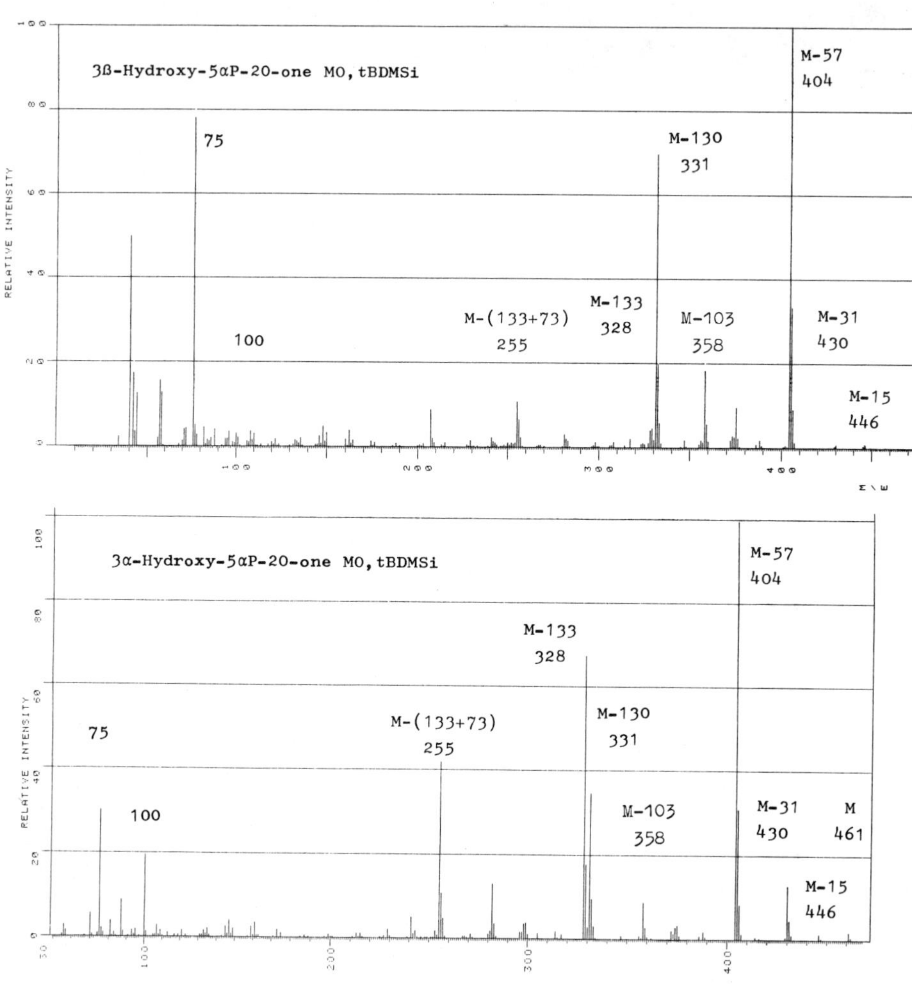

Fig. 6. Mass spectra of the methoxyimino-<u>t</u>.BDMSi ether derivatives of 3ß-hydroxy- and 3α-hydroxy-5α-pregnan-20-ones.

Important secondary ions of m/z 331 [M-(57+73)] and 255 [M-(57+73+76)] were found in the spectra of all four isomers. The secondary ion m/z 328 (poor in the spectrum of 3ß-hydroxy-5α-pregnan-20-one, as indicated in Table 4 and Fig. 6), resulting from losses of a fragment C_4H_9 and dimethylsilanol possesses diagnostic value. The ion at m/z 100 includes the side-chain and part of the ring D (C_{16}-C_{17}) and is also found in the spectra of methyloxime-trimethylsilyl ethers of 3-hydroxy-20-keto-C_{21}-steroids[23].

The 22.5 eV mass spectra of the methoxyimino-<u>t</u>.BDMSi ethers of 20α- and 20ß-dihydro-progesterones (20α-hydroxy- and 20ß-hydroxy-4-pregnen-3-ones) are shown in Fig. 7. The abundance of the molecular and significant fragment ions as well as their t_R-values to 5α-cholestane are reproduced in Table 5. The major fragmentations yield the ion at m/z 402 (M-57) or the secondary ions 326 [M-(57+76)] and 300 [M-(57+102)], resulting in losses of dimethylsilanol or side-chain breakdown together with the <u>t</u>.-butyl

Table 4. Abundance of Molecular and Sgnificant Fragment Ions, mean and (S.D.), found in the Electron-Impact Mass Spectra (22.5 eV) of the Isomeric 3-Hydroxy-Pregnan-20-ones after Methoxyimino-_tert_.Butyldimethylsilyl Ether Derivatization. n = Number of Spectra. t_R = Relative Retention Time to 5α-Cholestane (706s, 260C°)

	Steroids			
	3α/5α (n=10)	3ß/5α (n=10)	3α/5ß (n=14)	3ß/5ß (n=8)
m/z	t_R 1.35	t_R 1.71	t_R 1.41	t_R 1.46
461 (a)	1.6 (0.3)	0.3 (0.3)	1.2 (0.5)	0.8 (0.3)
446	1.3 (0.3)	0.9 (0.7)	1.5 (0.4)	1.0 (0.2)
430 (b)	7.2 (4.0)	2.8 (0.6)	7.2 (1.1)	5.2 (2.5)
404 (c)	(bp)	(bp)	99.2 (2.4) (bp)	99.5 (1.4) (bp)
375	5.1 (1.8)	8.2 (3.7)	9.1 (2.2)	5.6 (1.1)
358 (d)	11.7 (3.1)	18.8 (6.0)	25.0 (7.0)	10.7 (2.4)
331 (e)	38.2 (7.0)	51.2 (13.1)	36.2 (11.4)	36.8 (6.5)
328 (f)	67.5 (2.1)	3.7 (0.6)	26.4 (1.6)	17.4 (2.2)
298 (g)	4.4 (0.7)	1.8 (1.3)	8.4 (1.1)	4.5 (0.3)
255 (h)	46.6 (9.0)	8.1 (2.6)	37.9 (11.1)	30.6 (6.5)
100	21.3 (3.4)	6.0 (1.9)	19.8 (3.3)	14.5 (2.4)
.75	45.5 (18.1)	62.7 (23.1)	60.5 (19.1) *	63.8 (16.6) *

Molecular and fragment ions: (a) M (b) M-31 (c) M-57 (d) M-103 (e) M-(57+73) (f) M-133 (g) M-(132+31) (h) M-(133+73).
* base peak in some spectra.

moiety. The latter fragment is similar to the fragment M-117 found in the trimethylsilyl ether derivatives.

Besides a molecular ion of reasonable intensity, the spectra present a number of fragments the formation of which were directed by the presence of the methoxy-group, such as m/z 428 (M-31) and 268 [M-(57+32+102)]. The fragment ion at m/z 358 (M-101), a daughter ion of m/z 402, originates from a secondary loss of 44 daltons[3,6]. The fragment ion at m/z 159, which corresponds to the fragment of m/z 117 in the spectra of trimethylsilyl ether derivatives, is found together with the ion at m/z 103 [M-(57+44)] in the _t_.BDMSi ether derivatives of other 20-hydroxy-21-desoxy-steroids such as 5ß-pregnane-3α, 20α-diol[12], 20ß-hydroxy-4-pregnen-3-one[6] and its enol-form[3].

In several instances the base peak found in the 22.5 eV mass spectra of 20α- and 20ß-dihydro-progesterones as well as, to a lesser extent, in those of isomeric 20-hydroxy-pregnan-3-ones (Tables 4 and 5) was represented by the ion m/z 75 instead of M-57 (per cent of Σ_{50} measured for 20ß-dihydro-progesterone:14.7-18.5).

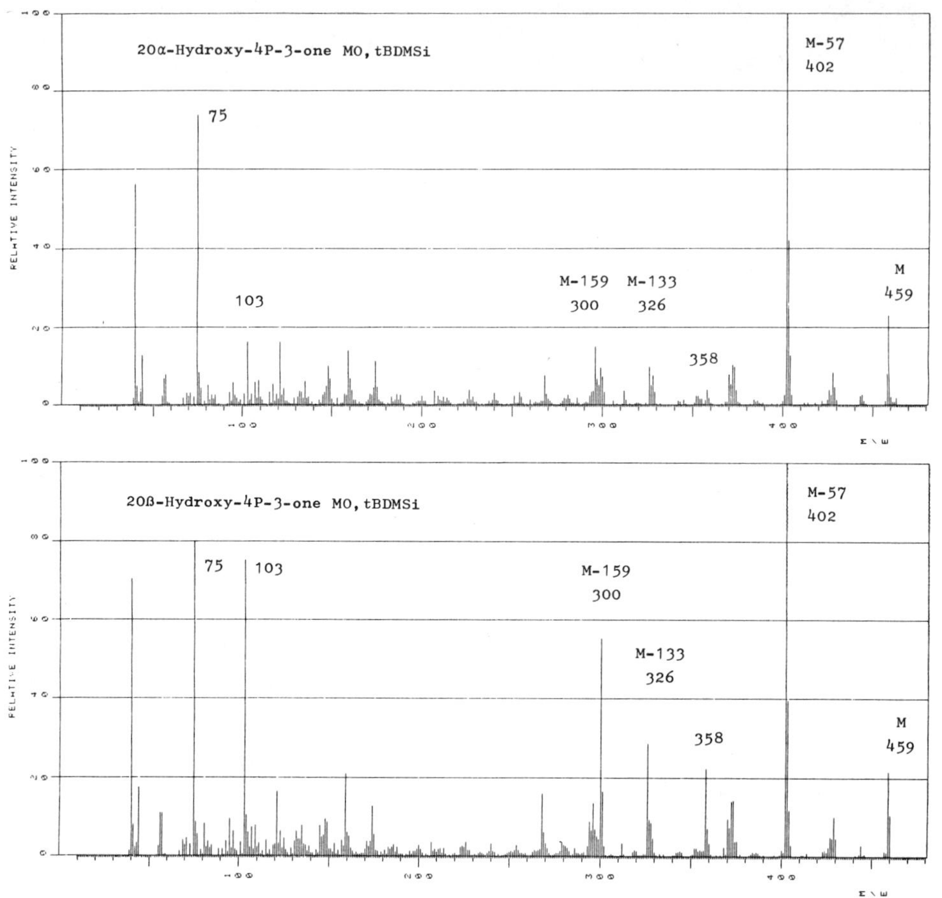

Fig. 7. Mass spectra of the methoxyimino-t.BDMSi ether derivatives of 20α- and 20ß-dihydro-progesterones.

DISCUSSION

The data presented in this investigation indicate an influence of the steroid stereochemistry on the fragmentation pattern of its methoxyimino-t.BDMSi ether derivatives, as observed by Quilliam and Westmore[12] and Lisboa and Ganschow[11] when single t.BDMSi ether derivatives were analysed.

The 22.5 eV spectra of 17-ketosteroids with an 3-axial hydroxyl group (3α/5α and 3ß/5ß-structure present fragment ions m/z 300 [M-(57+32)] and 268 [M-(133+32)] much more abundant than their 3-equatorial epimers. Higher intensity was observed for the fragment at m/z 128 in the spectra of the isomers with an axial 3-hydroxyl group. Fragment ions at m/z 300 [M-(57+32)] were more abundant in the spectra of 3-keto-5α-steroids than in the spectra of the corresponding 5ß-isomers. In the 5ß-series however, the ion m/z 270 [M-(132+31)] is much more pronounced (base peak for 5ß-dihydro-epitestosterone) than in the epimers with an 5α-configuration, as observed for the corresponding ion M-133 in the t.BDMSi ether derivatives[11]. The GC-MS behaviour of 17-ketosteroids as their methoxyimino-t.BDMSi ether derivatives investigated at

Table 5. Abundance of Molecular and Significant Fragment Ions
         Mean and (S.D.) found in the Electron-Impact Mass
         Spectra (22.5 eV) of 20ß-Hydroxy-4-Pregnen-3-one (n=8,
         t_R 1.22) and 20α-Hydroxy-4-Pregnen-3-one (n=14, t_R=1.31)
         after Methoxyimino-_tert_.Butyldimethylsilyl Ether
         Derivatization. n = Number of Spectra. t_R = Relative
         Retention Time to 5α-Cholestane (660s, 260°C)

| | Steroids | |
m/z	20ß-ol	20α-ol
459 (a)	16.8 (4.4)	18.2 (1.5)
444 (b)	2.6 (0.8)	2.7 (0.6)
428 (c)	6.3 (2.4)	3.9 (2.5)
402 (d)	95.8 (7.8)	98.5 (3.6)
373	12.2 (1.7)	10.7 (1.4)
358 (e)	19.3 (2.9)	4.5 (2.2)
328	8.2 (1.3)	6.8 (1.0)
327 (f)	8.3 (1.0)	4.2 (0.7)
326 (g)	23.4 (2.5)	11.5 (0.8)
300 (h)	51.9 (3.0)	9.4 (2.6)
299	3.1 (1.6)	10.1 (1.4)
296	12.5 (1.4)	13.0 (1.3)
295	5.7 (1.8)	3.7 (0.8)
268	14.5 (1.3)	8.0 (1.2)
174	10.4 (1.6)	10.9 (1.7)
159	18.5 (1.7)	15.4 (1.8)
148	9.5 (1.0)	9.2 (1.3)
121	18.1 (2.6)	14.3 (1.6)
103	73.2 (2.8)	16.8 (1.7)
75	90.0 (2.0) *	84.1 (10.2) *

Molecular and fragment ions: (a) M (b) M-15 (c) M-31 (d) M-57 (e)
M-101 (f) M-132 (g) M-(76+57) (h) M-(57+102).
* base peak in some spectra.

22.5 eV show interesting differences when compared with that found
when a higher ionization energy was used[15]. Under our experimental
conditions the ion at m/z 376 (M-57) has an abundance of about 30%
in the spectra of all isomers investigated and the base peak was
represented by m/z 344 (M-89), whereas at 70 eV[15] the ion M-57 was
very abundant (76-100%) and the relative intensities of the
fragment ion at m/z 344 were under 10%. Under both conditions the
ion m/z 300 was more pronounced in isomers with an axial configu-
ration. The spectra of 3-keto-steroids show similar fragmentation
at both 22.5 and 70 eV.

     It was observed that the 70 eV spectra of steroid-methoxime-
_t_.BDMSi ethers present more fragment ions and therefore less
pronounced base peaks[16] than the corresponding single _t_.BDMSi
ethers. For testosterone-methoxime-_t_.BDMSi ether, for instance,
the ion M-57 accounts for only 6.5% of the total ion current
measured at Σ_{50}, compared to 20.8% for the _t_.BDMSi ether derivative[3]. By using lower ionization energy as 22.5 eV more intensive

fragments were found in the higher regions of the spectra, an
important fact when mass fragmentography is used.

Under the present experimental çonditions intensity values
between 13.5 and 16.5 (%Σ_{50}) were obtained for the base peak (m/z
374) of testosterone. Even for the _bis-t_.BDMSi ether derivative of
11α-hydroxy-testosterone-methoxyimino the base peak m/z 75
accounted for values between 9.0 and 10.1 per cent of the total
ionization (%Σ_{50}). During investigations carried out at 70 eV with
2α,3α-cyclopropano-5α-androstanediols-_bis-t_.BDMSi ethers Quilliam
et al.[24] have obtained similar values (8.8-10.1% Σ_{50}) for its base
peak (m/z 75).

It can be concluded that the use of methoxyimino-_t_.BDMSi
ether derivatives carried out at 22.5 eV gives fragment ions of
pronounced intensity in the higher region of the spectra which
could be very advantageous for the qualitative or quantitative
analysis of steroids using single- or multiple-ion-detection
techniques.

REFERENCES

1. R.W. Kelly and P.L. Taylor, in: Summaries of the Proceedings
    of the 2nd International Symposium on Mass Spectrometry in
    Biochemistry and Medicine, Milan, Italy, 51, June (1974).
2. R.W. Kelly and P.L. Taylor, Anal. Chem. 48:465 (1976).
3. S.J. Gaskell and C.J.W. Brooks, Biochem. Soc. Trans., 560th
    Meet. 4:111 (1976).
4. D.S. Millington, J. Steroid Biochem. 6:239 (1975).
5. G. Phillipou, D.A. Bigham and R.F. Seamark, Steroids
    26:516 (1975).
6. A.G. Smith, S.J. Gaskell and C.J.W. Brooks, Biomed. Mass
    Spectrom. 3:161 (1976).
7. B.P. Lisboa and J.McK. Halket, in: A. Frigerio ed., Recent
    Developments in Mass Spectrometry in Biochemistry and
    Medicine, Vol.1, Plenum Publishing Corp., New York,
    441 (1978).
8. M.A. Quilliam and J.B. Westmore, Anal. Chem. 50:59 (1978).
9. B.P. Lisboa and J. McK Halket, in: A. Frigerio and L. Renoz
    eds., Recent Developments in Chromatography and
    Electrophoresis, Elsevier, Amsterdam, 141 (1979).
10. S.J. Gaskell, A.W. Pike and D.S. Millington, Biomed. Mass
    Spectrom. 6:78 (1979).
11. B.P. Lisboa and I. Ganschow, in: A. Frigerio ed.,
    Chromatography and Mass Spectrometry in Biomedical Sciences,
    Vol. 2, Elsevier, Amsterdam, 291 (1983).
12. M.A. Quilliam and J.B. Westmore, Steroids 29:579 (1977).
13. R.W. Kelly and P.L. Taylor, in: A. Frigerio and N. Castagnoli,
    eds., Advances in Mass Spectrometry in Biochemistry and
    Medicine, Vol. 1, Spectrum Publications, New York,
    449 (1976).
14. S.H.G. Andersson and J. Sjoevall, J. Chromatogr.289:195
    (1984).
15. S.J. Gaskell and A.W. Pike, Biomed. Mass Spectrom. 8:125
    (1981).
16. L. Ballhorn, W.F. Mueller and F. Korte, Steroids 33:379
    (1979).
17. A.G. Sharkey jr., R.A. Friedel and S.H. Langer, Anal. Chem.
    29:770 (1957).
18. J.A. Gustafsson, Arkiv foer Kemi 29:535 (1968).

19. B.P. Lisboa and J.A. Gustafsson , Eur. J. Biochem. 9:402 (1969).
20. J.A. Gustafsson and B.P. Lisboa, Acta Endocr. 65:84 (1970).
21. J.A. Gustafsson and B.P. Lisboa, Steroids 15:723 (1970).
22. S.L. Dale and J.C. Melby, Steroids 21:617 (1973).
23. C.J.W. Brooks and D.J. Harvey, Steroids 15:283 (1970).
24. M.A. Quilliam, J.F. Templeton and J.B. Westmore, Steroids 29:613 (1977).

OPTIMIZATION OF RADIORECEPTOR ASSAYS FOR ANTICHOLINERGIC DRUGS

IN BIOFLUIDS

K. Ensing

Department of Toxicology
State University
Groningen, The Netherlands

INTRODUCTION

In order to determine relationships between pharmacokinetics
and pharmacodynamics of some frequently used anticholinergic drugs
it was decided to develop an assay for these drugs in biofluids.
This assay should be generally applicable to anticholinergic
drugs, highly sensitive and stereoselective.

Because of their potency, therapeutic plasma concentrations
of the anticholinergics are generally in the lower ng/ml range and
difficult to analyze in conventional chromatographic systems.\
Besides this problem of sensitivity these methods actually do not
give any information about the biological activity of the parent
compound, its metabolites and degradation products or about
optical isomers.

For these reasons radioreceptor assays (RRA) for various
anticholinergics (e.g. oxyphenonium, ipratropium and scopolamine)
have been developed[1-3].

The principle of RRA is based on competition between a drug
and a radiolabelled ligand for binding to a certain receptor. When
a competitive drug is added to a mixture containing fixed amounts
of receptor and radiolabelled ligand, the drug will displace a
certain amount of labelled ligand depending on its concentration
and on its equilibrium dissociation constant. The development of
such an assay has to be undertaken with caution, and is dependent
on the binding characteristics of the drug and the presence of
active metabolites in biological samples[4].

When the pharmacological action of a drug resides in one
(parent) compound, the composition of the incubation medium and
the incubation conditions should be chosen so as to increase the
sensitivity of the assay. However, when more than one active
compound is present, equilibrium conditions and a near physiological incubation medium have to be used, otherwise no
meaningful quantitation of the total biological activity can be
made.

Despite structural similarities between the above mentioned
drugs, different assays were developed. Depending on the tertiary
or quaternary status of the nitrogen-atom, dexetimide or
N-methylscopolamine is used as labelled ligand, ensuring that
labelled and competitive ligand interact with the same binding
sites.

In this paper the different approaches will be discussed. The
sensitivity of a radioreceptor assay is primarily dependent on the
affinity of the drug towards the receptor, whereas the concentration of the labelled drug determines if a pre-incubation of the
drug with the receptors can be useful to improve the sensitivity
of the assay[5].

Another important factor is the accuracy of the assay. For
that reason we compare a filtration method and a centrifugation
method for the separation of free and bound labelled ligand. In
the first method, the concentration of competitive ligand is
inversely related to the bound fraction of labelled ligand whereas
in the latter method the concentration of competitive ligand is
related to the free fraction labelled ligand[6]. Finally the
applicability of the RRA for the bioanalysis of (anticholinergic)
drugs will be discussed.

MATERIALS AND METHODS

Chemicals

^{3}H-dexetimide (^{3}H-DEX) (15 Ci/mmol) was obtained from Janssen
Pharmaceutica (Beerse, Belgium). ^{3}H-N-methylscopolamine chloride
(^{3}H-NMS) (90 Ci/mmol) was supplied by NEN (Dreiech, FRG).
Ipratropium bromide was kindly supplied by Boehringer Ingelheim
(Ingelheim, FRG). Scopolamine and all other chemicals of analyti-
cal grade were from Merck (Amsterdam, The Netherlands).
Polyethylene tubes (10 ml) were obtained from Greiner (Alphen a.d.
Rijn, The Netherlands). The GF/B glassfiber filters were from
Whatman (Maidstone, U.K.). Plasmasol or Picofluor were used as
scintillation liquid and obtained from Packard Instruments
(Groningen, The Netherlands) in combination with 20 ml glass
vials, also from Packard. The 50 mol/l sodium phosphate buffer
(pH = 7.4) was composed from 4 volumes 50 mol/l sodium dihydrogen-
phosphate and 1 volume 50 mol/l disodiumhydrogenphosphate.

Preparation of Receptor Material

Calf brains without cerebellum, freshly prepared or stored at
-80°C were homogenized in 6 volumes ice-cold 0.32 mol/l sucrose
using a Teflon-glass Potter-Elvejehem homogenizer at 1200 rpm
(R.W. 18, Janke & Kunkel, Staufen i. Breisgau, FRG). The
homogenate was centrifuged, 10 min at 1000 g. The pellet was
discarded and the supernatant centrifuged, 60 min at 100,000 g.
The latter pellet was resuspended in buffer to the original volume
of the homogenate and centrifuged 30 min at 100,000 g. This
washing was repeated and finally the pellet was resuspended in
5 volumes buffer. The receptor preparation was lyophilized in
24 hours and the optimum tissue concentration for the RRA was
experimentally determined.

<u>Radioreceptor Assay for Ipratropium with the Centrifugation Method</u>

To duplicate polypropylene reaction vials (1.5 ml) were added
solutions of Ipbr in buffer, in a biofluid or in a suitably
prepared aqueous extraction solution, giving final concentrations
varying from 10^{-10} - 10^{-7} mol/l corresponding to about
40 pg - 40 ng/ml. Then 25 μl of radiolabelled methylscopolamine
(6.6×10^{-9}) was added giving a final assay concentration of
3×10^{-10} mol/l. Finally a volume of the receptor preparation was
added, containing 3 mg lyophilized receptors, in order to obtain a
550 μl incubation volume. The reaction vials were closed, shaken
and incubated 60 min at room temperature, then centrifuged 15 min
at 15,000 g in a Hereaus Christ Biofuge A (Osterode am Harz, FRG).
450 μl of the supernatants was transferred in the polyethylene
counting vials, 2 ml Picofluor 30 was added and vortexed for 5
seconds before counting in a Beckman 1800 liquid scintillation
counter (Irvine, CA, USA). The samples were counted for 40,000
counts with a five minutes counting time limit.

<u>Extraction and re-Extraction of Ipratropium from Plasma</u>

To 10 ml Sovirel tubes, 1 or 2 ml plasma containing 10^{-10} -
10^{-7} mol/l Ipbr and 100 or 200 μl 3×10^{-3} mol/l sodium picrate in
50 mmol/l sodium phosphate buffer were added, respectively, to
give a final picrate concentration of approximately 3×10^{-4} mol/l.
Five ml dichloroethane (DCE) was added, the tubes with 1 ml
aqueous solution were vortexed during 30 seconds whereas the 2 ml
aqueous samples were rolled during 60 min on a Denley Mixer 5
(Sussex, U.K.) in order to avoid the formation of inseparable
emulsions.
The tubes were centrifuged, 10 min at 5,000 g, the aqueous phase
was discarded, 4.5 ml of the DCE phase was transferred to another
Sovirel tube and 1.0 ml 2×10^{-6} mol/l tetrapentylammoniumhydroxide
(TPA) in buffer was added. The mixture was vortexed for 30 seconds
and centrifuged 10 min at 5,000 g. Two 0.25 ml aliquots of the
aqueous layer were transferred to 1.5 ml polypropylene reaction
vials and the RRA was carried out.

<u>Radioreceptor Assay for Scopolamine with pre-Incubation and the
Filtration Method</u>

To duplicate polyethylene tubes (10 ml) solutions of
scopolamine in buffer, in a biofluid or in a suitably prepared
aqueous extraction solution were added, giving final concentrations varying from 10^{-10} - 10^{-7} mol/l, corresponding to about 30 pg
- 30 ng/ml. Then a volume of the receptor preparation, containing
2 mg lyophilized receptors, was added in order to obtain a 500 μl
incubation volume. The tubes were mixed and incubated during 60
min at 0°C before 25 μl of radiolabelled ligand (4.2×10^{-8} mol/l
^{3}H-DEX) was added. The tubes were mixed again and incubated for
another hour at 0°C.

After the addition of 4 ml icecold buffer, the samples were
immediately filtered through Whatman GF/B glassfiber filters under
vacuum using a filtration apparatus (Multividor 40 S, Janssen
Scientific Instruments, Beerse, Belgium). The tubes were rinsed
with 4 ml icecold buffer, which was also filtered. The filters
were washed with 4 ml icecold buffer and then transferred to the
polyethylene counting vials, 3.5 ml Plasmasol was added and the
vials were shaken for 120 minutes. The total filtration rinsing

and washing process, taking place in approximately 15 seconds, was
carried out on each tube in turn. The vials were counted for
5 minutes or 40,000 counts in a liquid scintillation counter.

<u>Extraction and re-Extraction of Scopolamine from Plasma</u>

To 10 ml Sovirel tubes, 1 ml plasma containing 10^{-10} – 10^{-7}
mol/l scopolamine, 1 ml 0.1 mol/l Borax-buffer (pH = 9.75) and
5 ml DCE were added. The tubes were vortexed during 30 seconds,
cooled in ice, and centrifuged, 10 min at 5,000 g. The aqueous
phase was discarded, 4.5 ml of the DCE phase was transferred to
another Sovirel tube and 1.0 ml 0.1 mol/l phosphoric acid was
added. Then the samples were vortexed during 30 seconds, cooled in
ice and centrifuged, 10 min at 5,000 g. Two 250 µl aliquots of the
aqueous layer were transferred in polyethylene tubes, neutralized
with 50 µl 1 mol/l sodium hydroxide, frozen and lyophilized during
24 hours prior to the assay.

RESULTS

A prerequisite for the centrifugation mode of the RRA is that
in the absence of a competitive ligand the amount free labelled
ligand is smaller than the amount receptor bound labelled ligand.
The optimum tissue concentration for the inhibition curves was
estimated by measuring the differences in the concentration of
free labelled ligand, in the absence and presence of a fixed
concentration Ipbr giving approximately 50% inhibition of receptor
bound ^{3}H–NMS, for various tissue concentrations. These data are
plotted in Fig. 1. It can be concluded that the optimum tissue
concentration at the end of the linear part of the curves is about
6 mg/ml. Then the amount of displaced radiolabelled ligand is

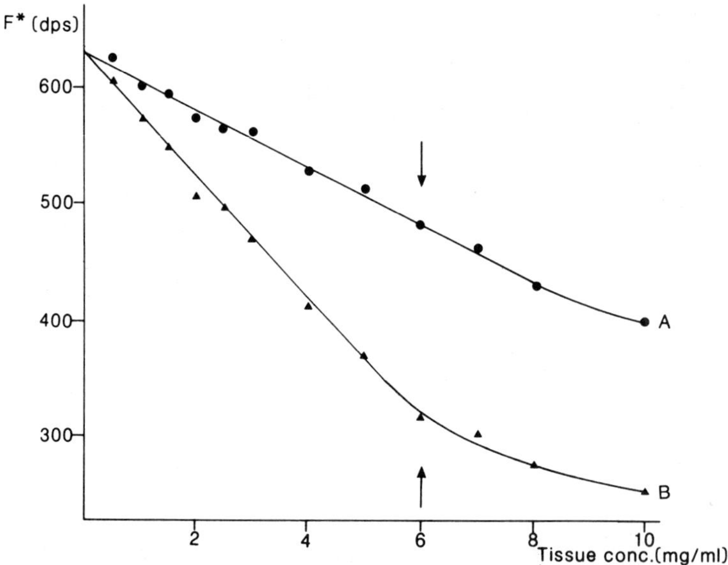

Fig. 1. Influence of the amount lyophilized receptor preparation
(tissue concentration) on the concentration free labelled
ligand (F*) in the presence (A) or absence (B) of
1 x 10^{-8} mol/l ipratropium bromide. The arrow indicates the
optimum tissue concentration.

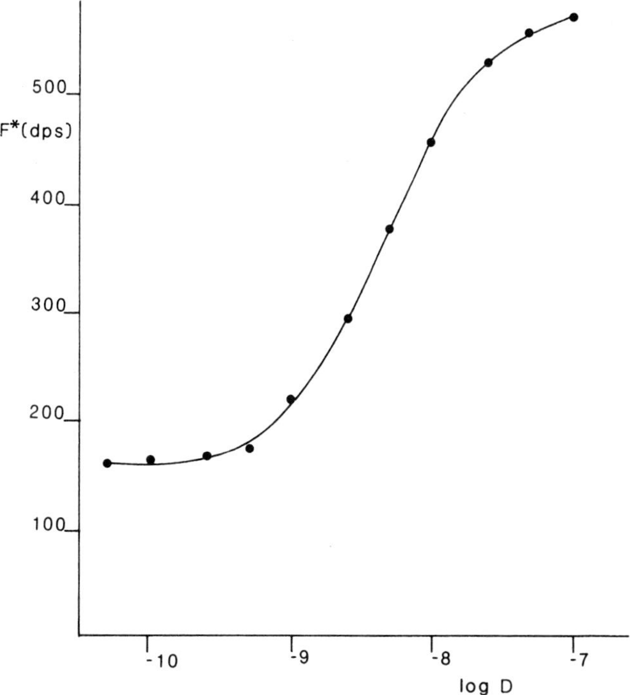

Fig. 2. Experimental inhibition curve for ipratropium bromide.
F* indicates the concentration free labelled ligand.
D indicates added drug concentration in mol/l.

maximal whereas the free fraction labelled ligand is rather small
as compared to the corresponding bound fraction.

In Fig. 2 a representative calibration curve for Ipbr is
given. The concentration Ipbr which causes 10% inhibition of
receptor bound ^{3}H-NMS is considered as determination limit of the
method. The obtained determination limit is 1 x 10^{-9} mol/l which
corresponds to 220 pg Ipbr in the assay.

When the RRA is to be carried out with urine, endogenous
urine compounds will interfere and reduce ^{3}H-NMS binding. However,
this inhibition can be made nearly constant and kept at an
acceptable level (<10%) when the urine is diluted about tenfold
with buffer. Because of the relatively high drug concentrations
that occur in urine after therapeutic dosing, a direct assay was
possible.

For plasma a sample clean-up and concentration step is
unavoidable. On the one hand, plasma proteins have a detrimental
effect on the sensitivity and accuracy of the assay because of
non-specific binding of ^{3}H-NMS and Ipbr. On the other hand, the
drug concentrations in plasma are close to the determination limit
of the assay, so that dilution of the samples as with urine will
result in undetectable levels.

Therefore, it was decided to use an ion-pair extraction with
picrate and re-extraction procedure with TPA, which transfers the

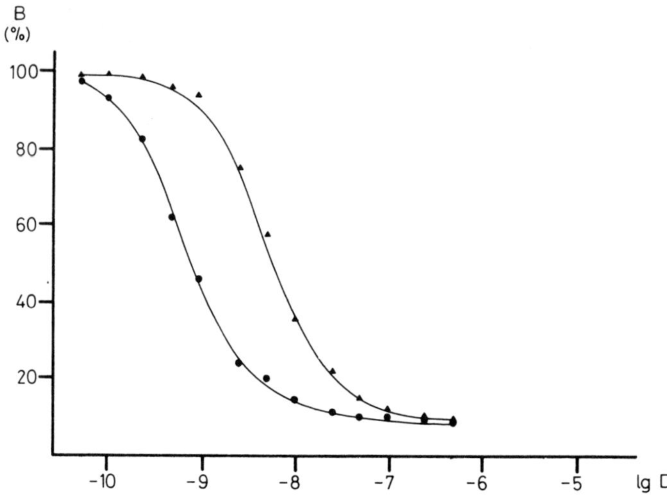

Fig. 3. Inhibition of ^{3}H-dexetimide binding by scopolamine under
equilibrium (▲) or non-equilibrium (●) conditions.
B = ^{3}H-dexetimide binding, 100% is total binding; D is
concentration of the drug in the RRA.

ipratropium back to the aqueous phase. This procedure is very
selective for quaternary anticholinergics and offers high
recoveries, even in the subnanogram range[7].

In the receptor assay for scopolamine a preincubation step
was introduced, creating non-equilibrium incubation conditions, in
order to improve the sensitivity of the assay. As can be deducted
from Fig. 3, a significant gain in sensitivity could be obtained.
The absolute detection limits of some anticholinergics under
equilibrium and non-equilibrium conditions are given in Table 1.

When the RRA for scopolamine was used for urine and plasma
samples, comparable observations were done with regard to interference: urine samples could be determined directly after
dilution, whereas for plasma samples a combined extraction and re-

Table 1.   Detection Limits for Some Anticholinergics under
           Equilibrium and non-Equilibrium Conditions in RRA

	equilibrium conditions (mol/l)	non-equilibrium conditions (mol/l)	gain in sensitivity
Atropine	5.5×10^{-13}	2.0×10^{-13}	2.8
Dexetimide	6.8×10^{-13}	7.5×10^{-14}	9.0
Promethazine	5.0×10^{-12}	2.0×10^{-12}	2.5
Scopolamine	5.0×10^{-13}	7.5×10^{-14}	6.7

extraction was required to eliminate plasma interferences. An
extra lyophilization step was introduced because trace amounts of
dichloroethane in the aqueous extract inhibited ^{3}H-dexetimide
binding[1].

Though both methods produce excellent inhibition curves, each
has its specific features. The centrifugation method for
ipratropium has a better precision for low concentrations of
competitive drug than the filtration method. On the other hand,
the opposite is true for higher concentrations.

The gain in sensitivity for scopolamine by the introduction
of a preincubation step has proven to be useful in the analysis of
scopolamine in plasma after transdermal drug delivery[3,5].

Quantitative radioreceptor assays, based on the competition
between a radiolabelled ligand and a drug for binding to a certain
class of receptors, can be employed for the determination of drugs
which induce a pharmacological effect via an interaction with this
type receptor.

The main advantage of the method, besides a fairly high
sensitivity, is the selective determination of biological active
compounds, e.g. optical isomers or metabolites. The RRA for
anticholinergics has been employed for the determination of the
quaternaries oxyphenonium, ipratropium and oxytropium in biofluids
after oral, i.v. and inhalation administration as well as for the
tertiary compounds scopolamine and atropine, administered via
transdermal or intrabronchial routes. Yet, despite its inherent
simplicity the RRA has to be optimized for every new compound.

Acknowledgements

This work was supported in part by grants of the Netherlands
Asthma Foundation, Boehringer Ingelheim (Alkmaar, The Netherlands)
and Ciba Geigy (Arnhem, The Netherlands). Besides that I want to
thank Prof. Dr. Rokus A. De Zeeuw, Willy G. in't Hout and Jolanda
Meindertsma for their contributions to this paper.

REFERENCES

1. K. Ensing, F. Kluivingh , T.K. Gerding and R.A. De Zeeuw,
    Development of a sensitive radioreceptor assay for
    oxyphenonium in plasma and urine, J. Pharm. Pharmacol.
    36:235 (1984).
2. K. Ensing, M. Pol and R.A. De Zeeuw, Radioreceptor assay of
    ipratropium bromide in plasma and urine, J. Pharm. Biomed.
    Anal. in press (1987).
3. K. Ensing, W.G.in't Hout, G.J. Ensing, P. Halma and R.A.
    De Zeeuw, Development and application of a radioreceptor assay
    for scopolamine, Arzneim. Forsch. (1988) in press.
4. K. Ensing and R.A. De Zeeuw, Radioreceptor assay - A tool for
    the bioanalysis of drugs, Trend Anal. Chem. 3:102 (1984).
5. K. Ensing and R.A. De Zeeuw, Radioreceptor assay of
    anticholinergic drugs in biological fluids, in: "Drug
    determination in therapeutic and forensic contexts", E. Reid,
    ed., Plenum Press, London (1984).

6. K. Ensing, K.G. Feitsma, D.A. Bloemhof, W.G. in't Hout and
    R.A. De Zeeuw, Centrifugation or filtration in quantitative
    radioreceptor assays, J. Biochem. Biophys. Meth. 13:85
    (1986).
7. J.E. Greving, J.H.G. Jonkman, F. Fiks, R.A. De Zeeuw, L.E. Van
    Bork and N.G.M. Orie, Determination of oxyphenonium
    bromide in plasma and urine by means of ion-pair extraction,
    derivatization and gas chromatography electron-capture
    detection, J. Chromatogr. 142:611 (1977).

COMPARISON OF AUTOMATED AND MANUAL METHODS OF LIQUID-SOLID SAMPLE

PREPARATION FOR DETERMINATION OF DRUGS IN PLASMA BY HPLC/UV

R.D. McDowall and J.C. Pearce

Department of Drug Analysis
Smith Kline and French Research Ltd.
Welwyn, Herts, U. K.

INTRODUCTION

Sample preparation schemes are often essential in the
analysis of drugs in biological materials in order to concentrate
the analyte within the detection limits of the instrument and
remove potential endogenous interfering compounds. The most common
method of sample preparation in the past has been liquid-liquid
extraction, however liquid-solid extraction using chemically
bonded silicas is becoming prominent[1]. Moreover, semi-automated
methods such as the Advanced Automated Sample Processor (AASP) can
give further gains in productivity in bio-analysis[2]. AASP methods
comprise a manual off-line sample preparation stage followed by
automated elution of the analytes by the mobile phase into a
liquid chromatograph.

We required the existing off-line manual sample preparation
stage to be automated in order to gain more benefit from the AASP
system. The reasons for this are two fold:
  a) greater increases in productivity as automated systems would
     be capable of operating overnight;
  b) the systems should be capable of the same or better
     precision and accuracy as the existing manual methods.

  The approaches followed were:
  a) the off-line automation of the manual stage using a Zymate
     robot;
  b) on line preparation and elution of the sample using a Gilson
     autosampler.

These were compared to the manual method of preparing the
AASP cassettes.

EXPERIMENTAL

The general assay technique for a family of inotropic agents
was to isolate the analyte from the biological sample using a C18
AASP cassette, followed by automated elution onto the liquid
chromatograph via the AASP-LC module (Analytichem Int., Harbor

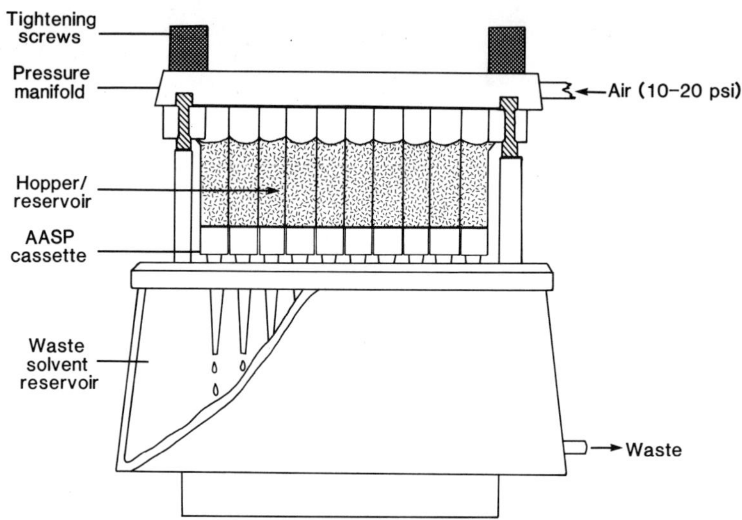

Fig. 1. Schematic representation of AASP VAC-ELUT.

City, Ca, USA). This sample preparation scheme is very quick and efficient but is limited by the capacity of the human operator. The heart of the off-line sample preparation scheme is the Vac-Elut manifold system.

<u>The Vac Elut</u>

This consists of an AASP Vac Elut vacuum box (Analytichem, Harbor City, Ca, USA) and pressure manifold which allows a cassette of ten extraction cartridges to be processed simultaneously at the bench (Fig.1). Solutions are dispensed to individual reservoirs above each cartridge and the liquid forced through by positive pressure (air or nitrogen, 15 psi). The vacuum box acts as a waste collector from which the unwanted fluid can be discarded. Once prepared the cassettes are transferred by hand to the AASP for analysis.

A number of problems have been experienced with this system and these include the cumbersome nature of the Vac Elut when changing solvents and the tendency not to make a good seal between cassette and reservoir. The procedure is also labour intensive and it was against this background that the decision was taken to investigate the potential for automation.

<u>Zymate-AASP System</u>

The Zymate Laboratory Automation System (Zymark, Hopkinton, Ma, USA) shown in Fig. 2 combines the use of a computer controlled robotic arm with laboratory work-stations which are strategically placed within the working envelope of the robot. The first stage of the procedure is to condition the AASP cassette. Methanol and water are delivered through separate lines from the Master Laboratory Station via the nozzle-hand. This hand also accomodates a supply of compressed air which is used to push both solvents and samples through each cartridge.

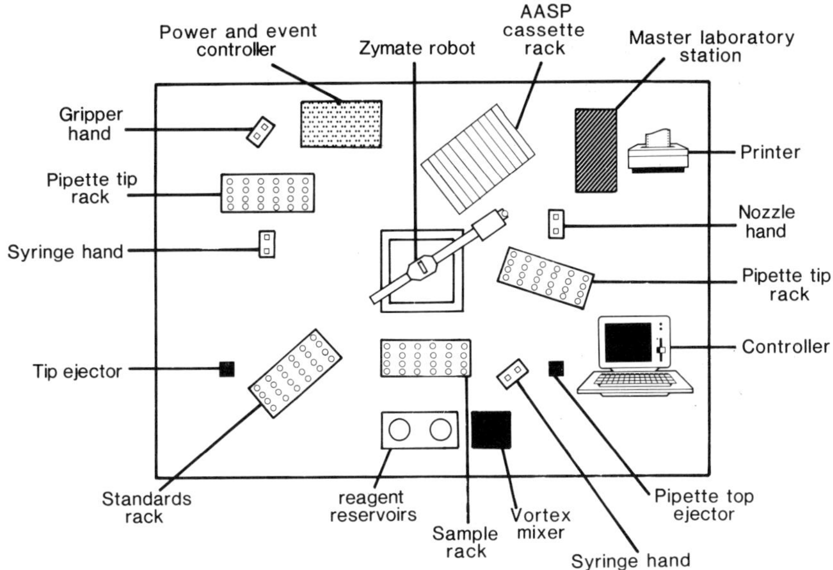

Fig. 2. Automated system for the extraction of inotropic agents from biological fluids using AASP cassettes.

Aliquots of plasma are presented to the system in polypropylene microfuge tubes located in the sample rack. Reagents such as internal standard and buffers are added to the sample using interchangeable syringe-hands in conjunction with disposable pipette tips. The diluted sample is mixed by vortex and then an aliquot is transferred to the appropriate cartridge in the AASP cassette rack. After a final wash step (typically 1 ml water) the cassette is ready for further processing in the AASP.

The system is capable of preparing up to sixty plasma samples during unattended overnight operation. There is however, still one point of manual intervention i.e. transfer of the cassettes from the rack on the robotic work station to the AASP-LCM. This, though not a serious limitation, does reduce total throughput of samples over a 24 h period.

Gilson 222/401-AASP System

Total automation of the procedure has been achieved using a Gilson 222 autosampler with a model 401 dilutor (Gilson, Villiers -le-Bel, France). This type of robotic unit has been described by Hawkes and Lindsay and will not be discussed further here[3]. The Gilson 222 assumes responsibility as the master controller of the system and communicates with the AASP via a 12 volt interface relay box. In this way the autosampler can remotely start the AASP and advance the cassette at a predetermined time. In addition, switching of the Valco valve is also effected by the controller.

The method of operation procedures are as follows:

1. _Sampling_ (cycle 1). Aliquots of plasma (50-100 μl) are presented to the system in test tubes in the autosampler rack. Internal standard solution is dispensed to sample n° 1 using

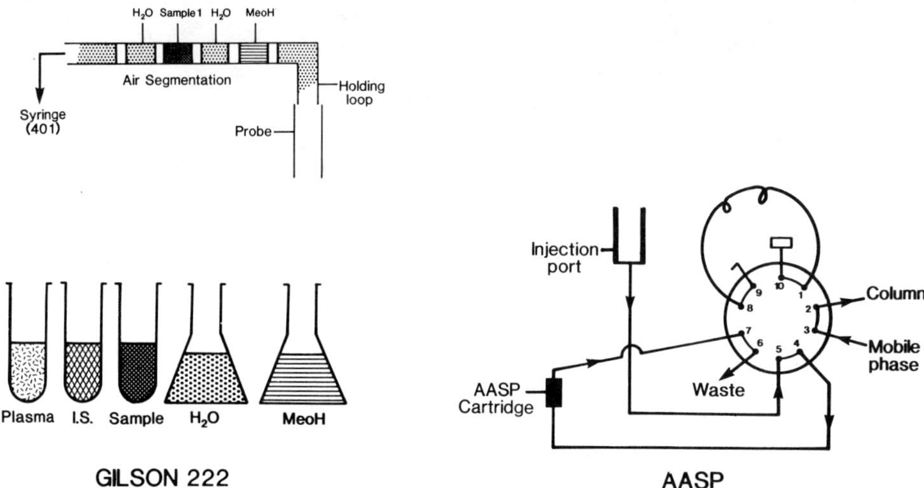

Fig. 3. Schematic of AASP-GILSON automated system.
Step 1: sampling.

the probe and the diluted sample mixed by bubbling air
through the liquid.
The 401 diluter in conjunction with the 222 then pulls up
the liquid phases, in reverse order, into a holding loop
with each solution separated by a small air gap (Fig. 3).
Whilst this is taking place the AASP has been instructed to
load the first cartridge in line with the 222 injection port
 which is connected to port 5 on the Valco valve. The HPLC
mobile phase at this time is simply routed to the analytical
column.

2. <u>Loading Cartridges</u> (cycle 1). The probe moves across to the
   injection port and the train of liquid is slowly pushed
   through the AASP cartridge and out to waste (Fig. 4). Thus,
   in this single operation the bonded sorbent is conditioned
   with methanol and water, the diluted sample applied and

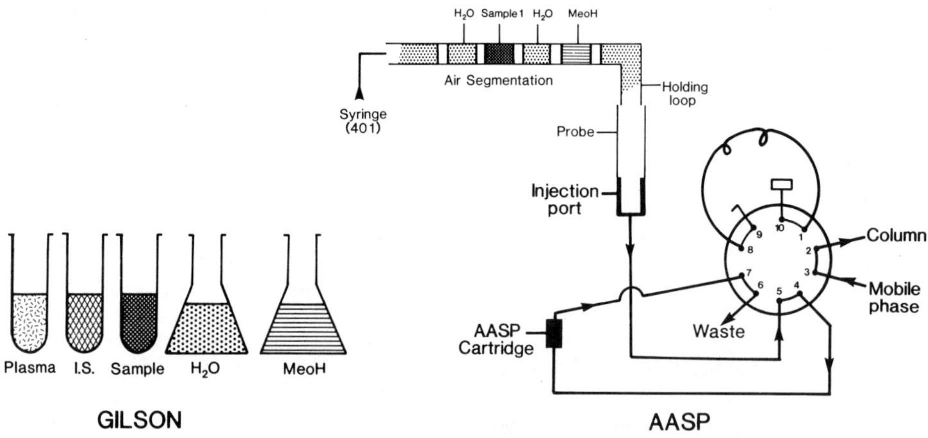

Fig. 4. Schematic of AASP-GILSON automated system.
Step 2: loading cartridge.

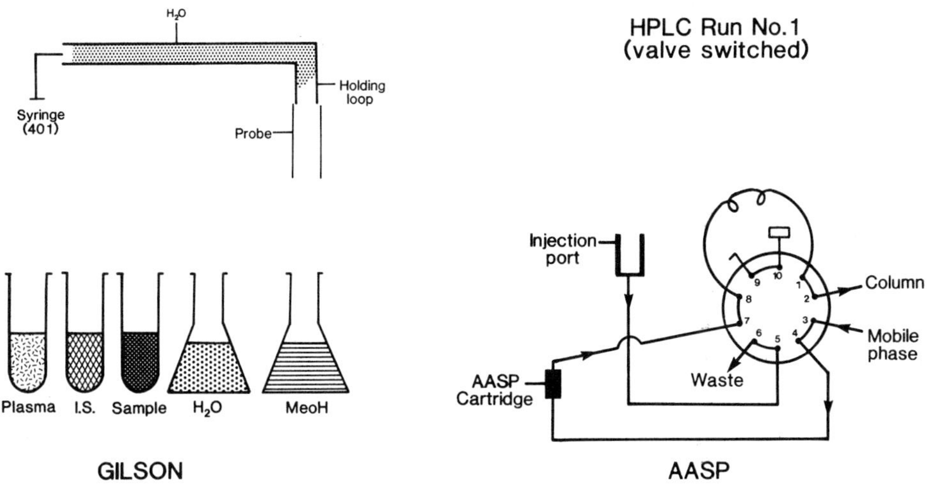

Fig. 5. Schematic of AASP-GILSON automated system.
Step 3: elution.

subsequently washed with an additional aliquot of water.
This sequence of course, could be modified to suit the
individual analysis.

3. <u>Elution</u> (sample 1). With loading now complete, the probe can
   return to its 'home' position where the holding loop is
   flushed with water prior to taking up the second sample. On
   command from the autosampler control program, the Valco
   valve is switched and the integrator run started. The mobile
   phase is directed through the cartridge, sweeping the
   analytes of interest onto the analytical column and HPLC run
   n° 1 is initiated (Fig. 5).

4. <u>Sampling</u> (sample 2). At a preset time (typically 0.5 min)

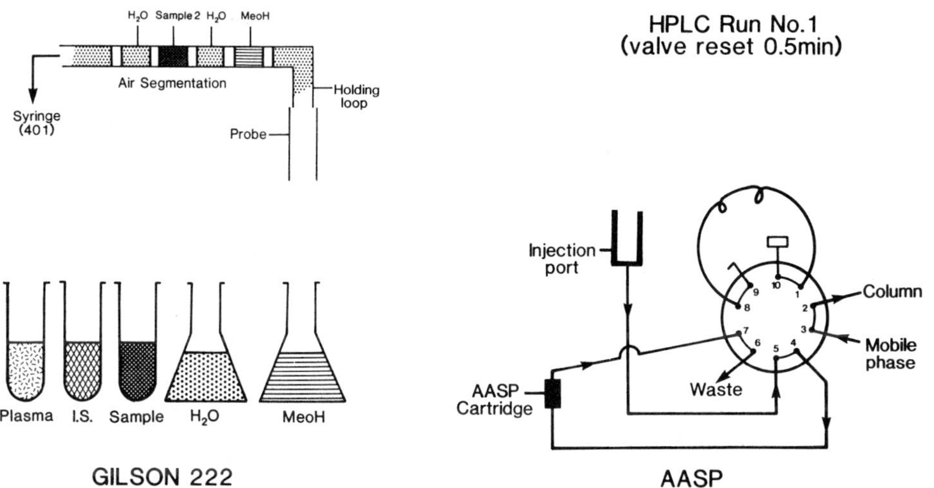

Fig. 6. Schematic of AASP-GILSON automated system.
Step 4: sampling-HPLC.

the valve is reset to the load position and the Gilson 222
proceeds to transfer sample n° 2 and associated liquid
phases into the holding loop (Fig. 6). The second AASP
cartridge is brought in line with the injection port ready
for loading.

5. <u>Loading Cartridge</u> (cycle 2). The probe, once again, moves
across to the injection port and the second sample is
processed through the cartridge (Fig. 7). On completion, the
probe can return 'home' ready for sample n° 3.
HPLC run n° 1 continues to the run end (6 min) and then the
valve is again switched to commence HPLC run n° 2. This
sequence of events will continue until all the plasma
samples in the rack have been processed and a complete set
of analytical reports have been generated.

RESULTS AND DISCUSSION

The manual and two automated techniques presented here were
compared against the following criteria:

- precision and accuracy of the assay of a cardiovascular drug in
  plasma;
- the number of samples assayed per 24 hours;
- cost and ease of implementation;
- human involvement in the assay scheme.

The precision and accuracy of the assay over a concentration
range of 25-500 ng/ml are shown in Table 1 and 2 respectively.
As can be seen the precision, as measured by the coefficient of
variation, of all three methods is better than 5%, which is acceptable at the concentrations measured. The corresponding accuracy,
as measured by bias, shows similar values with the exception of
the lowest concentration using the Gilson-AASP system. The reason
for this is unknown and requires further investigation.

Overall the quality of the data from either automated method
was comparable with the original manual method of sample preparation using the Vac Elut.

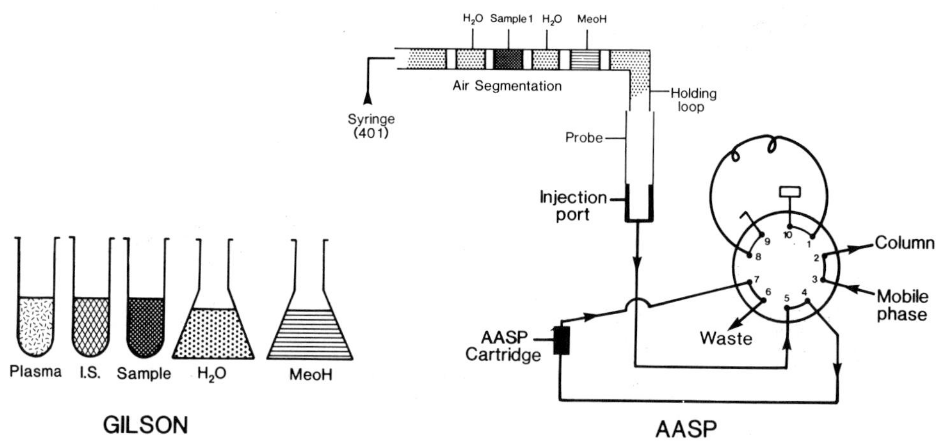

Fig. 7   Schematic of AASP-GILSON automated system.
Step 5: loading cartridge.

Table 1. Precision (as Coefficient of Variation) of an Inotrope
          Assay using the Manual Method with the two Automated
          Techniques

Drug Concentration ng/ml	Manual Vac Elut	Automated	
		Gilson-AASP	Zymate Robot
25	4	4	4
125	< 1	1	2
500	2	3	4

    Performance of the systems was then evaluated; the maximum
number of samples able to be processed per hour was 10 for the
manual system compared to six using the robot and 9 using the
autosampler. The manual approach is marginally faster, however it
is limited by the number of hours the analyst is able to extract
samples. When performance is assessed over a 24 h period the
manual system is only capable of approx 80 samples compared to 200
and 150 for the autosampler and the robot respectively. The
advantage of the automated systems is apparent from the numbers
processed in a full day; the difference is due to the ability of
automatic instrumentation to work outside of normal hours.
Comparison of the two types of automated systems shows that the
dedicated example (autosampler) is capable of a higher throughput
(30%) than the more flexible robotic system. This is due to the
fact that the Zymate system still requires human intervention to
transfer cassettes to the AASP whereas the Gilson approach is
totally automated.

    Reliability of automated systems is very important as
unattended operation is highly desirable, if not essential. The
robot arm is very robust, it has worked with 100% reliability; it
has been the peripheral devices such as the solenoid valves on the
master laboratory station that have presented a few minor
problems. Overall the robot has been very reliable. The Gilson-
AASP system is an unknown entity; it has great potential but a
suitable application has not been forthcoming in our laboratory.
Our present work is assessing, in depth, the reliability of this

Table 2. Accuracy (expressed as % bias) of the Manual and
          Automated Methods

Drug Concentration ng/ml	Manual Vac Elut	Automated	
		Gilson-AASP	Zymate Robot
25	+4	+12	+4
125	< -1	+ 1	-2
500	0	< - 1	+1

system. As mentioned previously, the manual Vac Elut system for off-line preparation can be unreliable especially after long periods of use as the seals wear; moreover, the cartridge and reservoir have to be positioned carefully.

Stability of the analytes on the solid-phase may be a factor to consider. The Gilson-AASP system had the advantage that the compounds of interest were only isolated on the phase for a very short time before analysis; this contrasted with the other two methods where the analytes may be isolated for up to 12 h before elution.

Cost is an interesting item: the Vac Elut costs £ 200 to purchase compared with £ 5,000 for the Gilson and £ 30,000 for the robot system. The speed of implementing automation also rises in a similar fashion: the manual system can be shown and implemented within two minutes in contrast to the Gilson which would take approximately two days to install and become operational. The robot is quite complex and would, in the opinion of the authors, require a minimum of two weeks to become familiar with the programming language and how to use the instrument. If a new function or use of the robot were required then re-engineering of the system would be necessary.

Human involvement is at its greatest with the Vac Elut where a trained analyst is required in all sample preparation stages. Less is needed with the robot where samples are laid out and the prepared cassettes are transferred manually to the chromatograph. The autosampler has the least human involvement as it can be set up and left to operate: all samples are prepared and analysed on-line.

These performance criteria are summarized in Table 3.

Table 3.   Summary of Performance of the Manual Method and the Automated Methods of Analysis

| | Manual Vac Elut | Automated | |
		Gilson-AASP	Robot System
Max per hour	10	9	6
Maximum No. Samples/24 h	80	200	150
Reliability	Poor	Unknown	Very good
Stability on Solid Phase	Stable Compounds	Any Compounds	Stable Compounds
Cost	£ 200	£ 5,000	£ 30,000
Implementation Time	2 minutes	2 days	2 weeks
Human Involvement	Labour Intensive	Minimal	Minimal

CONCLUSIONS

The three systems presented here can be used to analyse inotropic drugs in plasma; all are sufficiently precise and accurate. However the type, the degree of automation and the costs that a laboratory has to pay are varied. Flexible robotic systems are expensive and the most complex to implement and understand. According to Arndt[4], they are used to automate the "status quo" and do not provide an advance as such; essentially being tailored to a laboratory's needs.

The dedicated automation, outlined here, is more cost-effective and quicker to implement but has the disadvantage that it is limited to one task only. However, it has the potential to provide the greatest throughput but with the lowest cost for an automated system.

It is our belief that dedicated automation will find its greatest application in routine analysis whilst flexible systems will be used for development tasks.

REFERENCES

1. R.D. McDowall, J.C. Pearce and G.S. Murkitt, J. Pharm. Biomed. Anal. 4:3 (1986).
2. J.C. Pearce, J.A. Jelly, K.A. Fernandes, W.J. Leavens and R.D. McDowall, J. Chromatogr. 353:371 (1986).
3. D. Hawkes and M. Lindsay, Lab. Pract. September (1984).
4. R.W. Arndt, Chem. Brit. 22:974 (1986).

APPLICATION OF LIQUID-SOLID EXTRACTION: CONCOMITANT ANALYSIS OF
A CARDIOVASCULAR DRUG AND ITS METABOLITES IN PLASMA

R.D. McDowall and J.C. Pearce

Department of Drug Analysis
Smith Kline and French Research Ltd.
Welwyn, Herts, U. K.

INTRODUCTION

Sample preparation is usually a necessity for the analysis of
drugs in biological samples. In the past, this has been accomplished by liquid-liquid extraction which has the major disadvantage of emulsion formation and labour intensive liquid transfer
steps. The application of bonded-silica chemistry used for HPLC
column technology to sample preparation can avoid these disadvantages[1] whilst reducing the time and effort involved in the
transfer operations[2].

SK&F 94120 [5-(4-acetamidophenyl)pyrazin-2(1H)-one] is a
novel, potent, orally active inotropic agent with vasodilator
activity considered useful in the treatment of congestive heart
failure[3]. The compound is known to be extensively metabolised
after administration to the dog and the rat as shown in Fig. 1[4].
An assay involving the AASP (Analytichem Automated Sample
Processor) developed for the determination of the parent compound
in plasma, was also capable of quantifying metabolites II, III and
IV [5]. Chemical synthesis of the glucuronide metabolite (SKF 94120
MET I) gave us an opportunity to extend the assay to quantitate
all metabolites and the parent drug in a single plasma sample. The
simplest method was to modify the published assay to incorporate
the latter metabolite; a chromatogram of the standard reference
compounds eluting in the original solvent system is shown in
Fig. 2. It can be seen that the glucuronide metabolite elutes
immediately after the void volume. Interfering components of
plasma also eluted at this time and thus, to be successful in our
aim, we had to develop a new analytical strategy; this is shown in
Fig. 3. This assay is innovative in that sequential elution of the
AASP cassette with increasing strength mobile phases is used to
elute the analytes of interest. During the first elution the
glucuronide metabolite is measured, followed by a second elution
of the same cartridge to quantify the parent drug and the three
remaining metabolites. The use of the valve reset facility of the
AASP LC module is crucial for the success of this assay.

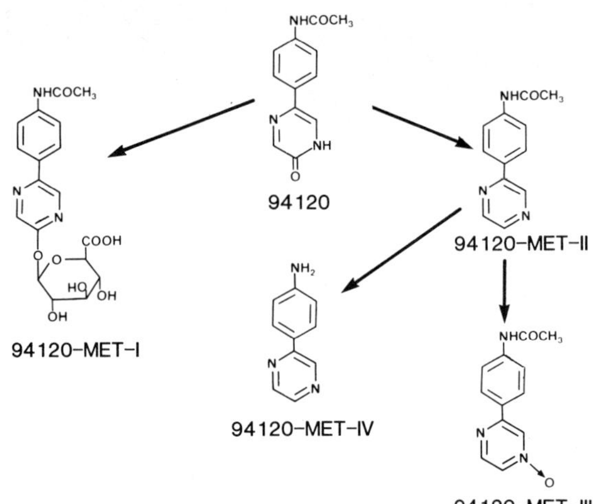

Fig. 2 . Chromatogram of SK&F 94120 and metabolite reference standards (HPLC-UV).

EXPERIMENTAL

Extraction of Plasma Samples

Plasma samples for assay were first thawed at ambient temperatures and then centrifuged at 2,000 g for 10 min to remove fibrous material. 100 μl was transferred to a 1.5 ml polypropylene centrifuge tube, then water (100 μl) and 500 μl of an internal standard solution (SKF 94857, the propyl analogue of SKF 94120) were added to the sample and mixed by vortex. The internal standard is used to quantify the parent drug and SKF 94120 MET II, III and IV. In contrast, quantification of the glucuronide metabolite (SKF 94120 MET I) used external standardisation.

The AASP C18 cassette was activated as normal[5] then 300 μl of each diluted plasma sample was transferred to a newly prepared cartridge and air applied until the reservoir was empty. Water (1.5 ml) adjusted to pH 3 with orthophosphoric acid was then used to wash off any highly polar material adsorbed to the sorbent. The cassette was removed from the manifold and transferred to the AASP HPLC systems for automated analysis.

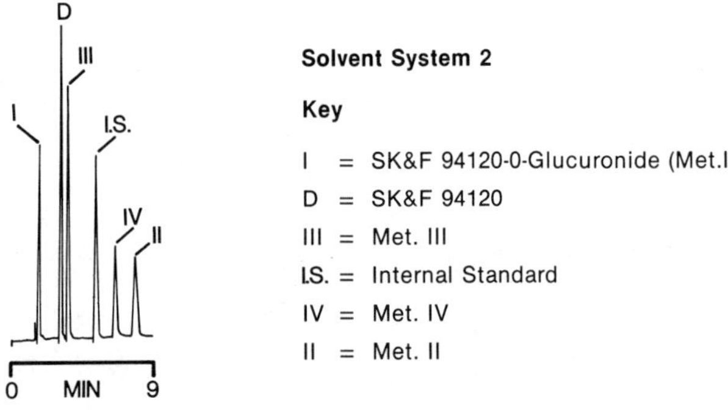

Fig. 1. SK&F 94120: Metabolic Pathway.

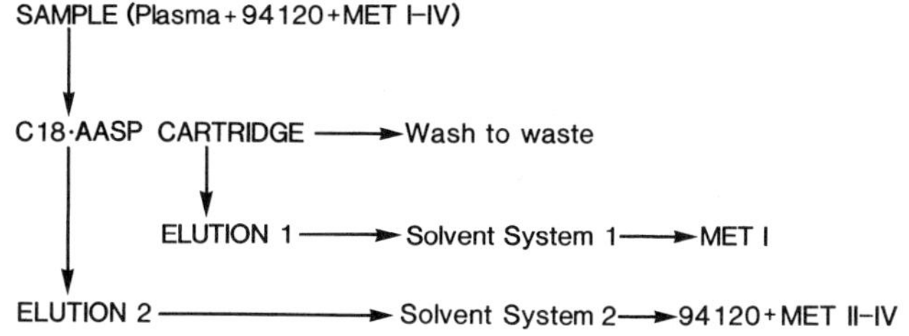

Fig. 3. Analytical strategy.

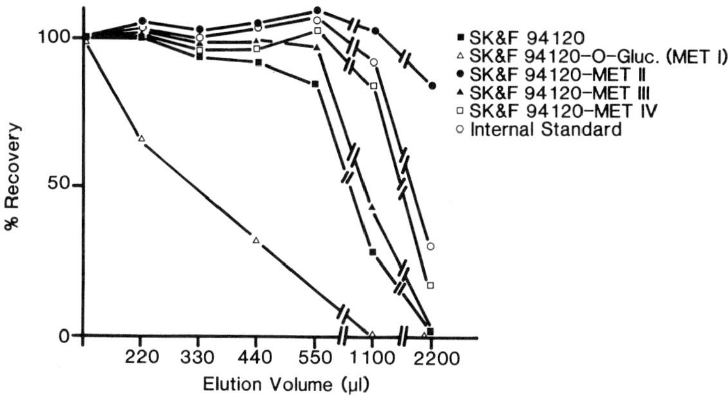

Fig. 4. Effect of cartridge wash volume (Solvent System 1) on
the recovery of SK&F 94120 and metabolites I-IV
(off-line).

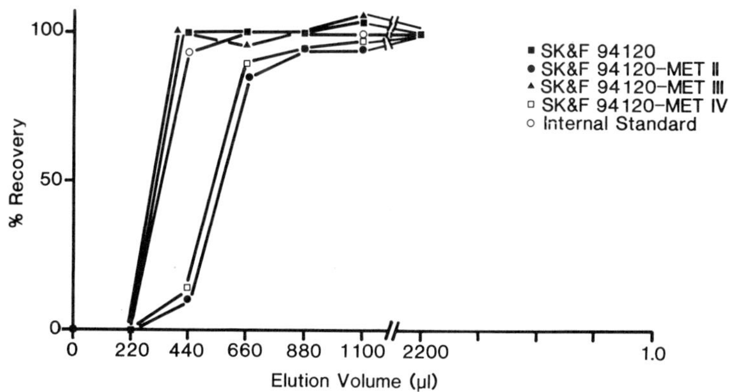

Fig. 5. Effect of valve reset time on the recovery of SK&F
94120 and metabolites II-IV (Solvent System 2).

High Performance Liquid Chromatography

Two chromatographs were set up. Both were identical with the
exception of the mobile phases and the instrument settings on the
AASP. Details of the AASP have been covered in other publications[1,5] and will not be repeated here. Each chromatograph
consisted of a Model 6000A (Waters, Milford, Ma, USA) pump, con-

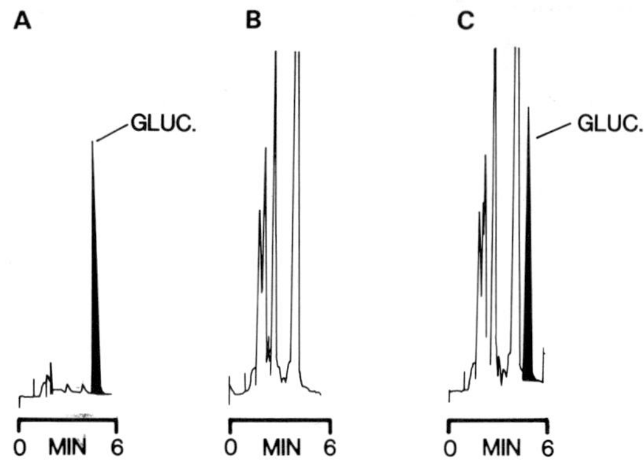

Fig. 6. Typical chromatograms (Solvent system 1) of
A: SK&F 94120-0-Glucuronide
B: Extract of control animal plasma
C: Extract of control animal plasma
spiked with Glucuronide (0.25 mg/l).

nected to an AASP-LC module with the following settings:

System 1:   the AASP run time was 6 min, cycle time was 7 min
              and the valve was reset after 0.3 min.
System 2:   the respective settings were 8,9 and 0.5 min.

Optimisation of the valve reset times was achieved by
monitoring the recovery of each analyte from the cartridge
following elution with varying volumes of the mobile phases. The
results are shown graphically for solvent system 1 and solvent
system 2 in Fig. 4 and 5 respectively.

The analytes were separed by a stainless steel column
300 x 3.9 mm I.D. packed with 10 µm C18-µBondpak (Waters) main-
tained at 35°C. The column effluent was monitored by a Kratos 783
variable wavelength UV detector, set at 280 nm and 0.005 absorbance units full scale. The signal from the detector was fed into
a LDC model 301 computing integrator.

The mobile phase of the first chromatograph (solvent system 1) consisted of 100 ml acetonitrile and 900 ml ammonium acetate buffer (0.01 mol/l); the pH value of the whole mixture was
adjusted between 4.5 and 5.0 with the addition of 300 µl
orthophosphoric acid. This was pumped through the system at
2.0 ml/min and under these conditions the retention time of
SKF 94120-MET I (glucuronide metabolite) was approximately 5 min.

The solvent system used in the second chromatograph (solvent
system 2) consisted of 200 ml acetonitrile, 800 ml 0.01 mol/l
ammonium acetate buffer and 300 µl orthophosphoric acid. At a flow
rate of 2 ml/min the approximate retention times of SKF 94120,
MET III, SKF 94857 (the internal standard), MET IV and MET II were
3, 3.5, 4.5, 6 and 7 min respectively.

RESULTS AND DISCUSSION

Typical chromatograms observed after the elution of analytes
from the AASP extraction cartridges are shown in Fig. 6 and 7.
Fig. 6 shows chromatograms of the standard solution, control

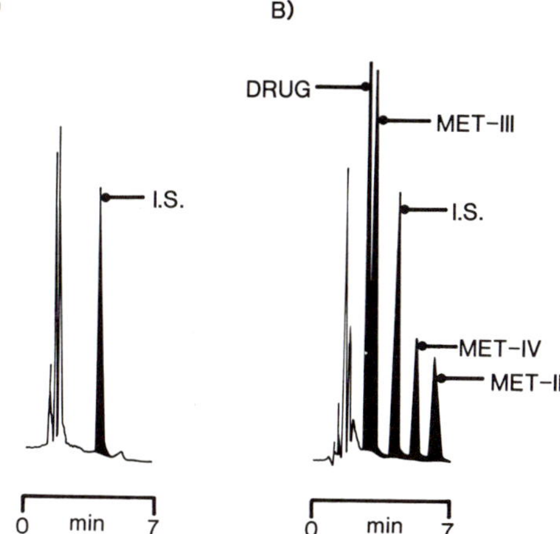

Fig. 7. Typical chromatograms of extracts of control animal
plasma spiked with:
A: Internal standard
B: Internal standard + Drug + 3 Metabolites.

plasma and plasma spiked with the glucuronide metabolite taken
through the procedure and eluted in solvent system 1. Fig. 7 shows
the elution of the remaining analytes from the cartridge in
solvent system 2. In both cases the peak shape of the analytes is
good and there is adequate resolution for accurate quantification.

The recovery of the parent drug and the glucuronide metabolite average 85% each whilst the recovery of the remaining
metabolites is quantitative. The importance of the pH value of the
initial water wash on the cartridge is clearly demonstrated in
Fig. 8. Rapid loss of the glucuronide metabolite occurs if the pH
value of this solution is not reduced to 3. Precision and accuracy
are within 10% for all analytes over the concentration range 125 -
500 ng/ml.

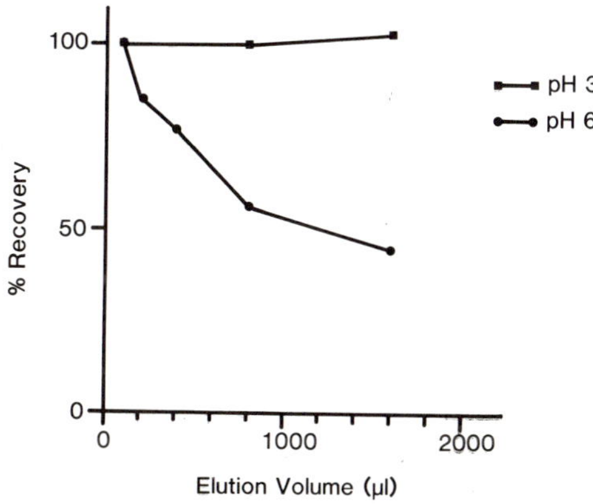

Fig. 8. Effect of water wash pH on the recovery of SK&F 94120-O-
Glucuronide from C_{18}AASP cassette [off line].

REFERENCES

1. R.D. McDowall, J.C. Pearce and G.S. Murkitt, J. Pharm. Biomed.
    Anal. 4:3 (1986).
2. R.D. McDowall, G.S. Murkitt and J.A. Walford, J. Chromatogr.
    317:471 (1984).
3. W.J. Coates, R.J. Eden, J.C. Emmett, R.W. Gristwood, D.A.A.
    Owen, R.A. Slater, E.M. Taylor and B.H. Warrington, Brit. J.
    Pharmacol. 84:22 (1985).
4. P.M. Osborne, S.H. Muth, R.M. Swallow, E.R. Franklin, M.A.
    Pue, T.J.A. Blake, E.S. Pepper, R.J. Chennery, D.A. Ross and
    R. Metcalf, presented at the International Society for the
    Study of Xenobiotics Meeting, Malta, 28th Febraury - 2nd March
    1985.
5. J.C. Pearce, J.A. Jelly, K.A. Fernandes, W.J. Leavens and
    R.D. McDowall, J. Chromatogr. 353:371 (1986).

DETERMINATION OF AMITRIPTYLINE, CHLORIMIPRAMINE AND THEIR
DEMETHYLATED METABOLITES IN PLASMA BY NORMAL PHASE AND UV
DETECTION

V. Ascalone and L. Dal Bo'

Pharmacokinetic Laboratory, Clinical Research Unit
L.E.R.S. - Synthèlabo
Milan, Italy

INTRODUCTION

Various GLC and HPLC methods for the determination of
tricyclic antidepressant drugs (TAD) (see Fig. 1 for chemical
structures) in biological fluids have been published in the recent years, these together with other methods: spectrophotometry,
spectrofluorimetry, RIA have been reported in excellent review
articles[1-2].

In our laboratory two methods are available for the determination of TAD in plasma and blood: capillary gas chromatography
with nitrogen-phosphorus selective detector (NPD) and HPLC on
silica column with UV detector. Capillary GC with NPD, despite the
high sensitivity, accuracy and specificity, is susceptible to
severe interference by residues from phosphate plasticizers
contained in plastic syringes and blood collection tubes[3,4]; for
such situations a method was developed, HPLC on silica column,
which enables the quantitation of TAD and metabolites also in
presence of plastic contaminants. This paper describes this new
HPLC method which consists of a single and rapid extraction,
chromatography on silica column and UV detection.

EXPERIMENTAL

Reagents and Solvents

NaOH 1 mol/l aqueous solution and NH4OH 25% Suprapur® were of
analytical grade from Merck (Darmstadt, FRG). Nanograde n-hexane
from Baker (Deventer, The Netherlands). Methanol, 2-propanol and
methylene chloride HPLC grade were from Baker, the water used for
the preparation of reagent solutions was HPLC grade produced by
the Milli Q-4 system (Millipore, Bedford, Ma, USA).

Amitriptyline (AT), chlorimipramine (CMI), nortriptyline
(NT), desmethylchlorimipramine (DCMI) and imipramine (IMI, inter-

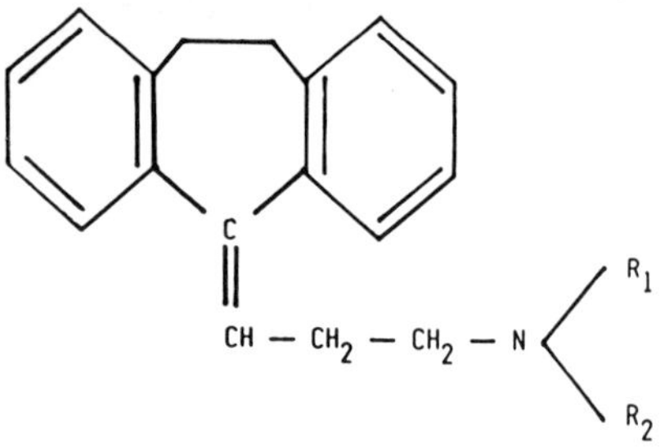

AMITRIPTYLINE   R$_1$ - CH$_3$ R$_2$ - CH$_3$

NORTRIPTYLINE   R$_1$ - H   R$_2$ - CH$_3$

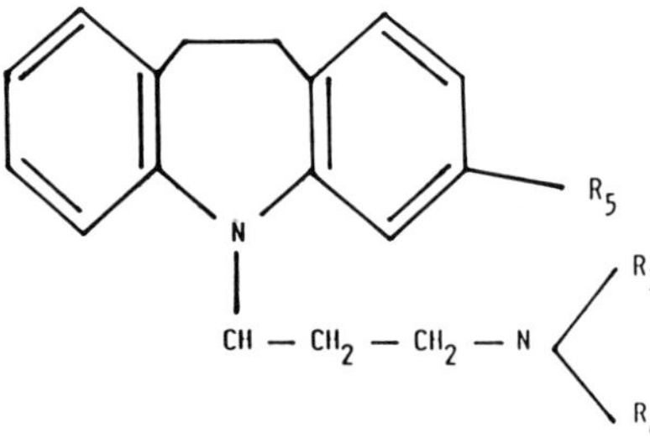

IMIPRAMINE   R$_3$-CH$_3$ R$_4$-CH$_3$ R$_5$-H
DESIPRAMINE   R$_3$-H   R$_4$-CH$_3$ R$_5$-H

CHLORIMIPRAMINE   R$_3$-CH$_3$ R$_4$-CH$_3$ R$_5$=Cl
DESMETHYLCHLORIMIPRAMINE R$_3$-H   R$_4$-CH$_3$ R$_5$=Cl

Fig. 1. Structures of amitriptyline, nortriptyline, imipramine, desipramine, chlorimipramine and desmethyl-chlorimipramine.

nal standard) were of pharmaceutical grade and provided by L.E.R.S. (Paris, France).

Standard Solutions

Stock solution of all the interested compounds were prepared by dissolving 5 mg of each compound in 10 ml of methanol. Standard solutions were prepared from stock solutions, by suitable dilution with methanol. Stock solutions are stable for at least 1 month if stored at 0-5°C.

HPLC Equipment and Operating Conditions

The chromatographic system consisted of: a model 414-T constant flow pump (Kontron, Zurich, Switzerland); a model PU 4020 UV spectrophotometric liquid chromatography detector (Pye Unicam, Cambridge, UK) operating at a wavelength of 240 nm and at a sensitivity of 0.02 AUFS; a Sedex 100 automatic sample injector

(Sedere, Vitry-sur-Seine, France) with an automatic injection
valve and a loop capacity of 50 µl; an analytical column
250 x 4.6 mm I.D. packed with silica, LC-Si 5 µm connected with a
guard column, 20 x 4.6 mm I.D., LC-Si 5 µm (Supelco, Bellefonte,
Pa, USA).

The mobile phase was methanol:aqueous ammonia 25%:methylene
chloride (4:0.2:95.8) and the flow rate was 1.5 ml/min. The UV
detector was coupled to an SP 4270 chromatographic computer
(Spectra Physics, San Jose, Ca, USA) for the determination of
peak-height and subsequent calculations using the internal
standard method and a multipoint linear calibration. Under these
conditions, the retention times for the interested substances
were: CMI about 5 min, AT about 5.2 min, IMI about 6.4 min, NT
about 13.9 min, DCMI about 14.7 min, as reported in Fig. 2a.

Procedure for Plasma Samples

A 20 µl volume of internal standard solution (generally
6 µg/ml) was transferred to a screw-topped test-tube (PTFE-lined-
caps, Sovirel 15) for each unknown sample, 1 or 2 ml of plasma was
added and the samples mixed well. A separate set of plasma standards, covering the range 5-200 ng/ml of plasma for all TAD, was
prepared by transferring 30 µl aliquots of standard solutions into
separate screw-topped test-tubes, pre-dose plasma (1 or 2 ml) was
added and mixed well (plasma standards used for calibration).

To all samples 0.5 ml of water and 0.2 ml of 1 mol/l NaOH was
added; then the samples were homogenized on a vortex mixer for a
few seconds, finally 6 ml of n-hexane was added and the tubes were
shaken on a tumble extractor (10 min at 30 rpm).

The tubes were then centrifuged at 2,500 g for 2 min at 4°C
on a refrigerated centrifuge.

After this operation the tubes were tipped into a cryogenic
bath at about -70°C (the aqueous phase being immediately frozen),
the organic phase was transferred into a conical glass tube
containing 50 µl of 2-propanol (to prevent glass adsorption during
the evaporation step).

All the organic extracts were evaporated to dryness under a
gentle stream of pure nitrogen in a thermostatted water-bath at
50°C.

The residues were redissolved in a suitable volume of HPLC
mobile phase (100-200 µl) and transferred to conical vials
(autosampler vials with conical inserts; Chrompack, Middelburg,
The Netherlands) set in a sample tray for automatic injection.

Calculation

Peak-height ratios substance/internal standard obtained from
plasma standards plotted versus concentration of the substances
were used to generate the linear least squares regression line.
The concentration of the interested substances in the unknowns
were determined by interpolation from the calibration curve using
the peak-height ratios substance/internal standard.

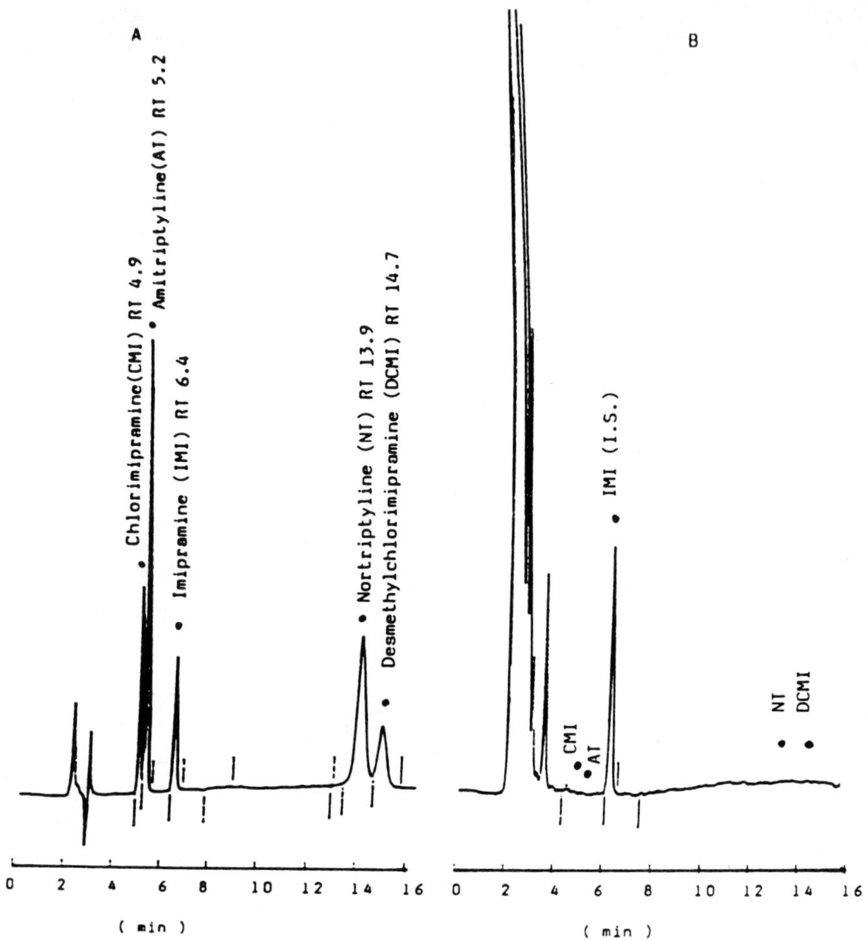

Fig. 2. a) Chromatogram of synthetic mixture containing CMI, AT, IMI, NT, DCMI.
b) Chromatogram of pre-dose plasma extract containing the internal standard.

RESULTS

Linearity

A linear correlation between the peak-height ratios AT:IMI, NT:IMI, CMI:IMI, DCMI:IMI and the concentration of AT, NT, CMI, DCMI respectively was found in the range 5-200 ng/ml of plasma; the linear least-squares regression calculation performed gave respectively the following correlation coefficients (r): 0.9988, 0.9992, 0.9989 and 0.9962 (Fig. 3), n=5 in all the cases.

Sensitivity

The detection limit for AT, NT, CMI and DCMI was respectively 1 ng/ml, 2 ng/ml, 2 ng/ml and 4 ng/ml in plasma with a signal-to-noise ratio of ca. 3:1.

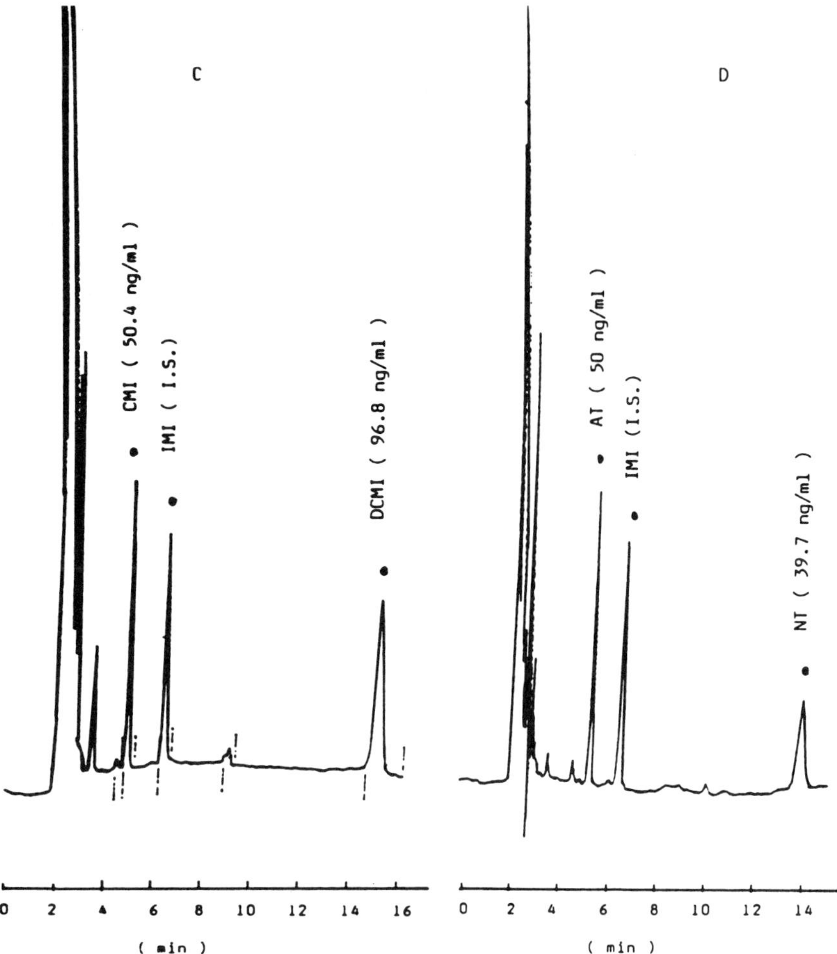

Fig. 2. c) Chromatogram of plasma extract from a patient administered orally with 50 mg chlorimipramine (single administration). Sample taken 2 hours after drug intake.
d) Chromatogram of plasma extract from a patient administered orally with 50 mg amitriptyline (single administration). Sample taken 2 hours after drug intake.

Selectivity

Several blank plasma samples from different humans were
tested for the absence of interfering endogenous components, the
Fig. 2b depicts a typical chromatogram of a drug-free plasma extract, spiked with the internal standard: there are no interfering
peaks at the retention times of the relevant compounds.

Interferences

The chromatographic interferences of other drugs that could
be co-administered were checked. No interferences were found when
control plasma samples spiked with largely prescribed drugs, such
as benzodiazepines, dextrometorphane, caffeine and cimetidine were
processed according to the described method.

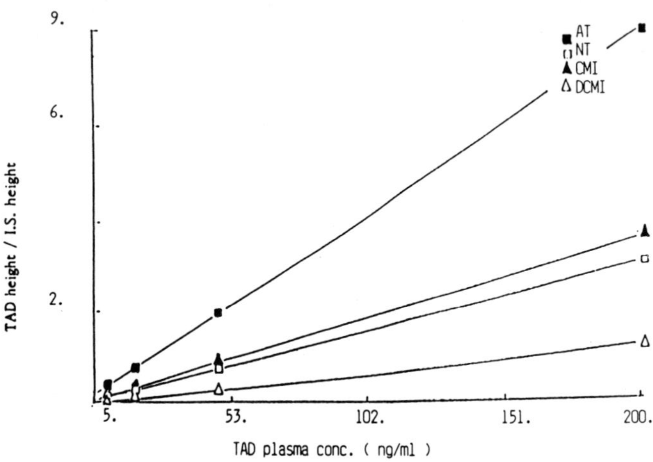

Fig. 3. Calibration curves for antidepressant drugs and their
metabolites.
Equations of the linear   (r)
regression curve
for AT   y= 21.9 x -0.53   0.9988
for NT   y= 55.8 x +0.09   0.9992
for CMI   y= 47.4 x -0.23   0.9989
for DCMI y= 139.2 x +0.3   0.9962

Table 1.   Reproducibility for Plasma Samples Spiked with AT, NT,
CMI, DCMI

	Amount added (ng/ml)	Amount found (Recovery ± C.V.%)	
		Within day (n=4)	Day-to-day (n=6)
AT	5	4.99 (99.8± 2.2)	5.08 (101.7±6.2)
	10	9.89 (98.9± 3.5)	10.10 (101.0±2.3)
	50	49.89 (99.8± 3.3)	50.04 (100.1±2.3)
	200	196.47 (98.2± 2.2)	193.61 (96.8±2.0)
NT	5	5.19 (103.8± 7.3)	4.75 (95.0±8.0)
	10	10.03 (100.3± 6.9)	9.77 (97.7±3.8)
	50	50.25 (100.5± 2.8)	50.39 (100.8±5.1)
	200	199.14 (99.6± 3.2)	208.55 (104.3±4.2)
CMI	5	5.26 (105.2± 3.0)	5.23 (104.6±4.3)
	10	10.15 (101.5± 3.5)	10.03 (100.3±3.2)
	50	50.31 (100.6± 3.3)	50.14 (100.3±2.5)
	200	195.25 (97.6± 4.2)	200.00 (100.0±4.1)
DCMI	5	5.24 (104.5±11.3)	5.09 (101.2±4.0)
	10	10.47 (104.7± 5.9)	10.03 (100.3±7.8)
	50	49.89 (99.7± 6.0)	50.13 (100.3±3.0)
	200	200.31 (100.2± 3.5)	200.76 (100.4±3.9)

<u>Statistical Validation of the Method</u>

Before performing the statistical validation of the method, by using the internal standard method, the absolute recoveries of all compounds were investigated from plasma spiked with the compounds. The overall recovery was about 90% for all the substances being independent on the concentration.

Intra-assay precision (within day) was performed on control plasma spiked with different amounts of all the interested compounds and processed during the same day. Inter-assay precision (day-to-day) was determined analysing the plasma samples on different days (about two week period).

The results are reported in Table 1 and demonstrate acceptable precision and reproducibility of the method over the ranges of concentrations investigated and during time.

<u>Application of the Method to Biological Specimens</u>

The assay was applied to the quantitative determination of AT, NT, CMI and DCMI in human plasma of patients treated orally with the drugs. Typical chromatograms of this experience are reported in Fig. 2c and 2d.

DISCUSSION

The method herein described is a fairly rapid and reliable HPLC method for the determination of two important TAD and their demethylated metabolites. The method, being automatized, has been successfully applied to analyse a large number of plasma samples arising from clinical and pharmacokinetic studies. The organic extracts chromatographically processed (n-hexane extracts) are rather clean and allow to work at high instrument sensitivity thus obtaining a detection limit very low and in the range 1-4 ng/ml.

Simplicity, sensitivity, accuracy and reproducibility, peculiar of this assay, make the method suitable for therapeutic drug monitoring and for pharmacokinetic studies when plasma concentrations are very low and high sensitivity is required.

REFERENCES

1. B.A. Scoggins, K.P. Maguire, T.R. Norman and G.D. Burrows, Measurement of tricyclic antidepressant. Part I. A review of methodology, Clin. Chem. 26:5 (1980).
2. R. Gupta and G. Molnar, Measurement of therapeutic concentrations of tricyclic antidepressant in serum, Drug Metab. Rev. 9:1 (1979).
3. J.D. Ramsey, T.D. Lee, M.D. Osselton and A.C. Moffat, Gas liquid chromatography retention indices of 296 non-drug substances on SE-30 or OV-1 likely to be encountered in toxicological analysis, J. Chromatogr. 184:185 (1980).
4. J.A.F. de Silva, Analytical strategies for therapeutic monitoring of drugs and metabolites in biological fluids, J. Chromatogr. 273:19 (1983).

DICLOFENAC IN PLASMA SAMPLES, A COMPARISON BETWEEN HPLC AND
CAPILLARY GLC

C. Giachetti, P. Poletti, and G. Zanolo
Biomedical Research Institute "A. Marxer" RBM
Ivrea, Italy

INTRODUCTION

2-[(2.6-dichlorophenyl)amino]phenyl acetate, Na salt,
diclofenac, is a potent steroidal analgesic and antiinflammatory
agent used in the treatment of rheumatoid arthritis.

In this work we compared HPLC and capillary GLC for the
determination of the drug in plasma samples. A detailed analysis
of the two methods (precision, accuracy and sensitivity) is
reported and discussed. Using the two techniques we analyzed 30
human plasma samples collected in a kinetics study of a marketed
formulation of diclofenac. The results obtained suggest that
capillary GLC as well as HPLC can be used to determine low plasma
concentrations of diclofenac in kinetics studies.

EXPERIMENTAL

Liquid Chromatography

Apparatus: Beckman 332 pumps
Detector :   Beckman 160 (Hg lamp)
Column: Prolabo (Paris, F) RP8, 5 μm, 250 x 4 mm I.D.
Wavelength: 280 nm
B%   70% acetonitrile in 0.01 mol/l H_3PO_4 adjusted to
                  pH 2.9 with triethylamine
Flow rate: 0.8 ml/min.

Sample Procedure for HPLC

A 0.5 ml plasma sample, to which 0.5 ml of 0.6 mol/l HCl and
250 ng of internal standard (mefenamic acid) had been added, was
extracted with 5 ml of benzene by shaking for 5 min. After
centrifugation the organic layer was transferred to another tube
and the drug extracted into 1 ml 0.1 mol/l NaOH by vortexing and
centrifuging for some minutes. The organic layer was discarded and
the aqueous solution reacidified with 50 μl of 6 mol/l HCl,
reextracted with 5 ml of benzene for 5 min and then centrifuged.
The supernatant was transferred to another tube and taken to
dryness under a stream of nitrogen. The residue was redissolved in
100 μl of the eluting solvent and 20 μl injected.

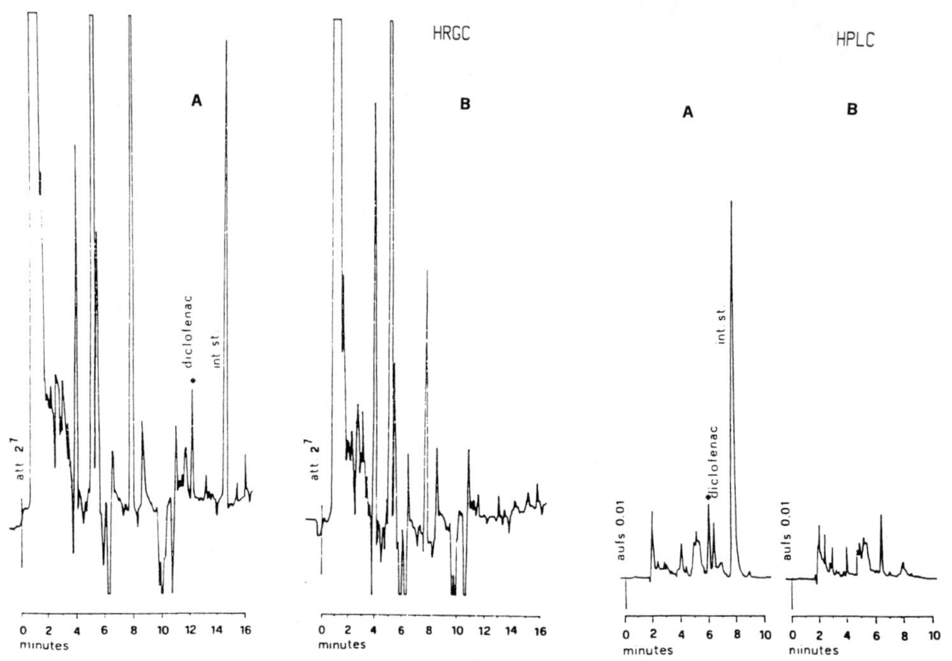

Fig. 1. Typical chromatograms of A) human plasma samples spiked
with 50 ng of Diclofenac and B) blank human plasma
samples, extracted and analyzed by both methods as
described.

Gas Chromatography (Capillary GLC)

Apparatus:	C. Erba Fractovap 2900 series (Milan, Italy)
Column:	OV 1 WCOT glass capillary column (Mega Milan)
	15 m x 0.32 mm I.D., film thickness 0.4:0.45 μm

$$180°C \xrightarrow[1\ min]{20°C/min} 200°C \xrightarrow[5\ min]{2.5°C/min} 240°C$$

Oven/temp : 180°C————> 200°C ——————> 240°C

Injector/Detector temperature	250°C/275°C
Detector:	C. Erba ECD HT 25 ^{63}Ni (10 mCi)
Voltage:	50 V, width 1 μs, frequency 100 μs
Excitation:	Pulsing type voltage
Procedure:	Constant period
Injection type:	Splitless, splitting ratio 1:20
Gas flow:	Carrier, He: 3 ml/min
	Make-up, N: 40 ml/min

Sample Procedure for GLC

A 30-200 μl plasma sample, H_2O up to 400 μl and 50 μl of
0.8 mol/l H_3PO_4 were mixed and extracted into 2 ml of a n.hexane:
dichloromethane 1:1 solution by shaking for 10 min. The organic
layer was taken up, dried under nitrogen and the residue dissolved
in 2 ml of 5 mol/l NaOH and left in a water bath at 40°C for 5
min. After cooling, 2 ml of dichloromethane, 200 μl of 0.05 mol/l
tetrabutyl ammonium hydrogen sulfate (in 0.1 mol/l NaOH) and 50 μl
of iodomethane were added to the sample and shaken for

Table 1. Calibration Curves

	Parameters		S.E.	95% C.I.		r
HRGC	Bx	0.0088	0.0001	0.0090	0.0085	0.9985
	C	0.0416	0.0139	0.0709	0.0122	
HPLC	Bx	0.0045	0.0003	0.0051	0.0039	0.9685
	C	0.0494	0.0337	0.1201	-0.0214	

Linear regression: y = Bx + C; Range: 10-25-50-100-250 ng/ml.
Mean values (4 replicates).

25 min. After centrifugation the organic layer was taken up and
dried under nitrogen. The residue, dissolved in 2 ml of distilled
water by vortexing, was extracted into 0.5 ml of hexane containig
40 ng of meclofenamic acid methyl ester as internal standard by
shaking for 5 min. After centrifugation the organic layer was
removed and 1 μl injected.

DISCUSSION

The quantification of diclofenac was evaluated by assaying a
set of plasma samples spiked with increasing amounts of drug using
both HPLC and GLC capillary methods (Fig. 1). In Table 1 the
calibration curves, obtained by plotting the peak height ratio
(drug/IS) versus the corresponding diclofenac concentrations, are
reported: these proved to be linear over the range considered.

The parallelism test confirmed that the two curves do not
differ significantly. In Table 2 the results on the accuracy and
precision of the methods are summarized. GLc appeared a precise
and accurate method at all concentrations above 5 ng/ml, while at
2 ng/ml the precision was not so good (CV > 20%) even if the
method still appeared to be accurate (recovery > 88%). The HPLC
method was imprecise but accurate enough at concentrations lower
than 10 ng/ml (CV > 29%). The other results were strongly
comparable with those obtained with the GLC. Therefore the
reproducibility of the two methods, as shown by the 95% confidence
interval, is very high.

Table 2. Reproducibility of the Assays. Mean Values (n=4)

	HRGC	HPLC	HRGC	HPLC	HRGC	HPLC
ng added	10.0	10.0	50.0	50.0	200.0	200.0
ng recovered	9.2	8.6	50.3	48.2	199.3	198.9
S.D.	1.5	2.5	1.6	2.9	2.8	2.8
recovery%	92.3	85.8	100.7	96.4	99.6	94.5
CV%	16.7	29.1	1.6	6.0	1.4	1.4
95% C.I.	11.4	12.5	52.2	51.6	200.9	202.3
	7.4	4.6	48.4	44.8	197.6	195.6

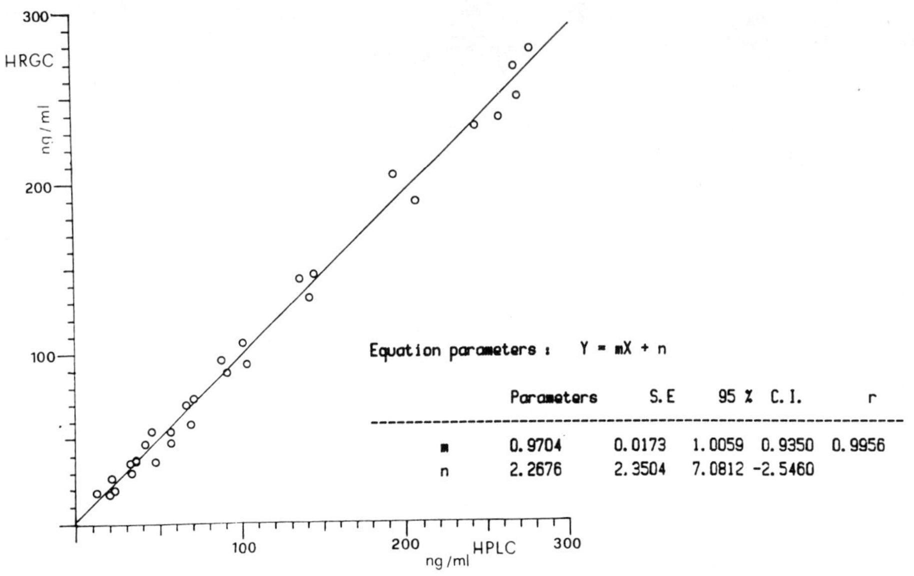

Fig. 2. Comparison of amount of Diclofenac in 30 plasma samples
taken from volunteers in a kinetics study of marketed
Diclofenac formulation.

The contemporary analysis using both techniques of 30
authentic plasma samples, taken from a kinetics study in man with
a marketed diclofenac formulation, furnished similar concentration
data. By plotting the HPLC results versus those of GLC and
applying least-square regression analysis, the Fig. 2 and the
related equation were obtained. In the 95% C.I. the regression
line obtained does not differ from a regression line with a 0
value as intercept and 1 value as slope and this fact confirms
that the results obtained by either GLC or HPLC do not differ
significantly.

Therefore both techniques are helpful in determining
diclofenac in biological fluids, but the HPLC, the sample extraction procedure, being simpler and faster than in GLC, should be
preferred in monitoring the drug, when sufficiently high concentrations are expected. On the other hand, when high sensitivity is
needed, due to the low range of concentration expected in the
biological samples, the GLC is the best choice even if a sophisticated sample extraction procedure is required.

REFERENCES

1. U.P. Geiger, P.H. Degen and A. Sioufi, J. Chromatogr. 111:239
    (1975).
2. P.J. Brombacher, H.M.H.G. Cremers, P.E. Verheesen and R.A.M.
    Quanjel-Schreurs, Arzeim.-Forsch 27:1597 (1977).
3. A. Schweizer, J.V. Willis, D.B. Jack and M.J. Kendall,
    J. Chromatogr. 195:421 (1980).
4. J.V. Willis, M.J. Kendall and D.B. Jack, Eur. J. Clin.
    Pharmacol. 18:415 (1980).
5. J.V. Willis, M.J. Kendall and D.B. Jack, Eur. J. Clin.
    Pharmacol. 19:34 (1981)

6. J.V. Willis, D.B. Jack, M.J. Kendall and V.A. John, Eur. J.
    Clin. Pharmacol. 19:39 (1981)
7. W. Schneider and P.H. Degen, J. Chromatogr. 217:263 (1981).
8. S.A.Said and A.A. Sharaf, Arzneim-Forsch, 31: 2089 (1981).

APPLICATION OF A SOLID PHASE AUTOSAMPLER TO THE HPLC DETERMINATION

OF DRUGS AND NATURAL COMPOUNDS IN BIOLOGICAL MATRICES

G. Grossi, A. Bargossi, R. Callivà, M.G. Salvatore,
R. Battistoni, and A. Lippi

Clinical Chemistry Laboratory
"St.Orsola" Hospital
Bologna, Italy

i. DETERMINATION OF XANTHINES

The use of a sample cleanup and extraction device and a
column switching technique allowed us to propose an automatized
method for the determination of plasma xanthines[1].

A study on the correlation of the results obtained by this
method with those ones that employ liquid-liquid extraction and
HPLC determination[2] and with fluorescence polarization immunoassay
(FPIA)[3], is here presented.

MATERIALS

Theobromine, theophylline, caffeine and the internal standards beta-hydroxy-propyltheophylline (AASP method) and beta
hydroxy-ethyltheophylline (micro-method) were obtained from Sigma
(St. Louis, Mo, USA).

For the AASP method, the extraction cartridges, silica bonded
with octadecylsilane (AASP cassette C 18) were purchesed from
Analytichem International (Harbor City, Ca, USA); the HPLC pump
mod. 2010, the solid-phase automatic sampler mod. AASP, the
ten-ports injection valve, the UV detector mod. 2050, the recorder
mod. 1201 were from Varian, (Palo Alto, Ca, USA); the C18 reverse-
phase analytical column, 150x4.6 mm, 5 μm (Erbasil C18) from Carlo
Erba (Milan, Italy); and the pre-column C18 40x4.6 mm from Bio-Rad
(Richmond, Ca, USA). The HPLC system used for the micro-method was
composed of a HPLC pump mod. Series 3B, a UV detector LC 75, an
autosampler LC 420 and a recorder mod. 56, from Perkin Elmer
(Norwalk, Ct, USA); the analytical column C18 7 μm, 250x4 mm
(Lichrosorb) was furnished by Merck (Darmstadt, FRG).

For the FPIA method, we utilized the TDX kit (Abbott, USA).

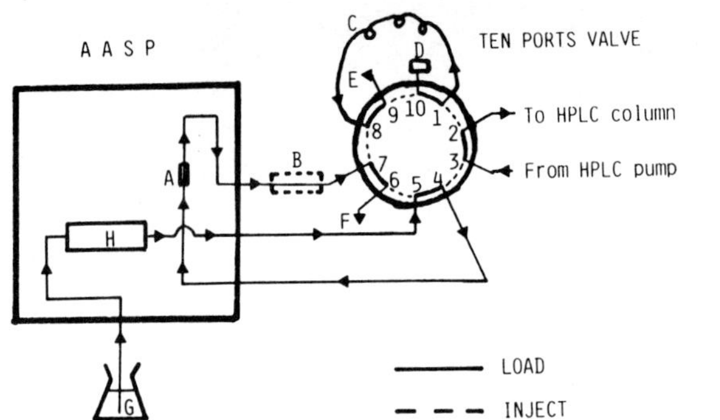

Fig. 1. AASP fluids diagram.

METHODS

Seventy three blood samples were drawn from patients treated
with theophylline and plasma was soon separated and stored at
-30°C till the day of the analysis.

The FPIA method was carried out according to the TDX kit
instruction manual.

The micro-method procedure can be summarized as follows: add
25 µl of plasma to 0.5 ml of $CHCL_3$:isopropyl alcohol (95:5)
(containing the internal standard); after having mixed by vortex
and having centrifuged, draw and evaporate 0.4 ml of the organic
phase; reconstitute with 0.2 ml of mobile phase and inject 50 µl
into the HPLC. Chromatography conditions were: mobile phase,
acetonitrile:0.2 mmol/l phosphate buffer (pH 4.0) (13.5:86.5);
flow, 1 ml/min; detector UV set at 272 nm.

The AASP method (Fig. 1) is here briefly reported:

- activate the cassette with 1 ml of methanol;
- equilibrate with 1 ml of 0.2 mmol/l phosphate
  buffer, pH 4.0;
- apply 25 µl of plasma sample and 50 µl of
  I.S. solution (20 mg/l);
- wash with 1 ml of buffer;
- wash with 1 ml of buffer, keeping the
  cartridge wet;
- put the cassette into the AASP autosampler;

autosampler program:
- pre-purge: 0 min;
- after purge: 0 min;
- run time: 5 min;
- cycle time: 5 min;
- valve reset time: 0.8 min.

HPLC conditions were: mobile phase, acetonitrile:0.2 mmol/l
phosphate buffer (pH 4.0) (13.5:86.5); flow, 2 ml/min; detector UV
set at 272 nm.

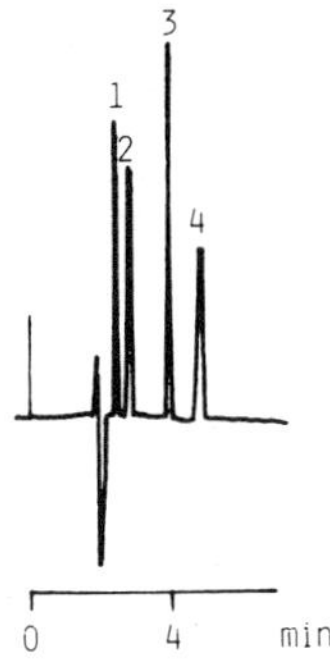

Fig. 2. AASP method, analytical profile obtained from a plasma
containing 10.0 mg/l of each xantine. 1 = theobromine,
2 = theophylline, 3 = I.S., 4 = caffeine.

RESULTS AND DISCUSSION

In Fig. 2 an analytical profile obtained with the AASP method
is shown. The retention times are:

        theobromine   2.0 min
        theophylline   2.5 min
        I.S.   4.0 min
        caffeine   4.8 min.

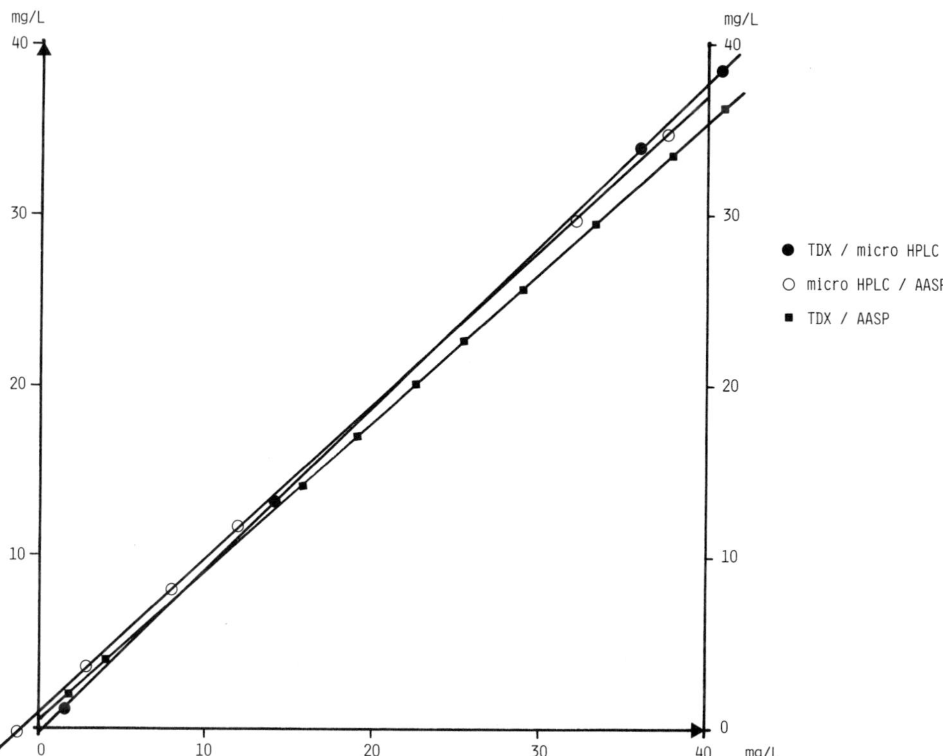

Fig. 3. Correlation study, graphic visualization of the results.

The mean recoveries (n=20) from plasma containing 5.0 mg/l of each xanthine are:

              theobromine   98.4% (CV 2.5%)
              theophylline   99.1% (CV 1.9%)
              caffeine   98.3% (CV.2.2%).

The mobile phase was stable at room temperature for more than 3 months. No interferences from other natural substances have been noted.

The results of the correlation study are summarized in Fig. 3. On the ground of the 73 samples tested, the correlation equations were:

    for Y = TDX and X = micro-HPLC ,
        $Y = 0.949 X - 0.268$ $(r = 0.987)$;

    for Y = micro-HPLC and X = AASP ,
        $Y = 0.900 X + 1.012$ $(r = 0.981)$;

    for Y = TDX and X = AASP,
        $Y = 0.866 X + 0.563$ $(r = 0.981)$.

The AASP solid phase autosampler resulted as a very useful tool for us: the results for xanthine analysis correlated well with other methods; the column switching adopted, changing the cartridge every sample, showed no problems of pressure enhancements and of change of the chromatographic behaviour. The cleanup procedure was fast, not needing any liquid transfer or vial labelling.

Finally the sample cleanup and extraction station can readily be fully automatized, utilizing only commercially available devices.

REFERENCES

1. G. Grossi, A. Bargossi, G. Righi, R. Pasquali, R. Battistoni
    and A. Lippi, Estrazione in fase solida ed analisi in HPLC
    delle xantine mediante un sistema completamente automatizzato,
    Biochim. Clin. 9:1037 (1985).
2. P.J. Naish and M. Cooke, Rapid assay for theophylline in
    clinical samples by reversed-phase high-performance liquid
    chromatography, J. Chromatogr. 163:363 (1979).
3. H.D. Hill, M.E. Jolley and C.H.J. Wang, Clin. Chem. 27:1086
    (1981).

ii.   APPLICATION NOTES ON HPLC OF ANTI-INFLAMMATORY DRUGS, CATECHOLAMINES AND THEIR METABOLITES

The determination of low levels of drugs or natural substances in plasma or urine usually involves extraction and sometimes sample concentration steps before HPLC analysis.

The column switching technique is the natural answer to the request for automation in HPLC[1]. Yet, the main problem is that, with the traditional column switching, the precolumn utilized for sample pretreatment often shows a very short life, so that to process a large number of biological samples, it needs to be changed many times. In this case it would be useful to perform the precolumn change automatically.

The solid phase autosampler previously described (AASP) can carry out sample preparation through disposable precolumns, and on-line sample elution and injection into HPLC[2].

In our experience, we have noticed that this autosampler is suitable to large routines and, changing the cartridge every sample, is free from problems of pressure enhancements and instability of the chromatographic behaviour.

Here we present preliminary data on its application to the determination of some drugs and endogenous compounds of clinical interest in biological fluids.

MATERIALS

For the AASP methods, the extraction cartridges, silica bonded (AASP cassette) were from Analytichem International (Harbor City, Ca, USA); the HPLC pump mod. 2010, the solid-phase automatic sampler mod. AASP, the ten-ports injection valve, the UV detector mod. 2050, the recorder mod. 1201 were from Varian (Palo Alto, Ca, USA); the coulometric electrochemical detector Coulochem 5100 A was from Environmental Sciences Assoc. (Bedford, Ma, USA).

For the traditional solid phase methods, the extraction cartridges, silica bonded (Bond Elut), and the vacuum manifold (Vac Elut) were from Analytichem; for the gradient elution of the catecholamine metabolites, we used an HPLC pump mod. Series 3B (Perkin Elmer, Norwalk, Ct, USA). All the standards were from Sigma (St. Louis, Mo, USA). The bidistilled water used for buffer preparation was purified using Norganic trace removal cartridges (Millipore, Bedford, Ma, USA). Acetonitrile and methanol were HPLC grade; all the other reagents were of analytical grade.

METHODS

<u>Anti-Inflammatory Drugs</u>

Sera spiked with diclofenac, naproxen, flunixin, ketoprofen and sulindac were used.

The HPLC conditions were: analytical column Erbasil C18 150 x 4.2 mm, 5 μm (Carlo Erba, Milan, Italy); precolumn: RP18

25 x 4 mm, 7 µm (Merck, Darmstadt, FRG); mobile phase:
acetonitrile:0.05 mol/l phosphate buffer pH 3 (1:1); flow:
2 ml/min; detector: UV, 254 nm.

Bond Elut C18 extraction conditions were: extraction column
C18, 500 mg; sample: 1 ml serum + 0.3 ml of 0.6 mol/l HCl.

          Method: - activate with methanol;
                       - equilibrate with 0.05 mol/l phosphate
                         buffer pH 3;
                       - apply sample;
                       - wash twice with buffer;
                       - elute with 2 ml of methanol.

Bond Elut SAX extraction conditions were: extraction column
Bond Elut SAX, 500 mg; sample: 1 ml serum, brought to pH 10.

          Method: - activate with methanol;
                       - equilibrate with water;
                       - apply sample;
                       - wash twice with water;
                      .- elute with 2 ml of 1.5 mol/l NaOH.

AASP C18 extraction and injection conditions were: column
AASP C18 cassette; sample: 50 µl serum + 15 µl of 0.6 mol/l HCl.

          Method: - activate with methanol;
                       - equilibrate with 0.05 mol/l phosphate
                         buffer pH 3;
                       - apply sample;
                       - wash with 1 ml buffer;
                       - wash with buffer, keeping the cartridge
                         wet.

          Program: - no purge;
          (AASP)   - run: 5 min;
                       - valve reset: 1 min.

AASP SAX extraction and injection conditions were: extraction
column AASP SAX cassette; sample: 50 µl serum + 10 µl NaOH 2%.

          Method: - activate with methanol;
                       - equilibrate with water;
                       - apply the sample;
                       - wash with water;
                       - wash with water, keeping the cartridge
                         wet.

          Program: - purge solvent: 1.5 mol/l NaOH;
          (AASP)   - pre-injection purge: 1 (25 µl);
                       - run: 5 min;
                       - valve reset: 2 min.

Catecholamine Metabolites Vanilmandelic Acid and Homovanillic Acid

Urines containing natural and spiked amounts of the sub-
stances.were assayed. The analytical column was an RP8 5 µm
125 x 4 mm (Merck) and the pre-column an RP8 7 µm 20 x 4 mm
(Merck). The other conditions were: mobile phase: B = 400 mg/l
sodium heptyl-sulphate buffer, pH 3; A = acetonitrile:buffer

(1:9); flow: 1.5 ml/min; detector: coulometric detector, det. 1:
+0.00 V; det. 2: +0.18 V.

                 Solvent program: - 1 min at 20% A;
                                    - from 20% to 99.9% A in 7 min
                                      with a convex gradient (curve 0.2);
                                    - 3 min at 99.9% A;
                                    - re-equilibration for 4 min at 20% A.

     Bond Elut extraction: extraction column SAX Bond Elut,
500 mg; sample: 0.25 ml urine + 0.25 ml water, at pH 7.5.

                 Method: - activate with methanol;
                              - equilibrate with water;
                              - apply sample;
                              - wash with water;
                              - elute with 1.5 ml of 1.5 mol/l NaOH.

     AASP SAX extraction: extraction column: AASP SAX cassette;
sample: 0.25 ml urine + 7.75 ml of 0.1 mol/l KH_2PO_4 pH 7.

                 Method: - activate with methanol;
                              - equilibrate with water;
                              - apply 0.2 ml sample;
                              - wash with water;
                              - wash with water, keeping the cartridge
                                wet.

                 Program: - purge solvent: 1.5 mol/l NaOH;
                  (AASP)   - pre-inject purge: 1 (25 μl);
                              - run: 14 min;
                              - valve reset: 2 min.

Urinary Catecholamines

     Urines containing natural and spiked amounts of norepine-
phrine, epinephrine and dopamine were assayed.

     Analytical column: Bio-Sil ODS 150 x 4 mm, 5 μm, with pre-
column ODS 40 x 4.6 mm, 5 μm (Bio-Rad, Richmond, Ca, USA); mobile
phase: 50 mmol/l phosphate buffer, pH 2.8, containing 200 mg/l of
sodium dodecyl sulphate and 20% acetonitrile; flow rate: 2 ml/min;
detector: coulometric; conditioning cell : +0.43 V; det. 1:+0.15
V; det. 2:-0.30 V.

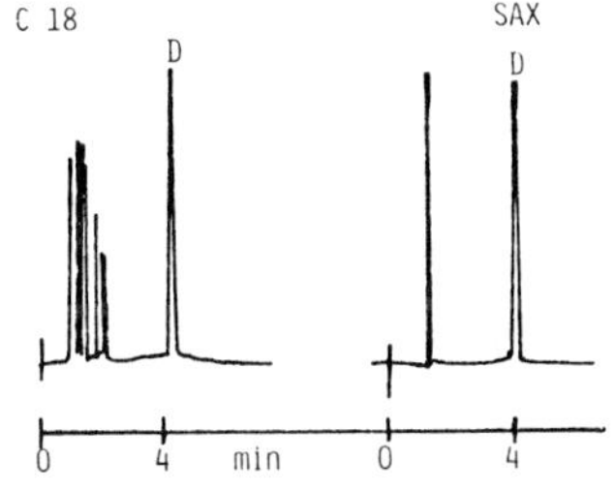

Fig. 1. Anti-inflammatory drugs, AASP extraction and injection:
        serum spiked with Diclofenac
        (Diclofenac 10 μg/ml, 0.16 AUFS).

Bond Elut extraction: extraction column: Bond Elut CBA (carboxylic acid), 500 mg; sample: 5 ml urine + 14 ml EDTA 0.1%, brought to pH 6.5.

Method:	- activate with methanol; - equilibrate with water; - apply 6 ml of sample; - wash with water; - elute with 2 ml of 4% boric acid; - inject 50 µl into the HPLC.

AASP CBA extraction: extraction column: AASP CBA cassette; sample: 5 ml urine + 14 ml of 0.1% EDTA; brought to pH 6.5.

Method:	- activate with methanol; - equilibrate with water; - apply 200 µl of sample; - wash with water; - wash with water, keeping the cartridge wet.
Program: (AASP)	- purge solvent: 4% boric acid; - pre-injection purge: 6 (150 µl); - time: 5 min; - valve reset: 3 min.

RESULTS

Anty-Inflammatory Drugs

For all the substances, the recoveries were more than 90% with Bond Elut C18, more than 80% for Bond Elut SAX, more than 95% for AASP C18 and more than 90% for AASP SAX. In Fig. 1 the analytical profiles of serum spiked with diclofenac and extracted with AASP C18 and AASP SAX are compared. It is clear that the SAX extraction is cleaner and more selective.

As a matter of fact, the problem was keeping the peak volume coming from the extraction cartridge as little as possible; so every time that the HPLC mobile phase could not elute the substance easily, it was necessary to use a stronger eluent. In these case it was better to protect the analytical column, inserting a precolumn between the ten ports valve and the analytical column. Actually, all the precolumns used in the present work had this aim.

Another problem for the AASP was that the cassette should have been wetted, when injection took place; otherwise, an air peak, that could not be removed by the purge pump, arose in the chromatograms. In Fig. 2 an example of this peak, in the case of xantine analysis, is shown.

Catecholamine Metabolites

The Bond Elut SAX extraction has been previously discussed[3]. For the AASP SAX extraction, the recoveries were: VMA 86% and HVA 92%. The analytical profiles obtained by Bond Elut and by AASP extraction were quite similar.

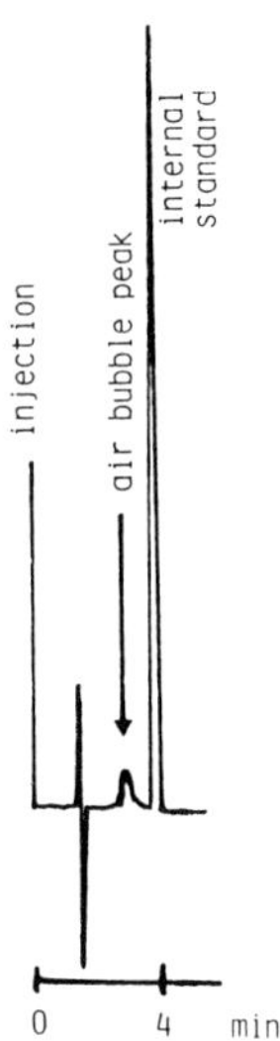

Fig. 2. The "air bubble" interferent peak (xanthine analysis).

Urinary Catecholamines

The selectivity of the extraction was given by the use of boronic acid, as previously reported[4]. The recovery for all the catecholamines was more than 90% with both Bond Elut CBA and AASP CBA extraction. The selectivity of the HPLC separation and of the coulometric detector contributed to obtaining very clean analytical profiles, with practically no peaks but the injection peak and the catecholamine ones.

CONCLUSION

Table 1 summarizes the sample preparation approaches adopted to the analytical problems we dealt with in this paper.

According to our data, it is possible to utilize a mixed mode of analysis, i.e. to extract the sample on an ion-exchange cartridge, and to perform the analysis on a reverse phase column.

Finally, we would point out that, to use the column switching technique with the AASP autosampler, it is necessary that the substances to be determined show column capacity ratio values (k') in the order:

$$(\text{cartridge}) \leq (\text{precolumn}) \leq (\text{analytical column})$$

Since these k' values are not reported in literature, we tested many C18 sorbents for the xanthine analysis, and the comparison showed that k', when evaluated in the condition: acetonitrile:0.2 mol/l phosphate buffer pH 4.0 (13.5:86.5) increases in the order:

RP18 Lichrosorb (Merck) < C18 (Bio-Rad) < C18 (Varian) < AASP C18 cassette (Analytichem) < C 18 Erbasil (C. Erba).

Table 1.   Bond Elut and AASP Extraction Procedures

Substances	Matrix	Reverse-phase Bond Elut	Ion-exchange Bond Elut	Reverse-phase AASP	Ion-exchange AASP
xanthines	serum	yes	no	yes	no
anti-inflam-matory drugs	serum	yes	yes	yes	yes
catechola-mine metab-olites	urine	no	yes	no	yes
catechola-mines	urine	no	yes	no	yes

REFERENCES

1. R.W. Frei, New sample handling strategies in HPLC, Swiss. Chem. 6:55 (1984).
2. G. Grossi, A. Bargossi, Estrazione in fase solida ed analisi in HPLC delle xantine mediante un sistema completamente automatizzato, Biochim. Clin. 9:1037 (1985).
3. G. Grossi, Urinary homovanillic acid and vanilmandelic acid: solid phase extraction and determination by high-performance liquid chromatography with electrochemical detection, in: Proceedings of the Conference "La Cromatografia Liquida ad Elevata Risoluzione (HPLC) in Analitica Clinica: Situazione Attuale e Prospettive", Verona, July 1-2, Cortina , Verona, in press.
4. G. Grossi, Determination of catecholamines in plasma and urine by high-performance liquid chromatography: a comparison among different extraction, separation and detection methods, in: Proceedings of the Conference "La Cromatografia Liquida ad Elevata Risoluzione (HPLC) in Analitica Clinica: Situazione Attuale e Prospettive", Verona, July 1-2, Cortina, Verona, in press.

MEASUREMENT OF PLASMA CATECHOLAMINES - STUDY AT BASAL AND DURING

INSULIN-INDUCED HYPOGLYCEMIA IN NORMAL AND DIABETIC SUBJECTS

G. Piemonte, A. Bolner, P. Moghetti*, E. Bonora*,
V. Cacciatori*, M. Querena*, and M. Muggeo*

Laboratory of Medical Research
* Chair of Metabolic Diseases
University of Verona
Verona, Italy

INTRODUCTION

Many studies demonstrated that catecholamines (CA) play an important role in the regulation of a number of physiological processes involved in body homeostasis[1-5]. In particular, noradrenaline (NA) and adrenaline (A) contribute to the regulation of several metabolic pathways, such as lipolysis and glycogenolysis. In stress conditions (starvation, physical exercise, disease, etc.), catecholamines favour the prompt availability of substrates, as glucose from liver and free fatty acids from adipose tissue[6-15].

Thus, it is of great importance to have informations about the plasma levels of catecholamines in physiological and pathological conditions. Unfortunately, the methods used for catecholamine assay were not always completely satisfactory[16,17]. This was mainly due either to the low concentration of these hormones in blood, or to the difficulty in the clinical interpretation of analytical results. To overcome these problems, many different analytical methods and a number of tests to evaluate catecholamine response in standardized dynamic conditions have been proposed[8,10,17].

CATECHOLAMINES ASSAYS

Several methods were proposed to measure with sensitivity, accuracy and reliability the CA content in biological samples. Here we briefly review some different assays so far employed, which still suffer from problems linked maily to matrix interferences and to the very low concentrations at which these amines are present, particularly in the plasma.

Fluorimetry of Purified Extracts

Fluorimetry (FL) of purified extracts was the earliest assay developed for CA determination, and many attempts have been made to increase both sensitivity and specificity of these methods. CA can be measured[18] either by direct fluorimetry of the unmodified molecules or by fluorescence determination of the CA derivatives obtained by chemical reactions. All these methods need extensive clean-up procedures associated with concentration steps in order to obtain valid results. Fig. 1 shows the basis of the fluorimetric methods most largely used.

At low pH values the catechol ring shows native fluorescence, however several compounds with phenolic structure has the same feature; moreover the sensitivity is poor for clinical applications. By using derivative formation, it is possible to increase both sensitivity and specificity: condensation with ethylenediamine and especially CA-oxidation to form related trihydroxyindoles (THI) by means of manganese oxide, iodine or ferricyanide are the most commonly used reactions. The last reaction has also been utilized in an automatic assay[19] and in HPLC post-column reactors[20].

Fig. 2 shows the flow diagram of the extraction and purification procedure needed to perform a NA and A plasma assay through the reaction to trihydroxyindole derivatives. Several steps including chromatographic purifications on alumina and/or cation exchange resins are needed to remove interferent compounds. NA and A are then individually determined at specific excitation and emission wavelengths or by selective oxidation of the two molecules at different pH values.

<u>Radioimmunoassay</u>

Few attempts have been made to develop radioimmunoassays (RIA) of CA because of the problems arising in the production of

REACTION	FLUOROPHORE	E_X — E_M nm	SENSITIVITY ng
NATIVE FORM PHENOLIC STRUCTURE	LOW pH	285-325	1.0-5.0
ETHYLENEDIAMINE CONDENSATION		420-520	0.2-1.0
OXIDATION -MnO$_2$ -Iodine -Ferricyanide (R$_2$ = H, CH$_3$)		415-500	0.1-0.5

Fig. 1. Fluorimetry features of catecholamines by direct assay or through derivatization.

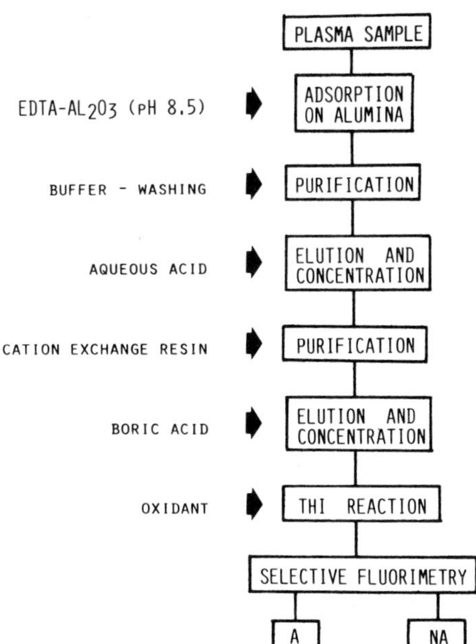

Fig. 2. Flow diagram of adrenaline and noradrenaline fluorimetric
assay in plasma samples, through derivatization to
trihydroxyindoles.

specific and reliable antibodies. These small molecules are not
immunogenic compounds and, consequently, they must be conjugated.
Other problems come from the instability of the catechol moiety
and finally from the low specific activity of tracers, with
consequent poor sensitivity of the assays.

These difficulties were in part circunvented by enzymatic
conversion of CA to more stable compounds, such as metanephrines,
and by using antibodies specific to metanephrine[21]; the scheme of
this procedure is reported in Fig. 3. Nevertheless this method
involves two enzymatic methylations with S-adenosyl-L-methionine
(SAM) catalyzed by phenylethanolamine-N-methyltransferase (PNMT)
and catechol-O-methyltransferase (COMT) before the final
immunoreaction. Moreover in this method NA can be measured only
indirectly by difference between total CA and A concentrations.

Radioenzymatic Assay

Several radioenzymatic assays (REA) for the measurement of
plasma CA have been developed in the recent years. Some of these
methods are based on the enzymatic conversion of NA to A by SAM
and PNMT[22], as shown in Fig. 4. Other assays[23] use methoxyamine
formation from NA, A and dopamine (DA), by SAM and COMT; both the
enzymatic reactions introduce a methyl radioactive group which
labels the molecules. The labelled compounds are then purified
with various treatments, including extractive and chromatographic
steps. Finally CA are measured on the basis of the recovery of
radioactivity.

A scheme of a REA based on COMT is reported in Fig. 5. NA, A
and DA are firstly converted to methoxyamine derivatives and then

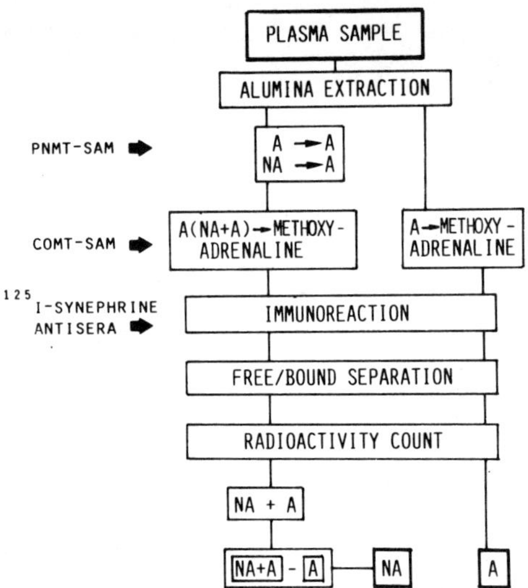

Fig. 3. Flow diagram of adrenaline and noradrenaline radioimmunoassay in plasma samples.

extracted with toluene/isoamyl alcohol. Further purification steps
are then needed, including thin layer chromatography, oxidation to
vanillin and liquid/liquid extraction. As compared to the other
methods, REA allows to detect the lowest amounts of CA, but this
potential advantage is hampered by the need to use small plasma
volumes, because of the presence of enzymatic inhibitors.

Chromatographic Methods

All the methods so far discussed require seriate purification
steps to achieve an acceptable degree of selectivity, with some

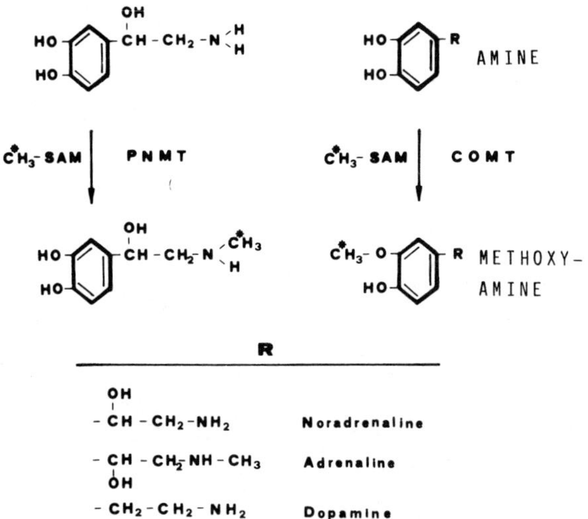

Fig. 4. Radioenzymatic assay of catecholamines in plasma samples
by different enzymatic pathways.

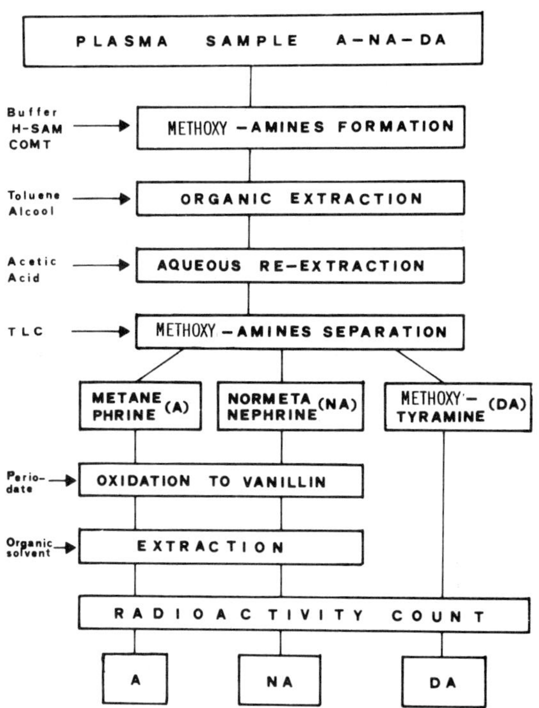

Fig. 5. Flow diagram of adrenaline, noradrenaline and dopamine radioenzymatic assay through the conversion of catechol-amines to their labelled O-methyl derivatives by means of catechol-O-methyl transferase.

drawbacks affecting reliability. To overcome these problems, highly efficient gas liquid chromatographic (GLC) and high performance liquid chromatographic (HPLC) procedures with on line detection have been proposed by different Authors[20,24-27]. In these cases, less restrictive conditions are requested for sample preparation. GLC methods use flame ionization (FID), electron capture and mass spectrometric (MS) detection of previously derivatized CA. HPLC methods employ different detection principles, such as direct determination with photometric, fluorimetric and electrochemical detectors (ECD), or pre- and post-column derivatization; on line clean-up, to simplify or eliminate sample treatment, is employed too.

The main features of the above mentioned analytical procedures are summarized in Fig. 6. In synthesis: direct FL and RIA cannot be applied to routine determinations of plasma CA because of the poor sensitivity of the first method and the scarce reliability of the latter. GLC methods still suffer from problems related to volatile derivative formation, furthermore, only the very expensive MS detection allows acceptable sensitivity. Only REA and HPLC with ECD or post-column derivatization meet the needs of sensitivity and reliability. However, being REA affected by longer analysis time, tedious purification steps, and need of radioactive reactants, HPLC seems to be the method of choice for CA determination in clinical chemistry laboratories.

ASSAY	TIME (H)	SPECIFICITY	SENSITIVITY (PG)	RECOVERY (%)	COST	RELIABILITY
FL	3-8	L	100 - 1000	50 - 70	L	L
RIA	12	L	10 - 30	35 - 50	M	L
REA	48	M	3 - 10	40 - 60	H	L-M
GLC (FID-ECD)	24	M-H	50 - 200	70 - 90	L	L
GLC (MS)	24	H	20 - 40	70 - 90	H	L
HPLC (ECD)	2-3	M	5 - 10	70 - 95	L	M-H
HPLC (PRE-COL. DERIVAT.)	2-8	M-H	50 - 200	70 - 90	L	M-H
HPLC (POST-COL. DERIVAT.)	2-3	M-H	10 - 200	70 - 90	L	H

H = HIGH   M = MEDIUM   L = LOW

Fig. 6. Comparison among the different methods proposed for the determination of catecholamines.

AIM OF THE STUDY

In the present study we have verified the clinical usefulness and the practicability in a clinical chemistry laboratory of the NA and A determination in plasma through an improved HPLC-ECD method[28]. Therefore, we measured plasma concentrations of NA and A in the fasting state and during insulin-induced hypoglycemia in a group of healthy subjects and in a group of patients with type 2 diabetes mellitus. The plasma CA content was also measured in a patient with pheocromocytoma before and after the surgical removal of the tumour.

SUBJECTS AND METHODS

Subjects

Plasma CA were measured in six nonobese healthy subjects (3 males and 3 females, age 31 ± 5 years (mean ± S.D.), body mass index = 23 ± 3 Kg/m^2), five obese type 2 diabetics with mild hyperglycemia (5 males, age 46 ± 12 years, body mass index = 31 ± 5 Kg/m^2) and one man with adrenal pheocromocytoma (50 years old, body mass index = 23 Kg/m^2).

Experimental Design

In healthy subjects and diabetic patients plasma CA were measured in the fasting state and during an insulin tolerance test (ITT). The ITT was performed at 9.00 a.m. and consisted of a bolus injection of 0.1 U/Kg of regular insulin. Blood samples were drawn in the basal state and after 3, 6, 9, 12, 15, 20, 30, 40, 50, and 60 minutes after insulin administration. In the patient with pheocromocytoma plasma CA content was measured in the fasting state only, either before or three months after the surgical removal of the tumour.

Method

The principles of the CA assay we used[28] are illustrated in Fig. 7. Purification and concentration of CA from plasma samples were performed in polypropylene test tubes by liquid-liquid solvent extraction. At alkaline pH CA form stable anionic com-

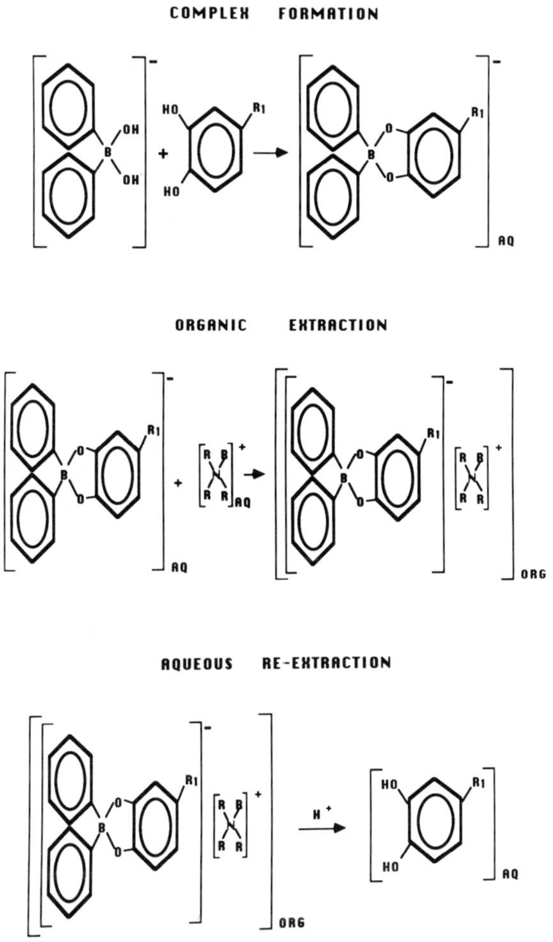

Fig. 7. Reactions involved in solvent extraction of catecholamines from biological fluids.

plexes with diphenylborate (DPB). In the presence of hydrophobic
quaternary amines, such as tetraoctylammonium bromide (TOA), the
CA-DPB complexes are easily and selectively extracted in n-heptane
containing little quantities of n-octanol, via ion-pair formation.
CA are finally concentrated by acid aqueous reextraction and
analyzed by HPLC. Fig. 8 shows the flow diagram of the entire
procedure, which results very simple and fast, requiring only few
analitycal steps easily achievable in any clinical chemistry
laboratory.

In comparison with the original method, improvements of
analytical performances have been gained by the following
modifications:

- readjustment of the volume ratios of organic and aqueous phases
  in the extraction steps;
- standardization of the extraction times and control of
  temperature;
- use of N-isopropyl-noradrenaline, wich presents an analytical
  behaviour quite similar to that of NA and A, as an internal

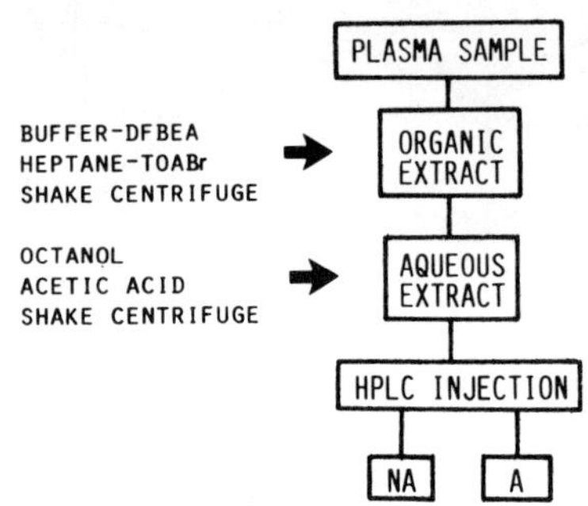

Fig. 8. Flow diagram of the proposed solvent extraction
procedure.

standard;
- use of a mobile phase containing formic acid and very low
  amounts of inorganic ions, which lowers the background current;
- use of a dual electrode cell, oxidizing interferent substances
  at the former electrode (coulometric mode) and detecting CA
  with improved selectivity at the latter one (amperometric mode).

<u>Reagents</u>

Noradrenaline, adrenaline and N-isopropyl-noradrenaline (IS)
were obtained from Sigma (St. Louis, Mo, USA), diphenylboric acid
ethanolamine complex (DPB-EA) from Janssen (Beerse, Belgium),
tetraoctylammonium bromide (TOA-Br) from Fluka (Buchs, Swisse) and
n-octylsulfate sodium salt (OSA) from Merck (Darmstadt, FRG).
Ammonium chloride, ammonium hydroxide, hydrochloric acid, sodium
metabisulfite, EDTA, n-octanol, n-heptane, acetic acid, formic
acid, citric acid, diethylamine, acetonitrile and sodium azide
were supplied from C. Erba (Milan, Italy). All the chemicals were
of analytical grade and were used without further purification.

Solutions were prepared in deionized water and filtered
before use through a 0.22 µm pore filter obtained from Gelman (Ann
Arbor, Mi, USA). 1.0 mg/ml stock solutions of CA in 0.1 mol/l HCl
were prepared and stored at 4°C, working standards were obtained
by making suitable dilutions of the stock solutions in 0.01 mol/l
HCl.

<u>Sample Preparation</u>

Blood was collected in chilled lithium heparin tubes, containing 2.0 µg/ml sodium metabisulfite as an antioxidant, and
readily centrifuged at 4°C within half an hour since sampling. The
plasma stored at -20°C showed stable CA content for three months
at least.

In a polypropylene tube, 50.0 µl of 60.0.ng/ml IS solution,
1.0 ml of 2.0 mol/l NH4Cl-NH4OH buffer pH 8.7 containing 0.2% (w/v)
DFB-EA and 0.5% (w/v) EDTA, 5.0 ml of a n-heptane solution
containing 0.9% (v/v) n-octanol and 0.25% (w/v) TOA-Br were in
sequence added to 2.0 ml of plasma. The mixture was shaken for
5 min at 60 rpm on a roller-shaker and centrifuged for 1.0 min at
3,000 g. 4.5 ml of the organic phase was transferred to a second
tube, 2 ml of n-octanol and 150.0 µl of 0.1 mol/l acetic acid

freshly prepared were then added and the mixture was shaken for
5 min. After a short centrifugation, 100.0 µl of the aqueous phase
was injected into the HPLC.

<u>Cromatography</u>

A HPLC pump model 420 from Kontron (Zurich, Swisse) connected
to a sample injection valve model 7125 (with a 200 µl loop) from
Rheodyne (Cotati, Ca, USA) was used as the solvent delivery
system. Chromatographic separation was performed on a Chromspher
C8 5µm 200x3 mm I.D. column from Chrompack, (Middelburg, The
Netherlands) in series with a reverse phase Chromsep guard column
10.0x2.1 mm I.D.

The mobile phase consisted of an aqueous solution of 0.1
mol/l formic acid containing 0.5 mmol/l EDTA, 1.0 mmol/l citric
acid, 25.0 mmol/l diethylamine and 8.0% (v/v) acetonitrile
adjusted to pH 3.2; the eluent solution was filtered through a
0.22 µm Gelman membrane and delivered by the pump at a constant
flow rate of 0.4 ml/min.

Detection of CA was achieved using a Coulochem model 5100 A
electrochemical detector equipped with an analytical cell model
5011 (ESA, Bedford, Ma, USA). The adopted working conditions were:
first electrode +170 mV, second electrode +480 mV, gain 100x50
corresponding to a full scale current of 20 nA.

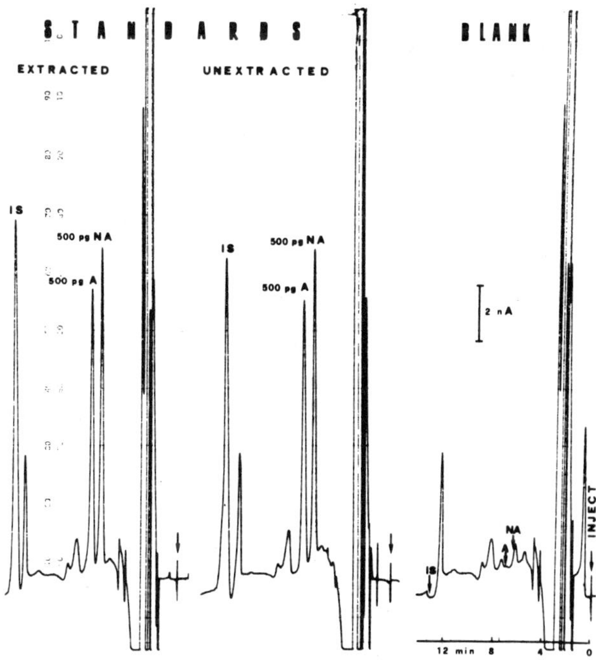

Fig. 9. Chromatograms of a standard solution before and after
extraction and a blank extraction. Amount injected:
noradrenaline 500 pg, adrenaline 500 pg, N-isopropyl-
noradrenaline (internal standard) 1500 pg.

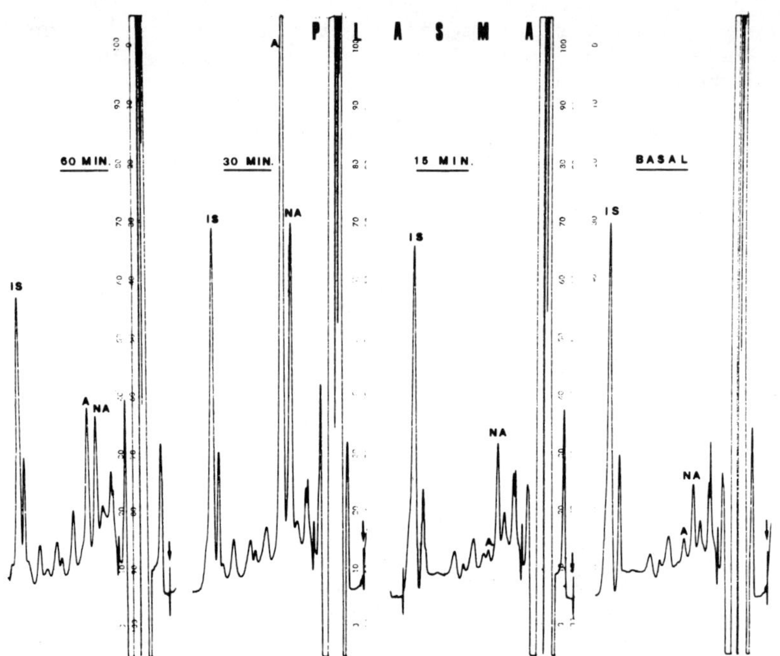

Fig. 10. Chromatograms of plasma samples drawn at different times
after insulin injection, in a healthy subject.

RESULTS

The chromatograms of a standard solution of NA, A and IS
before and after the extraction steps are depicted in Fig. 9. The
two chromatographic patterns were superimposible, showing the high
selectivity and recovery of the solvent extraction procedure;
moreover, the few peaks observed in the chromatogram of a "blank"
solution submitted to the same procedure were well separated and
did not interfere with CA peaks.

Fig. 10 shows the chromatograms of plasma extracts in basal
conditions and during ITT in a healthy subject. The profiles
result quite similar to those of the standards, confirming the
selectivity of the assay. Furthermore, the "cleanness" of the
chromatograms together with the high response of the electrochemical detection, allow to measure NA and A in plasma at levels
down to 5÷10 pg/ml.

Other main characteristics of the method are:
a) the recovery of CA added to plasma samples at physiological and
   pathological levels is ranging from 95 to 100%;
b) the method is able to discriminate plasma CA concentrations
   ranging from 10 to more than 1500 pg/ml;
c) the overall time from blood drawing to the results is less than
   two hours;
d) a single technician can process 12 samples at the same time up
   to the injection into the chromatograph.

The CA response to insulin-induced glucose fall in normal
subjects and diabetic patients are shown in Fig. 11 and 12,

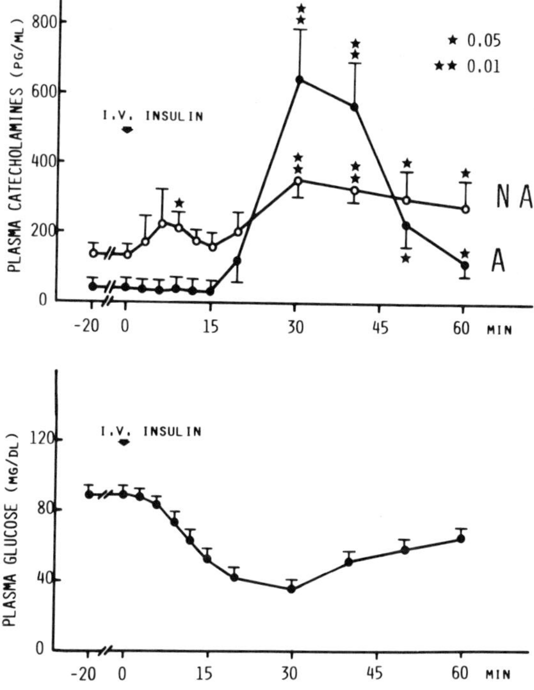

Fig. 11. Mean ± S.E.M. catecholamine response (top) to insulin-
induced glucose fall (bottom) in normal subjects.

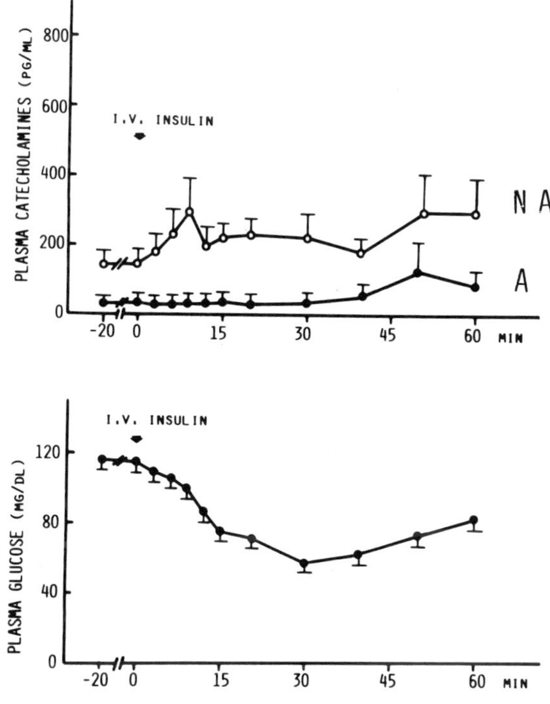

Fig. 12. Mean ± S.E.M. catecholamine response (top) to insulin-
induced glucose fall (bottom) in diabetic patients.

asterisks show the statistical difference vs. basal state. CA basal levels were similar in the two groups. In normal subjects A remained at the basal level of approximately 30 pg/ml until the 15th minute of the test and then rose to a peak of approximately 650 pg/ml at the 30th minute (a value about 20÷22 times higher than the basal). NA increase was much less and its profile showed a double peak at the 6th and the 30th minute after insulin injection. In diabetic patients, where the plasma glucose fall was lower than in normal subjects, NA profile was quite similar to that observed in normals, while A showed a smaller and delayed increase. This phenomenon was conceivably related to the lower and delayed fall of plasma glucose concentration. These data are consistent with those reported from other Authors[29,30].

In the patient with pheocromocytoma, during the week before the operation, a fasting blood sample was drawn each morning for CA determination. The mean ± S.D. of the 7 samples was 1396 ± 144 pg/ml for NA and 745 ± 85 pg/ml for A, respectively. Three months after removing the tumour, the values were 554 ± 47 pg/ml and 36 ± 10 pg/ml (n=5), respectively.

CONCLUSIONS

The results here presented demonstrate that the method we used is sensitive, specific and fast. Other main features of the assay are: ease of the analytical steps, stability of the reactants, low cost per analysis. These features, together with the verified reliability, make it really suitable for a wide application in clinical chemistry.

REFERENCES

1. L.J. De Groot, ed., "Endocrinology," Volume 2, Grune and Stratton, New York (1979).
2. H.U. Buhler, M. Da Prada and W. Hafely, Plasma adrenaline, noradrenaline and dopamine in man and different animal species, J. Physiol. 276:311 (1978).
3. A.J. Garber, P.E. Cryer, J.V. Santiago, M.V. Haymand, A.S. Pagliara and D.M. Kipnis, The role of adrenergic mechanism in the substrate and hormonal response to insulin-induced hypoglycemia in man, J. Clin. Invest. 58:7 (1976).
4. H.C. Guldberg and C.A. Marsden, Catechol-O-methyl-transferase: pharmacological aspects and physiological role, Pharmacol. Rev. 27:135 (1975).
5. N. Ben-Jonathan, J.M. Bahr and R.I. Weiner, eds., "Catecholamines as hormone regulators," Raven Press, New York (1985).
6. P.M. Vanhoutte and T.F. Luscher, Peripheral mechanisms in cardiovascular regulation: transmitters, receptors and the endothelium, in: "Handbook of Hypertension, Pathophysiology of Hypertension-Regulatory Mechanisms," Volume 8, A. Zanchetti and R.C. Tarazi, eds., Elsevier, Amsterdam (1986).
7. J. deChamplain, L. Farley, D. Cousineau and M.R. van Ameringen, Circulating catecholamine levels in human and experimental hypertension, Circ. Res. 38:109 (1976).
8. H. Galbo, J.J. Holst and N.J. Christensen, Glucagon and plasma catecholamine response to graded and prolonged exercise in man, J. Appl. Physiol. 38:70 (1975).

9. S.T. Mason, ed., "Catecholamines and behaviour," Cambridge University Press, Cambridge, (1984).
10. N.J. Cristensen, The role of catecholamines in clinical medicine, Acta Med. Scand. 624:9 (1979).
11. N.J. Christensen, Catecholamines and diabetes mellitus, Diabetologia 16:211 (1979).
12. N.J. Cristensen and V. Jorgen, Plasma catecholamines and carbohydrate metabolism in patients with acute myocardial infarction, J. Clin. Invest. 54:278 (1974).
13. S.G. Ball, Pheochromocytoma, in: "Handbook of Hypertension, Clinical Aspects of secondary Hypertension," Volume 2, J.I.S. Robertson, ed., Elsevier, Amsterdam (1983).
14. K. Haegnevik and P. Belfrage, Fetal and maternal plasma catecholamine levels at elective cesarean section under general or epidural anesthesia versus vaginal delivery, Am. J. Obstet. Gynecol. 142:1004 (1982).
15. J. Jaerhult, V. Angeras, L.O. Farnebo, H. Graffner, B. Hamberger and K. Uvnas-Moberg, The possible role of circulating catecholamines in the control of gastric function in health and duodenal ulcer disease, Scand. J. Gastroenterol. 19:137 (1984).
16. E.L. Bravo and R.C. Tarazi, Plasma catecholamines in clinical investigation: a useful index or a meaningles number?, J. Lab. Clin. Med. 100:155 (1982).
17. J.M.P. Holly and H.L.J. Makin, The estimation of catecholamines in human plasma, Anal. Biochem. 128:257 (1983).
18. S. Udenfriend, ed., "Fluorescence assay in biology and medicine," Academic Press, New York (1969).
19. R.J. Merrills, An autoanalytical method for the estimation of adrenaline and noradrenaline, Nature 193:988 (1962).
20. Y. Yui, T. Fujita, T. Yamamoto, Y. Itokava and C. Kawai, Liquid-chromatographic determination of norepinephrine and epinephrine in human plasma, Clin. Chem. 26:194 (1980).
21. W.J. Raum and R.S. Swerdloff, A radioimmunoassay for epinephrine in tissues and plasma, Life Sci. 28:2819 (1981).
22. D.P. Henry, B.S. Starman, D.J. Johnson and R.H. Williams, A sensitive radioenzymatic assay for norepinephrine in tissues and plasma, Life Sci. 19:375 (1976).
23. M. Da Prada and G. Zuercher, Simultaneous radioenzymatic determination of plasma and tissue adrenaline, noradrenaline and dopamine within the femtomole range, Life Sci. 19:1161 (1976).
24. U.E.G. Bock and P.G. Waser, Gas chromatographic determination of some biogenic amines as their pentafluorobenzoyl derivatives in the biological materials, J. Chromatogr. 213:413 (1981).
25. I. Molnar and C. Horvath, Reverse-phase chromatography of polar biological substances: separation of catechol compounds by high-performance liquid chromatography, Clin.Chem. 22:1497 (1976).
26. D.S. Goldstein, G. Fewerstein, J.L. Izzo, I.J. Kopin and H.R. Keiser, Validity and reliability of liquid chromatography with electrochemical detection for measuring plasma levels of norepinephrine and epinephrine in man, Life Sci. 28:467 (1981).
27. E. Gerlo and R. Malfait, High-performance liquid chromatographic assay of free norepinephrine, epinephrine, dopamine, vanillylmandelic acid and homovanillic acid, J. Chromatogr. 343:9 (1985).

28. F. Smedes, J.C. Kraak and H. Poppe, Simple and fast
    quantitative isolation of adrenaline, noradrenaline and
    dopamine from plasma and urine, J. Chromatogr. 231:25 (1982).
29. U. Lilavivathana, R.G. Brodows, P.D. Woolt and R.G. Campbell,
    Counterregulatory hormonal responses to rapid glucose lowering
    in diabetic man, Diabetes 28:873 (1979).
30. G. Boden, M. Soriano, R.D. Holdtke and O.E. Owen,
    Counterregulatory hormone release and glucose recovery after
    hypoglycemia in non-insulin-dependent diabetic patients,
    Diabetes 32:1055 (1983).

DETERMINATION OF FREE 3-METHOXY-4-HYDROXYPHENYLETHYLENE GLYCOL IN

HUMAN PLASMA AND CEREBROSPINAL FLUID BY HPLC WITH ELECTROCHEMICAL

DETECTION

V. Rizzo and G.V. Melzi d'Eril

Headache Center, "C.Mondino" Foundation
University of Pavia,
Pavia, Italy

INTRODUCTION

3-Methoxy-4-hydroxyphenylethylene glycol (MHPG) is the major
metabolite of norepinephrine in the central nervous system[1].

MHPG levels have previously been measured with electron-
capture gas chromatography[2] or gas chromatography-mass spectrometry[3,4]. Although the latter technique is highly sensitive and
very specific, it cannot be employed in routine work because it
requires equipment that is expensive to purchase and to maintain,
and considerable sample preparation and technical expertise.

Recently, High-Performance Liquid Chromatography (HPLC) with
amperometric detection has been developed to measure MHPG in
plasma[5-9], urine[10,11] and cerebrospinal fluid[12,13]. This kind of instrumentation is less expensive to purchase and less troublesome
to operate than that used in previous methods.

In most published works ethylacetate has been used for the
extraction of MHPG, but this sample pretreatment gives rise to
many non desirable peaks on the chromatogram: then some investigators have recently used disposable micro-columns for the
enrichment before HPLC separation[8,14,15]. In general, the methods
are expensive, time consuming, and require rather large sample
volumes.

The use of the coulometric electrochemical detector with the
electrodes in series provides a tool for MHPG determination in the
low picogram level with excellent specificity.

This paper describes an enhancement of HPLC methodology for
quantitation of free MHPG in both plasma and cerebrospinal fluid
(CSF): the method requires minimal sample pretreatment coupling
reversed-phase HPLC with coulometric electrochemical detection.

MATERIALS AND METHODS

<u>Materials</u>

MHPG hemipiperazine sodium acetate, acetic acid, EDTA, and heptanesulfonic acid were purchased from Sigma (St. Louis, Mo, USA). PM-10 filters were purchased from Amicon (Danvers, Ma, USA).

<u>Equipment</u>

The Liquid Chromatography system consisted of a model 5700 pump (ESA, Bedford, Ma,USA), a model 7125 injection valve fitted with a 50 µl loop (Rheodyne Inc., Berkeley, Ca, USA) and a 5 µm Spheri-5 RP18 (100 x 4.6 mm) column (Brownlee Labs, Santa Clara, Ca, USA). The ESA coulometric detector model 5100 was equipped with two cells (models 5011 and 5021, ESA) containing the electrodes through which the mobile phase and the sample flow. The first cell, called the conditioning cell, is set at a potential to transform the substance to be measured in its reduced or oxidized form; the second cell, called the analytical cell, contains two electrodes: the potential of the first is set to eliminate all the possible interfering substances; the second detector is set at a potential to get the highest signal from the molecule that we wish to measure.

<u>HPLC conditions</u>

The mobile phase consisted of 25 mmol/l sodium acetate, adjusted to pH 4.0 with 3 mol/l acetic acid, containing 100 mg/l EDTA and 150 mg/l heptanesulfonic acid. It was filtered though a 0.20 µm filter, and degassed under vacuum prior to use. The flow rate was 1.2 ml/min. After the injection of a certain number of plasma samples, a shift in the current/voltage curve for the MHPG

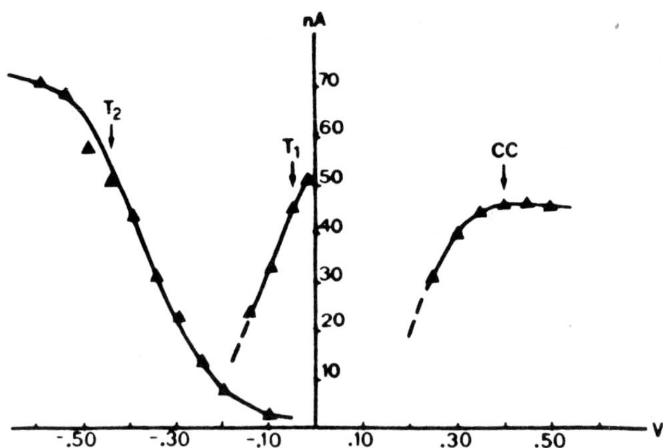

Fig. 1. Voltammogram of MHPG from three detectors at different electrode potentials (V) showing on the y axis the absolute value of the current (nA).
The voltammogram of T_2 was determined whilst the CC and T_1 were set at +0.50 V.The voltammogram of T_1 was determined whilst CC was set at +0.50 V, and that of the CC was determined whilst T_1 and T_2 were at -0.05 V, and -0.45 V respectively.

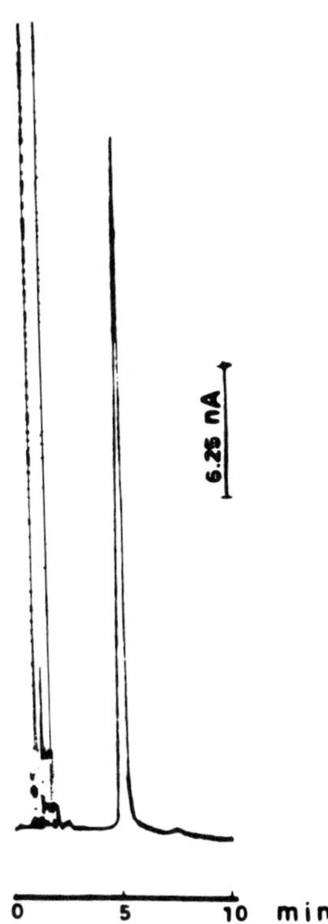

Fig. 2. Chromatogram of MHPG standard (20 ng/ml concentration).

was observed. This is due to a layer of lipidic materials building
up on the reference electrode surface which is causing a shift in
its potential. To remove this problem we flushed the cells every
100 injections with the following sequence: 100 ml H_2O, 50 ml
CH_3CN, 50 ml Dioxane, 150 ml Hexane, 50 ml CH_3CN, 100 ml H_2O.

Procedure

Prior to the analysis all samples (plasma and CSF) were
ultrafiltered through an Amicon PM 10 membrane and then 20 μl were
injected into the HPLC system. Samples were eluted isocratically
and quantitated by comparing the peak height with the standard
solution. The potentials of the conditioning cell (CC) and
electrode 1 (T_1) and 2 (T_2) were +0.40, -0.05 and +0.45 V respectively; the response time was 10 s.

RESULTS

The potentials of the three electrodes were selected after
injection of fixed amounts of MHPG (8 ng/ml, Fig. 1). The choice
of -0.45 V for T_2 was a good compromise between a high sensitivity
and a good selectivity. The potentials of the other two detectors
were chosen to get the best signal.

A chromatogram of MHPG standard is shown in Fig. 2. The
retention time of MHPG at a flow rate of 1.2 ml/min, was 5 min;
limits of detection at signal/noise ratio of 2 were 3.0 pg per
injection.

The effect of concentration and pH of the mobile phase on the retention time is shown in Fig. 3.

Fig. 4 exemplifies the application of the present technique to the determination of MHPG in plasma and CSF.

The identity of chromatographic peak resulting from the analysis of plasma and CSF samples was demonstrated in several ways. There was a complete correspondence between the retention time of the sample and that of the standard. Varying the acetonitrile concentration and pH value of the mobile phase the retention times of standard and sample varied in the same manner. Finally, the ratio of the peak areas of standard and sample measured at different potentials was the same.

The analytical recovery was tested by adding 4, 8, 10, 20 ng of MHPG to 1 ml of plasma sample that was processed as described above. The actual recovery of the added MHPG was 100±1 %, making the use of an internal standard unnecessary.

The reproducibility of the method was tested by repeated injections of the samples: the within-day and between-days coefficients of variation of the assay are shown in Table 1.

DISCUSSION

The recent improvement in the coulometric detectors in comparison with the traditional amperometric detectors has given the chromatographer increased versatility. At the appropriate voltage setting, 100% of an electroactive substance will react passing through the cells, while in an amperometric system only 1-5% will react. In this procedure the potential of the CC was set at +0.40 V to oxidize most of the MHPG present in the sample. By setting the potential of T_1 at -0.05 V, it reduced the response of T_2 by only 10% but many possible interfering substances were eliminated. The potential of T_2 was set at -0.45 V to obtain the highest signal from the MHPG reduction together with the lowest noise: this yielded a good baseline stability.

The applicability in practice of this simple isocratic proce-

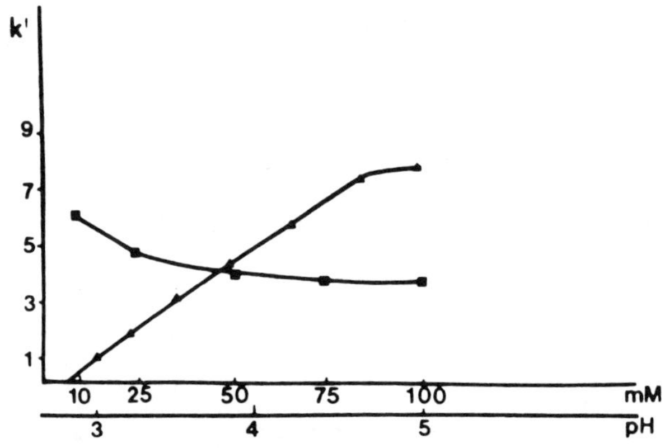

Fig. 3. Effect of sodium acetate (▲) molar concentration and
pH (■) on capacity ratios (k') of MHPG.

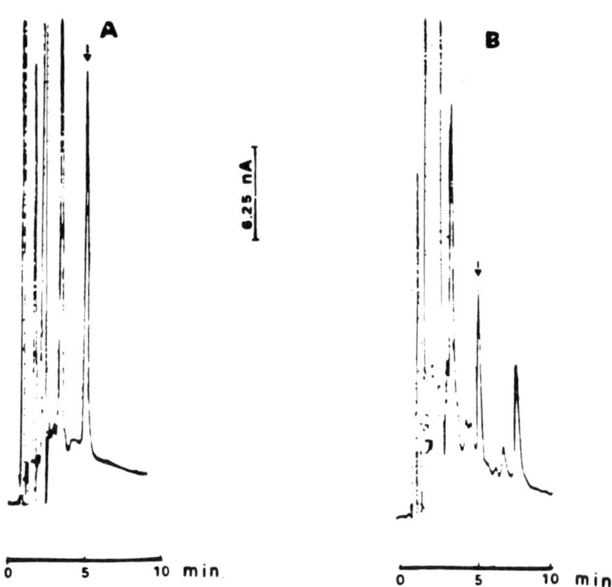

Fig. 4. Chromatogram of a 20 µl ultrafiltered CSF sample (A).
MHPG concentration is 15 ng/ml.
Chromatogram of a 20 µl ultrafiltered plasma sample (B).
Plasma MHPG concentration is 5 ng/ml.

dure has been demonstrated by the analysis of both plasma and CSF samples. The retention of the compounds was simply controlled by pH and concentration adjustments of the mobile phase: there was no need for the addition of organic modifiers.

A reversed phase system not requiring organic modifiers is advantageous, as degassing leads to gradual evaporation of the organic modifier resulting in a change of mobile phase concentration. EDTA, included in the mobile phase to protect the electrodes from heavy metals contamination, do not affect the retention or the resolution of MHPG.

Table 1. Within-Day and Between-Days CV% Values of MHPG Assay

	Mean Value (ng/ml)	CV%		n
		within-day	between-days	
Plasma	5	3.0	6.9	10
CSF	15	3.8	5.5	10

The linearity of the procedure and of the detector response was verified up to 250 ng/ml. Response was linear over the range investigated.

Increasing the concentration of sodium acetate, the capacity
factor (k') decreases. This is probably due to the more effective
masking of residual silanol groups by the buffer, when higher
concentrations of sodium acetate were used. Many peaks appear in
the plasma sample chromatogram: however, the selectivity seems
assured by the combined effect of chromatography and applied
potential of the electrodes.

The advantages of the HPLC with coulometric detection on-line
are cost, selectivity and sensitivity. An isocratic HPLC with
electrochemical detection is much less expensive to purchase and
maintain than either a gas chromatography-mass spectrometry or a
electron capture gas chromatography system. Furthermore, the
coulometric detector has three in series electrodes which provide
further selectivity. The ratio of the signal from T1 to the signal
T2 would be constant for a specific compound, and would give more
selectivity to the system.

In conclusion this assay appears adequate to resolve and
measure MHPG. Because no preliminary sample purification other
than ultrafiltration is needed, the possibilities for technical
errors are reduced. Thus the speed and the reproducibility make
this MHPG assay highly suitable for clinical use.

REFERENCES

1. J.W. Maas and D.H. Landis, In vivo studies of the metabolism
   of norepinephrine in the central nervous system, J. Pharmacol
   Exp. Ther. 163:147 (1968).
2. H. Dekirmenjian and J.W. Maas, 3-methoxy-4-hydroxy phenyl-
   ethlene glycol in plasma, Clin. Chim. Acta 52:203 (1974).
3. S. Takahashi, D.D. Godse, J.J. Warsh and N.C. Stancer, A gas
   chromatographic mass spectrometric (GC-MS) assay for
   3-methoxy-4- hydroxyphenylethylene glycol and vanilmandelic
   acid in human serum, Clin. Chim. Acta 81:183 (1977).
4. J.J. Warsh, D.D. Godse, S.W. Cheung and P.P. Li, Rat brain
   and plasma norepinephrine glycol metabolites determined by
   gas chromatography-mass fragmentography, J. Neurochem.
   36:893 (1981).
5. H. Ong, F. Capet Antonini, N. Yamaguchi and D. Lamontagne,
   Simultaneous determination of free 3-methoxy-4-hydroxymandeli
   acid and free 3-methoxy-4-hydroxyphenylethylene glycol in
   plasma by liquid chromatography with electrochemical
   detection, J. Chromatogr. 233:97 (1982).
6. M. Scheinin, W.H. Chang, D.C. Jimerson and M. Linnoila,
   Measurement of 3-methoxy-4-hydroxyphenylethylene glycol in
   human plasma with high-performance liquid chromatography
   using electrochemical detection, Anal. Biochem. 132:165
   (1983).
7. P.A. Shea and J.B. Howell, High-performance liquid
   chromatographic method for determining plasma and urine
   3-methoxy-4-hydroxyphenylethylene glycol by amperometric
   detection, J.Chromatogr. 306:858 (1984).
8. A. Minegishi and T. Ishizaki, Determination of free 3-methoxy
   4-hydroxyphenylethylene glycol with several other monoamine
   metabolites in plasma by high-performance liquid chroma-
   ography with amperometric detection, J. Chromatogr. 311:51
   (1984).

9. S. Schinelli, G. Santagostino, P. Frattini, M.L. Cucchi, and G.L. Corona, Assay of 3-methoxy-4-hydroxyphenylethylene glycol in human plasma using high performance liquid chromatography with amperometric detection, J. Chromatogr. 338:396 (1985).

10. J.R. Shipe, J. Savory and M.R. Wills, Improved liquid chromatographic determination of 3-methoxy-4-hydroxyphenylethylene glycol in urine with electrochemical detection, Clin. Chem. 30:140 (1984).

11. A.M. Krustulovic, C.T. Matzura and L. Bertani-Dziedzic, Endogenous levels of free and conjugates urinary 3-methoxy-4-hydroxyphenylethylene glycol in control subjects and patients with pheochromocytoma by reversed-phase liquid chromatography with electrochemical detection, Clin. Chim. Acta 103:109 (1980).

12. P.J. Langlais, W.G. Mc Entee and E.D. Bird, Rapid liquid chromatographic measurement of 3-methoxy-4-hydroxyphenyl glycol and other monoamine metabolites in human cerebrospinal fluid, Clin. Chem. 26:786 (1980).

13. M.G. Anderson, J.G. Young, D.Y. Cohen, B. Shaywitz and D.K. Batter, Amperometric determination of 3-methoxy-4-hydroxyphenylethylene glycol in human cerebrospinal fluid, J. Chromatogr. 222:112 (1981).

14. F. Karege, Method for total 3-methoxy-4-hydroxyphenyl glycol extraction from urine, plasma and brain tissue using bonded-phase materials: comparison with the ethyl acetate extraction method, J.Chromatogr. 311:361 (1984).

15. J. Semba, A. Watanabe and R. Takahashi, Determination of free and total 3-methoxy-4-hydroxyphenylethyl glycol in human plasma by high performance liquid chromatography with electrochemical detection, Clin. Chim. Acta 152:185 (1985).

SIZE EXCLUSION HPLC COUPLED TO RIA IN THE STUDY OF HUMAN CHORIONIC

GONADOTROPIN (hCG) IN PHYSIOLOGICAL AND PATHOLOGICAL CONDITIONS

R. Dorizzi, M. Pradella, M. Giavarina, and F. Rigolin

Laboratory of Clinical Chemistry and Microbiology
Civic Hospital
Legnago, Verona, Italy

INTRODUCTION

Human chorionic gonadotropin (hCG) is a glycoprotein hormone
normally secreted by throphoblastic cells of the placenta during
pregnancy; the peak production is attained in the latter part of
the first trimester of pregnancy. During the first 4-6 weeks of
pregnancy, hCG mantains the corpus luteum and stimulates steroid
hormone secretion[1]. Similar to the pituitary gonadotropins,
luteinizing hormone (LH), follicle stimulating hormone (FSH) and
thyroid stimulating hormone (TSH), hCG comprises a carbohydrate
part and a peptide part. The molecular mass of the complete molecule is about 40,000 Daltons. The carbohydrate moiety, comprising
30-33% of the hCG molecule, is essential for the transport and for
the biologic effect of the hormone and, if removed partly or
totally, the biological activity of hCG is reduced correspon-
dingly[2]. hCG is composed of two dissimilar subunits: α and ß linked
by not covalent bonds which do not possess intrinsic biological
activity[3]. The α-subunits of all four human glycoprotein hormones
are nearly identical; they consist of 89-92 aminoacids in the same
sequence but are significantly different in the carbohydrate
moieties[4]. hCGß contains 145 aminoacids and six carbohydrate
chains and differs from TSHß and FSHß but is very similar to LHß.
Approximately 80% of the first 115 aminoacids of hCGß are in
identical positions to those of LHß. The elucidation of the amino
acids sequence of hCG revealed that hCGß possesses an extended,
unique COOH-terminal glycopeptide of about 30 aminoacid residues
not present in other glycoprotein hormones[5].

hGC is secreted also by neoplastic and nonneoplastic tissues[6].
According to Rosen[7] the highest incidence of this secretion is
among patients with gonadal tumours: all the patients with testis
choriocarcinoma (8 out of 8), 2 of 4 patients with teratocar-
cinoma, 6 of 16 patients with seminoma, 30 of 52 patients with
embryonal cell carcinoma and 5 of 12 patients with adenocarcinoma
of ovary had detectable hCG. It has been reported that there is
not a selective elevation of hCG-subunits without an increment of
the hormone[8].

Some Authors recently reported the presence of hCG in the plasma of patients without tumours but with a variety of benign diseases and also in normal non pregnant subjects[9,10].

Several Authors previously investigated hCG through traditional chromatographic techniques[11-15]. Yet, these methods require large samples and long analytical times, with poor selectivity and reliability.

We report here our preliminary results concerning the development of a size exclusion HPLC method that, coupled to RIA, seems to be quite promising for the study of hCG molecule with improved accuracy.

CASES AND METHODS

<u>Cases</u>

Samples of serum from 20 normal pregnant women in the first trimester of pregnancy, from a surgically confirmed ectopic pregnancy and from 10 male patients suffering from neoplasms, in which hCG was detectable, were investigated. In 3 of the male patients a testicular carcinoma was surgically demonstrated; in the other 7 a carcinoma in the lung (3) or in the colon (4) was demonstrated. In our investigation only samples without hemolysis, that causes a decrease of hCG of 20-30%[16], were used; all the samples were frozen at -20°C and assayed within 2 months.

<u>Methods</u>

We used a Beckman (Berkeley, Ca, USA) Liquid Chromatograph (Model 421 System Controller; Model 210 Sample Injector and Model 112 Pumps) fitted with a Bio-Gel TSK 30 7.5 x 300 mm column (Bio-Rad, Richmond, Ca, USA). The mobile phase was 0.1 mol/l phosphate buffer containing 0.1 mol/l NaCl pH 7.4 isocratically delivered at a flow of 1 ml/min. 20 µl of the sample of the women and 200 µl of the samples of the males previously filtered through 0.2 µm disposable filters (FlowPore D, Flow, FRG) were injected. 120 fractions of 250 µl were manually collected and then assayed by a commercial RIA kit furnished by Biodata (Rome, Italy), employing an antibody directed against ß-subunit. When defined as the apparent concentration at two standard deviations from the maximum binding, the sensitivity of the kit is 6 mIU/ml (1° IRP WHO 75/537). According to the manufacturer, the interferences of the following substances, evaluated by the Abraham method (X/Y x 100), where X and Y are respectively the weight of hCG and of the interferent that causes a 50% decrease of binding, are:

hCG	100.00 %
hLH	0.53 %
hFSH	0.84 %
hTSH	0.04 %
hPRL	0.00 %

RESULTS

In all the samples of pregnant women the chromatographic profile was very similar to a standard WHO of hCGß; hCG immunoreactivity was measured in the fractions 25-31 with the maximum immunoreactivity in the fractions 25-26 (Figure 1A). In 1 out of 7

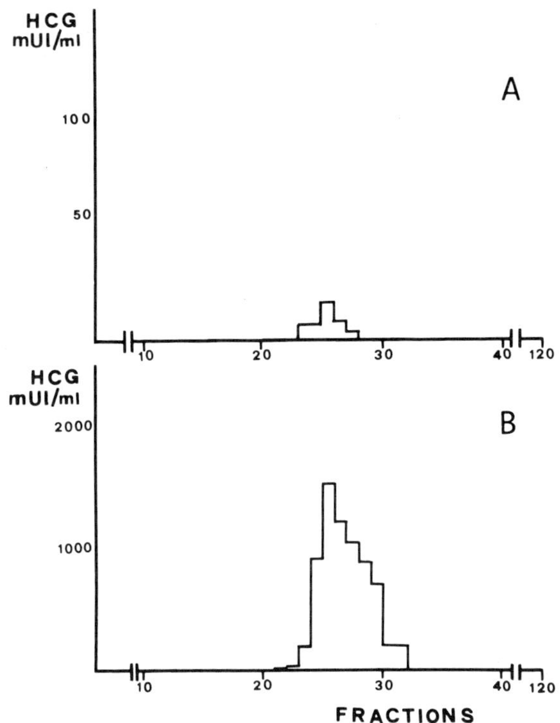

Fig. 1. Typical chromatograms of samples of a male with a
testicular carcinoma (A) and of a normal pregnant
woman (B).

samples of patients with extratesticular carcinoma in which immunoreactivity was detectable after chromatography and in all the
samples of patients with testicular carcinoma the hCG-like chromatographic profile was very similar to the standard: Fig. 1B shows
the chromatographic profile of the sample with the highest hCG
level. In all the cases, immunoreactivity was never detected after
the fraction n. 35.

DISCUSSION

The biological activity and the structure of hCG is very similar to that of LH, differing mainly by the 30 additional C-termi-
nal residues of its ß-subunit and by carbohydrate differences.
Previous reserches[10] showed that hCG-like substances found in the
plasma of normal non pregnant subjects and in women suffering from
breast cancer was similar to standard hCG as far as molecular size
(identical elution from Sephadex G-100), ionic strength (identical
elution from DEAE-Sephadex A-50), and immunoreactivity (different
elution profiles of hCG and hLH on Sephadex G-100 and Sephadex
DEAE A-50) are concerned. This material is similar to that found
by Chen[6] and Robertson[9] in the urine of normal postmenopausal
women. Serum hCG or hCG-ß have been measured in subjects with
different tumors, expecially gonadal and gastrointestinal. Indeed,
material with hCG immunoactivity has been extracted from normal
human testis, liver, colon, pituitary, lung, kidney and other tissues such as fibroblasts[17,18]. The secretion of hCG-like substance

by nonmalignant cells can be viewed as a primitive paracrine secretion persisting through evolution[19].

The entire molecule of hCG and its subunits have been extensively studied with chromatographic and immunological techniques and nowadays there are satisfactory separation and quantitative methods. Nevertheless, we think that the chromatographic methods actually used for study of the molecule, are scarcely useful in the routine since they are cumbersome and require many hours of analysis. HPLC investigation of hCG molecule assures clear advantages:

A) samples of very little volume (20-200 μl), compared to traditional Sephadex chromatography (10-40 ml), may be injected;
B) very short time of analysis is required; in our study the entire hCG immunoreactivity was collected in no more than 10 minutes.

Our researches show that the hCG molecule secreted in normal pregnancy, in ectopic pregnancy, in testicular and extratesticular carcinoma yield the same profile if investigated in Size Exclusion HPLC coupled to RIA.

The greatest limit of our method is the incapacity to detect hCG immunoreactivity in samples containing hCG levels of less than 1000 mIU (1° IRP WHO 75/537) since it is not advisable to inject volumes greater than 200 μl in the column actually used by us. Nevertheless, we think that the use of recently developed columns and of an antibody directed against α-hCG will allow a more accurate study of hGC molecule in physiological and pathological conditions.

REFERENCES

1. R.O. Hussa, Biosynthesis of human chorionic gonadotropin, Endocr. Rev. 1:268 (1980).
2. Anonymous, Chorionic gonadotropin (hCG), in: "Quantitation of reproductive hormones. When, why and how," A. Kallner, ed., IFCC Education Committee, Farmos Diagnostica, Stockholm (1985).
3. H.R. Masure, W.L. Jaffee, M.A. Sickel, S. Birken, R.E. Canfield and J.L. Vaitukaitis, Characterization of a small size urinary immunoreactive human chorionic gonadotropin (hCG)-like substance produced by normal placenta and by hCG-secreting neoplasms, J. Clin. Endocrinol. Metab. 53:1014 (1984).
4. G.T. Ross, Clinical relevance of research on the structure of human chorionic gonadotropin, Am. J. Obstet. Gynecol. 129:795 (1977).
5. F.J. Morgan, S. Birken and R.E. Canfield, The amino acid sequence of human chorionic gonadotropin: the α subunit and ß subunit, J. Biol. Chem. 250:5274 (1975).
6. H.C. Chen, G.D. Hodgen, L. Matsuura, L.J. Lin, E. Gross, L.E. Reichert, S. Birken, R.E. Canfield and G.T. Ross, Evidence for a gonadotropin from nonpregnant subjects that has physical immunological and biological similiarities to human chorionic gonadotropins, Proc. Natl. Acad. Sci. USA 73:2885 (1979).
7. S.W. Rosen, B.D. Weintraub, J.L. Vaitukaitis, H.H. Sussman, J.M. Hershman and F.M. Muggia, Placental proteins and their subunits as tumor marker, Ann. Int. Med. 82:71 (1975).

8. K. Mann and H.J. Karl, Molecular heterogeneity of human
    chorionic gonadotropin and its subnits in testicular cancer,
    Cancer 52:654 (1983).

9. D.M. Robertson, H. Suginami, H.H. Montes, C.P. Puri, S.K. Choi
    and E. Diczfalusi, Studies on a human chorionic gonadotropin-
    like material present in non-pregnant subjects, Acta
    Endocrinol. 89:492 (1978).

10. A. Borkowski, V. Puttaert, M. Gyling, C. Muquardt and J.J.
    Body, Human chorionic gonadotropin-like substance in plasma
    of normal nonpregnant subjects and women with breast cancer,
    J. Clin. Endocrinol. Metab. 58:1171 (1984).

11. J.L. Vaitukaitis and E. Ebersole, Evidence for altered
    synthesis of human chronic gonadotropin in gestional
    trophoblastic tumours, J. Clin. Endrocrinol. Metab. 42:1048
    (1975).

12. I.A. Kourides, B.D. Weintraub, S.W. Rosen, E.C. Ridgway, B.
    Kliman and F. Maloof, Secretion of alpha subunit of
    glycoprotein hormones by pituitary adenomas, J. Clin.
    Endocrinol. Metab. 43:97 (1975).

13. C. Hagen, E.D. Gilby, A.S. McNeilly, K. Olgard, P.K. Bondy and
    L.H. Rees, Comparison of circulating glycoprotein hormones and
    their subunits in patients with oat cell carcinoma of the lung
    and uraemic patients on chronic dialysis, Acta Endocrinol.
    83:26 (1976).

14. S.A. Amr, R.E. Wehmann, S. Birken, R.E. Canfield and B.C.
    Nisula, Characterization of a carboxiterminal peptide
    fragment of the human choriogonadotropin ß-subunit excreted
    in the urine of a woman with choriocarcinoma, J. Clin. Invest.
    71:329 (1983).

15. R.J. Norman, C. Lowings, T. Oliver and T. Chard, Human
    chorionic gonadotropin and subunits-heterogeneity in serum of
    male patients with tumors of the genital tract, Clin.
    Endocrinol. 23:25 (1985).

16. C.V. Rao, R.O. Hussa, F.R. Carman, M.L. Rinke , C.L. Cook and
    M.A. Yussman, Stability of human chorionic gonadotropin and
    its alpha subunit in human blood, Am. J. Obstet. Gynecol.
    146:65 (1982).

17. G.D. Braunstein, V. Kamdar, J. Rasor, N. Swaminanathen and
    M.E. Wade, Widespread distribution of chorionic gonadotropin-
    like substance in normal human tissues, J. Clin. Endocrinol.
    Metab. 49:917 (1979).

18. Y. Yoshimoto, A.R. Wolfsen, F. Hirosa and W.D. Odell, Human
    chorionic gonadotropin-like material: presence in normal
    tissues, Am. J. Obstet. Gynecol. 134:729 (1979).

19. J. Roth, D. Le Roith, J. Shiloah, J. Rosenzweic, M. Lesniak
    and J. Havrankova, The evolutionary origins of hormones,
    neurotransmitters, and other extracellular messengers:
    implications for mammalian biology, N. Engl. J. Med. 306:523
    (1982).

HPLC-RIA DETERMINATION OF URINARY 2,3 DINOR - 6 KETO -

PROSTAGLANDIN F1α IN HUMANS

P. Minuz, M. Degan, F. Paluani, G.P. Velo*,
and A. Lechi

Institute of Clinical Medicine
* Institute of Pharmacology
University of Verona
Verona, Italy

2,3 dinor-6 keto-prostaglandin (PG)F1α has recently been identified as the main urinary metabolite of systemic prostacyclin (PGI2) in humans. It derives from beta-oxidation of 6 keto-PGF1α, the non enzymatic hydrolitic derivative of PGI2[1].

Urinary-2,3 dinor-6 keto-PGF1α could therefore be considered a reliable although not specific index of the "in vivo" systemic production PGI2[2]. This compound is currently assayed by using gas-chromatography/mass-spectrometry.

We have developed a radioimmunoassay (RIA) method for the detection of this compound. A HPLC technique was utilized in order to improve the specificity of the RIA.

METHODS

Samples from 24-hour urine collections were stored at -20°C until the time of analysis. After thawing, 20 ml samples were centrifuged (10 min, 1,000 g) and acidified with citric acid (pH 3.0-3.4); about 6,000 DPM of [3]HTxB (180 Ci/mmol, Amersham, Buckinghamshire, U.K.) were then added to each sample.

As a first chromatographic step, the urine samples were passed through Sep-Pak C18 Cartridges (Water, Milford, Ma USA), previously washed with ethanol (20 ml) and water (20 ml).

Cartridges were then eluted with water (20 ml), ethanol:water (15:85, 20 ml), petroleum ether (20 ml), ethyl-acetate (10 ml). This last fraction, containing PG and Tx metabolites, was collected. (All solvents were RPE grade, C. Erba, Milan, Italy).

The ethyl-acetate fractions were immediately eluted through silica disposable columns (Baker, Phillisburg, USA), pretreated with ethyl-acetate:toluene (20:80, 5 ml).

The columns were then rinsed with: ethyl-acetate:toluene (40:60, 5 ml), ethyl-acetate:toluene:methanol (40:60:20, 10 ml). The last fraction was dried "in vacuo" and suspended in acetonitrile:water (30:70, 300 µl).

The HPLC apparatus used was a Beckman 112 Solvent Delivery Module (Beckman, Berkeley, Ca, USA). A reverse-phase ODS column was used (Hyperchrome ODS2 150 x 4,6 mm, 5 µm, Bishoff, Leonberg, FRG).

PGs were eluted isocratically with acetonitrile:water (30:70, HPLC grade, C. Erba, Milan, Italy) acidified with trifluoroacetic acid to pH 3. Flow rate was 1 ml/min. Fractions were collected using an LKB Ultrorac collector. Fractions containing 2,3-dinor-6 keto-PGF1α were identified according to the K[1] of authentic 2,3-dinor-6 keto-PGF1α kindly provided by Upjohn Co. (Kalamazoo, Mich, USA), and later extracted with a volume of ethyl-acetate 3 times greater.

The dried extract, suspended in phosphate buffer (0.02 mol/l, pH 7.4, 2.5 ml), was used for RIA. Each fraction contained both 2,3 dinor-6 keto-PGF1α and TxB2, and therefore ^{3}HTxB2 added to the samples was used to determine the percentage of recovery.

2,3 dinor-6 keto-PGF1α assay was carried out as follows. Incubated in a polystyrene tube for 24 hours at 4°C were: 100 µl of phosphate buffer (0.02 mol/l, pH 7.4) containing 2,3 dinor-6 keto-PGF1α (standard curve) or 50-200 µl of sample, 100 µl of phosphate buffer containing about 5,000 DPM of ^{3}H-6 keto-PGF1α Amersham (150 Ci/mmol), 500 µl of phosphate buffer containing

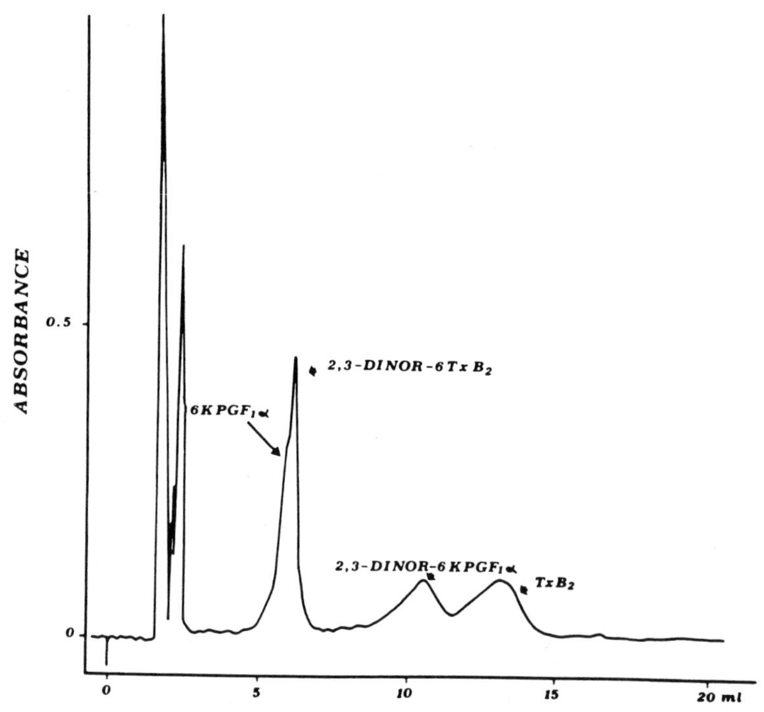

Fig. 1. HPLC chromatogram of PGI$_2$ and TXA$_2$ metabolites.

Table 1. Percentage of Recovery after HPLC

Sample	2,3 dinor-6 keto-PGF1α	^{3}HTxB$_2$
1	37.8	37.8
2	46.8	40.2
3	9.3	8.4
4	47.3	54.9
5	13.5	18.0

bovine albumin (1.5 g/l) and specific antiserum to
2,3 dinor-6 keto-PGF1α; the phosphate buffer was added to achieve
a final volume of 1 ml. Samples were assayed in duplicate. Two
different volumes of each sample were used. Antiserum to
2,3 dinor-6 keto-PGF1α was obtained in our laboratory by immu-
nizing rabbits to a conjugate 6 keto-PGF1α-human albumin.

The affinity constant to 2,3 dinor-6 keto-PGF1α of the antiserum used was 0.83 l/mol. The sensitivity (IC 50) was 56 pg/ml
(B = 36%) at a final dilution of anti-serum 1:20,000 in the assay.

Cross reactivities of the antiserum to 2,3 dinor-6 keto-PGF1α
were: TxB$_2$:0.15%, PGE$_2$:0.30%, PGF2α:1.50%, 6 keto-PGF1α:100.00%.

RESULTS AND CONCLUSIONS

A typical chromatogram of PGI$_2$ and TxA$_2$ metabolites, obtained
using 10 µg of each compound and a Beckman UV detector (wavelength
214 nm), is shown in Fig. 1.

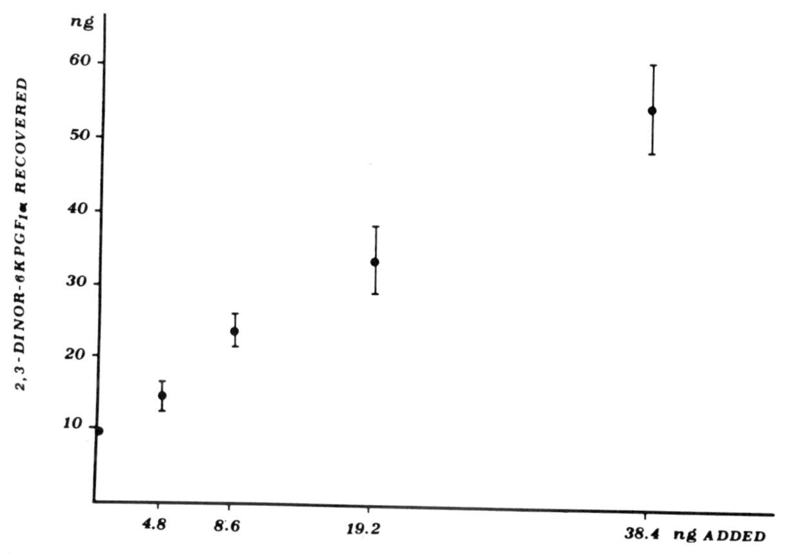

Fig. 2. Recovery by RIA of 2,3 dinor-6keto-PGF$_{1α}$ added to urine
before the purification procedures.

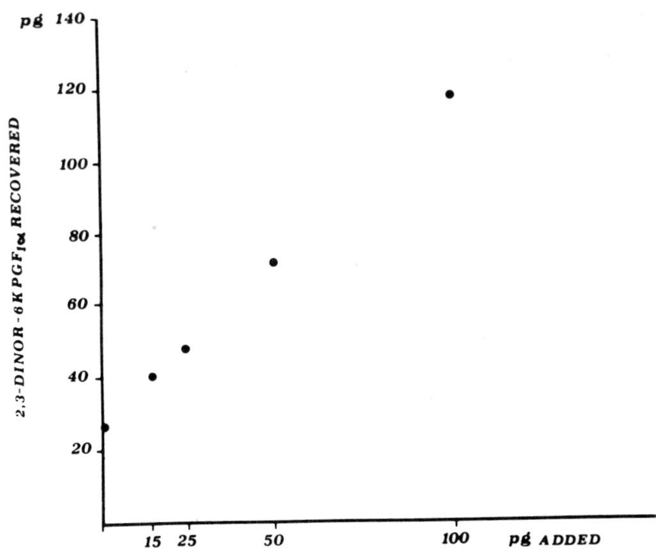

Fig. 3. Recovery by RIA of 2,3 dinor-6keto-PGF$_{1\alpha}$ added to urine
before RIA.

Recoveries of ^{3}HTxB$_2$ added to each sample of urine after
silica gel chromatography (n = 28) and after HPLC (n = 32)
(mean ± S.D.) were: 70.2 ± 8.1%, and 53.2 ± 23.8%, respectively.

The recovery by RIA of 2,3 dinor-6 keto-PGF1 α after HPLC
(n = 5) was similar to, and correlates with that of the added
^{3}HTxB$_2$ (Table 1). The recovery by RIA of different amounts of 2,3
dinor-6 keto-PGF1α added to aliquotes of a sample of urine before
the purification procedures (mean ± S.D.) is shown in Fig. 2.
Recovery by RIA of 2,3 dinor-6 keto-PGF1α added to urine after
chromatographic procedures is shown in Fig. 3. 24-hour urinary
excretion of 2,3 dinor-6 keto-PGF1α determined in a group of
6 healthy volunteers was: 153 ± 73.6 ng (mean ± S.D.), range
47 - 245 ng.

RIA is a sensitive, simple and relatively inexpensive
alternative to gas-chromatography/mass-spectrometry for the
determination of PGs in human biological fluids. Since no specific
antisera nor specific radioactive tracers for 2,3 dinor-6 keto-
PGF1α RIA are as yet available, we made use of the high cross-
reactivity with 6 keto-PGF1α of an antiserum produced in our
laboratory. The lack of specificity of the antiserum used was
overcome by a HPLC preliminary purification. This method permits a
good resolution (R > 1.5) between 2,3 dinor-6 keto-PGF1α and
6 keto-PGF1α.

The main criteria for the validation of RIA were satisfied.

The data obtained by measuring urinary 2,3 dinor-6 keto-PGF1α
in the studied subjects are similar to those observed by other
authors using gas-chromatography/mass-spectrometry[3].

This method therefore seems to provide a useful and reliable means of specific detection of 2,3 dinor-6 keto-PGF1α in urine.

REFERENCES

1. B. Rosenkranz, C. Fischer, K.E. Weimer and J.C. Froelich, Metabolism of prostacyclin and 6 keto-prostaglandin F1α in man J. Biol. Chem. 225:10194 (1980).
2. G.A. Fitzgerald, A.K. Pedersen and C. Patrono, Analysis of prostacyclin and thromboxane biosynthesis in cardiovascular disease, Circulation 67:1174 (1983).
3. O. Vesterqvist and K. Green, Development of a GC-MS method for quantitation of 2,3 dinor-6 keto-PGF1α and determination of the urinary excretion rates in healthy humans under normal conditions and following drugs, Prostaglandins 28:139 (1984).

COMPARISON OF DIFFERENT METHODOLOGICAL APPROACHES TO THE STUDY OF
THE ERYTHROCYTE $NA^+ - K^+$ PUMP

P. Delva, M. Degan, C. Lechi*, and A. Lechi

Institute of Clinical Medicine
* Institute of Clinical Chemistry
University of Verona
Verona, Italy

INTRODUCTION

The $Na^+ - K^+$ pump is a complex membrane transport system which
catalyzes a simultaneous Na^+_i (internal sodium) $- K^+_o$ (external
potassium) exchange characterized by an asymmetric ($3\ Na^+:2\ K^+$)
stoichiometry, responsible for cell polarization. The energy for
this active process is supplied by the coupled hydrolisis of ATP.
The erythrocyte is the commonest cell model utilized for investigations in this field, mainly because of its easy availability.
Furthermore, differently from other cells, $Na^+ - K^+$ ATPase from
erythrocyte homogenates is exclusively of plasma membrane origin.

In the last few years the widespread attempt to investigate
the biochemical basis of many pathological conditions has led to
the characterization of disturbances of a number of membrane ion
transport systems.

Alterations of the erythrocyte $Na^+ - K^+$ pump have been
associated with arterial hypertension[1], renal failure[2], obesity[3],
hyperthyroidism[4], and some hypokalemic conditions including
hyperaldosteronism[5].

The erythrocyte $Na^+ - K^+$ pump can be studied by several
different techniques, three of which are currently applied in our
laboratory.

METHODS

Enzymatic Method

The $Na^+ - K^+$ ATPase activity of the erythrocyte membrane is
measured utilizing a bioluminescence technique[6] on the basis of
the extent of ATP hydrolysis before and after incubation at 37°C.

The oxidative decarboxylation of luciferin leads to the formation of a product which emits light and is directly proportional

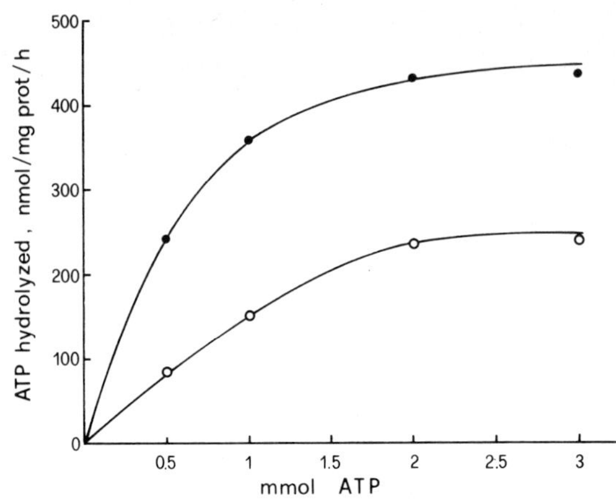

Fig. 1. Enzyme activity as a function of ATP concentration:
total ATPase (●) and ouabain-inhibited ATPase (o).

to the ATP concentration. In the incubation conditions used, with
a 2 mmol/l ATP concentration, we are in the presence of a zero-
order kinetics (Fig. 1).

Total ATPase activity is measured in the absence of ouabain,
while ouabain-sensitive ATPase ($Na^+ - K^+$ ATPase) is calculated on
the basis of the difference between the enzyme activities in the
absence and in the presence of ouabain respectively.

Cation Fluxes

Two main technical approaches, differing with regard to the
ionic composition of both the intra- and extracellular environment, must be considered: (i) the "fresh" erythrocyte method, in
which the red cell mantains its physiological intracellular sodium
concentration (approximately 8-9 mmol/l cells); and (ii) the
"loaded" erythrocyte method, characterized by the utilization of
an ionophore (2,5-chloromercuribenzene sulfonate) which allows
changes in the intracellular sodium concentration (up to 80 mmol/l
cells). In both cases we measure the activity of the $Na^+ - K^+$ pump,
as the ouabain inhibitable fraction of either Na^+ efflux or K^+
influx.

It is worth noting that the possibility of varying both intra
and extracellular sodium and potassium concentrations enables us
to study the kinetics of the $Na^+ - K^+$ pump (see Fig. 2 and 3). The
linearization of the different ion curves with appropriate mathematical models makes it possible to calculate the maximal
velocity (Vmax) and the apparent affinity (K_{Nai} or K_{Ko}) for both
outward and inward cation fluxes.

Red Cell Ouabain Binding

Extensive investigations demonstrated that ouabain binds
avidly to the $Na^+ - K^+$ pump with a 1:1 molar stoichiometry, and it
is generally accepted that the maximal number of molecules of
ouabain bound per cell closely reflects the number of $Na^+ - K^+$ pump
units.

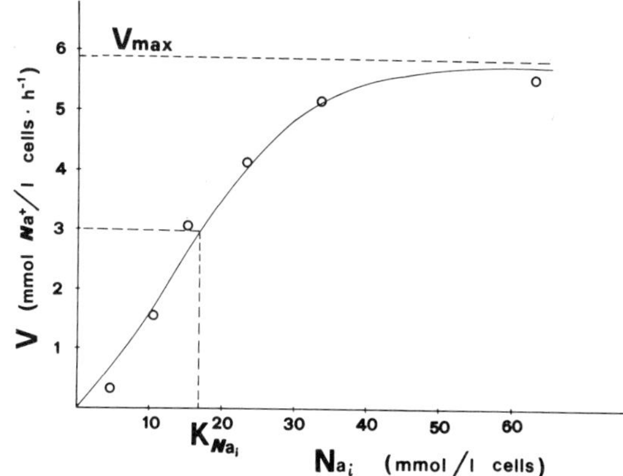

Fig. 2. Ouabain-sensitive Na⁺ efflux at different intracellular
sodium (Na_i) concentrations.

The erythrocytes are incubated at 37°C for 5 hours with
different ³H-ouabain concentrations (from 1.5 to 40 nmol/l) and
are then separated from the incubation solution by filtration
through glass microfiber filters. Specific ouabain binding is
calculated on the basis of the difference between total binding
and both binding in the presence of an excess of unlabelled
ouabain (non-specific ouabain binding) and the binding to the
filters themselves.

Erythrocyte maximal capacity together with the apparent
dissociation constant for ouabain (K_d) are calculated on the basis
of Scatchard analysis of equilibrium binding data (Fig. 4).

DISCUSSION

We have compared these methods in terms of three parameters:
(i) type of information provided, (ii) reproducibility: and (iii)
technical complexity.

The enzymatic method is relatively easy, is not time consuming, shows good reproducibility, the coefficient of variation
being very low (<1%), and is very sensitive. Actually it enables
us to measure ATP concentrations in the order of nanomoles and
very small substrate variations. However, it does not allow direct
measurement of ion exchange, and, furthermore, cellular integrity
is not preserved.

Measurement of cation fluxes has the obvious advantage of
direct assessment of the ions movements catalyzed by the Na⁺ - K⁺
pump. Many important parameters are available, such as apparent
affinity for both intracellular sodium and extracellular potassium
as well as maximum velocity for both potassium influx and sodium
efflux. It is, however, a complex technical approach and sometimes
is very far from reflecting physiological conditions. What is
more, optimal reaction conditions are still a debatable issue.

The ouabain-binding technique is both time consuming and
complex. It quantifies the number of pump units, but this is probably relatively poor information, because it depends on protein

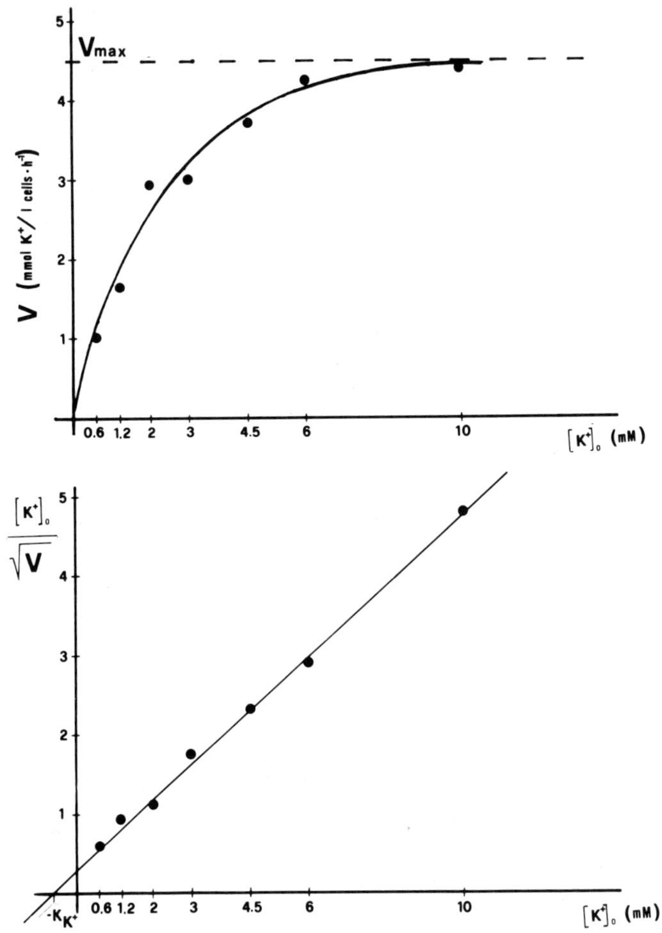

Fig. 3. Top panel: ouabain-sensitive K$^+$ influx at different
extracellular potassium (Ko) concentrations.
Bottom panel: linearization by Hanes plot.

synthesis, which is uncapable of undergoing rapid variations, and
the method is thus unsuitable for detecting possible acute pump
variations. Moreover, this methodological approach provides no
information on cation traslocation.

In agreement with Blaustein[7], one may expect the sodium
pumping rate to be increased or decreased depending on whether or
not: a) the inhibitor is partially or totally dissociated from the
pumps during preparation of cells subjected to repeated washing;
b) intracellular sodium is increased as a result of pump inhibition with consequent changes in efflux rate; and c) there is a
compensatory increase in the number of pump sites when the pumps
are chronically inhibited, as observed in patients on therapy with
digitalis cardiokinetic agents. Point a), in particular, presents
a thorny problem because all the methods used for measuring either
pump activity or membrane ATPase activity require washing of
erythrocytes.

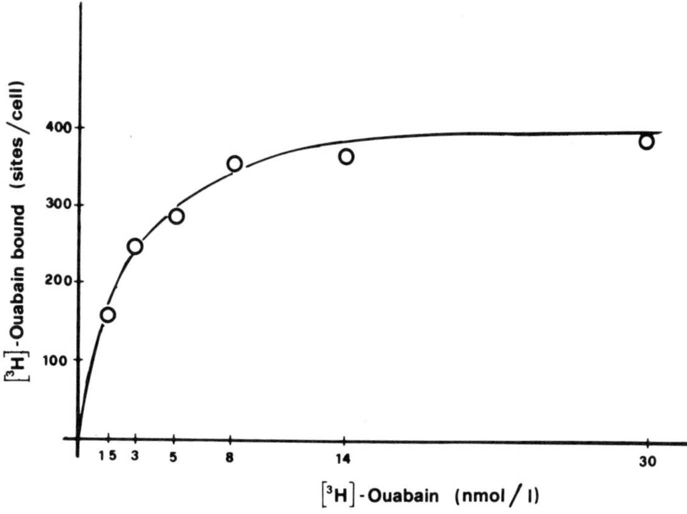

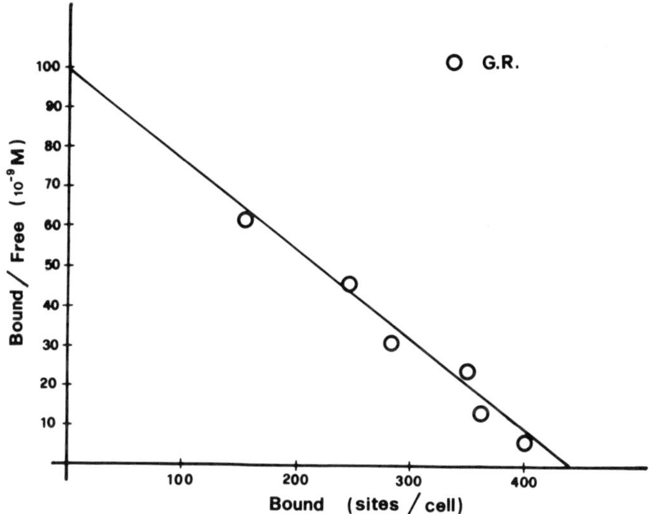

Fig. 4. Top panel: 3H-ouabain binding to human erythrocytes at different concentrations.
Bottom panel: Scatchard plot of the same experiment.

In conclusion, because of the inherent drawbacks of each method, the choice of the most suitable method of testing the activity of the $Na^+ - K^+$ pump for clinical purposes also depends on both the degree of simplicity and precision required. However, in view of the complementary data provided by the different methods, it may be necessary, in some cases, to study the $Na^+ - K^+$ pump from more than one point of view.

REFERENCES

1. A.F. Aderounmu and L.A. Salako, Abnormal cation composition
    and transport in erythrocytes from hypertensive patients,
    Eur. J. Clin. Invest. 9:369 (1979).
2. C.H. Cole, Decreased ouabain-sensitive adenosine triphosphate
    activity in erythrocyte membrane of patients with chronic
    renal disease, Clin. Sci. Mol. Med. 45:775 (1973).
3. M. De Luise, G.L. Blackburn and J.S. Flier, Reduced activity
    of red cell sodium-potassium pump in human obesity, N. Engl.
    J. Med. 303:1017 (1980).
4. M. De Luise and J.S. Flier, Status of the red cell $Na^+ - K^+$
    pump in hyper-and hypothyroidism, Metabolism 32:25 (1983).
5. J. Duhm and J. Behr, Role of exogenous factors in alterations
    of red cells $Na^+ - K^+$ cotransport in essential hypertension,
    primary hyperaldosteronism and hypokalemia, Scand. J. Clin.
    Lab. Invest. 46:82 (1986).
6. E. Soderling, Y. Le Bell, J. Laikko and M. Larmas, Conditions
    for assay of ATPase from biological materials using a
    bioluminescence technique, Clin. Chim. Acta 111:33 (1981).
7. M.P. Blaustein, Sodium transport and hypertension,
    Hypertension 6:445 (1984).

INHIBITION OF CHEMOTACTIC ACTIVITY OF PMN BY ARSENIC: A MEMBRANE

MEDIATED EFFECT

M. Governa, M. Valentino, M. Rocco, E. Bertoli*,
and G. Zolese*

Institute of Occupational Medicine and
* Institute of Biochemistry
University of Ancona
Ancona, Italy

INTRODUCTION

Arsenic is a ubiquitous element which is widely distributed
in minerals of the earth crust. The dissemination of arsenic
compounds into the environment has been facilited by the metal
smelting and industrialization. These compounds, where the arsenic
is found in relatively soluble forms, can be adsorbed into the
body chiefly by inhalation and ingestion. The rate of arsenic
metabolism in organisms changes in relation to the chemical form
of the administered compound and to the animal species. Actually,
arsenite, the trivalent compound, causes higher tissue concentrations than arsenate, and the same ingested dose of arsenite is
excreted in different percentages after 48 hours in man (60%),
rabbits and mice (90%)[1]. Owing to this great variation in arsenic
metabolism among species, it is difficult to find a single animal
model which can explain the mechanism involved in arsenic toxicity. For this reason several studies in vitro have employed
cellular models such as lymphocytes[2,3], fibroblasts[4] and alveolar
macrophages[5].

It has been found that the macrophagic cells are highly
sensitive to cytotoxic arsenic action in vitro. Therefore, in
order to clarify the mechanism involved in cytotoxicity we have
studied the chemotactic activity, which is a typical activity of
phagocytic cells, in polymorphonuclear leukocytes (PMN). PMN are
blood phagic cells which are involved in the response to infection
mainly by protecting the organism against pyogenic infection.

Since our findings are indicative of an action of arsenic on
the chemotactic receptor placed on the plasma membrane, we have
also studied, in a second series of experiments, the effect of
arsenic on the membrane of erytrocyte ghosts, which is a well
known membrane model.

MATERIALS AND METHODS

<u>PMN Locomotion</u>

PMN were isolated from venous blood of three healthy laboratory staff members using Ficoll-Hypaque gradient centrifugation.
They were washed twice in Hanks' (Flow Labs., McLean, Va, USA) and
suspended $2x10^6$ in RPMI 1640 (Flow Labs.) medium supplemented with
sodium arsenite (Merck, Darmstadt,FRG) in arsenic concentrations
ranging from 0 to 10^{-4} mol/l.

No significant decrease in cell viability was observed with
trypan blue exclusion test after incubation for 3 hours at 37°C of
PMN suspensions containing arsenic at the concentration of
10^{-4} mol/l.

Five chemotaxis experiments were performed using the modified
Boyden technique, as previously described[6]. The upper compartment
of Boyden chamber was filled with arsenic (at concentrations from
0 to 10^{-4} mol/l) in RPMI 1640, while the lower compartment was
filled with Zymosan (Sigma Chemical Co., St. Louis, Mo, USA) activated serum, diluted 1:10 in RPMI 1640, prepared according to
Maderazo and Woronich[7]. Chemotactic activity was expressed as
chemotactic index calculated according to Hill[8].

Three experiments were also carried out in order to evaluate
the effect of arsenic on the random migration, which is a non
oriented locomotion of cells. In these tests, the lower compartment of Boyden chambers was filled only with RPMI 1640 without any
chemotactic agent. Since there was an incomplete migration of
cells through the filter, the leading front of Zigmond and Hirsch[9]
rather than the chemotactic index of Hill[8] was calculated.

<u>Erythrocyte Membrane Preparation and Incubation with DPH</u>

Erythrocyte membranes were prepared according to Steck and
Kant[10] from freshly collected human blood drawn into heparinized
tubes. Erythrocytes were washed three times with isotonic saline
pH 7.4 and lised with the ipo-osmotic 5 mmol/l phosphate buffer
pH 8. One volume of a diluted erythrocyte membranes approximately
100 µg/ml was mixed with a volume of 2 µmol/l DPH (Molecular
Probes Inc., Junction City, Or, USA) dispersion and incubated for
1 hour at 37°C.

<u>Fluorescnce Measurements</u>

The fluorescence polarization measurements were done on a
Perkin Elmer (Norwalk, Ct, USA) spectrofluorimeter MPF 44 A,
equipped with two glass prism polarizers. Suspension of unlabelled
membranes at the same concentration were used as reference blanks
to evaluate light scattering contribution.

The steady state fluorescence polarization (P) was calculated
using the equation[11]:

$$P = \frac{I^= - I^+}{I^= + I^+}$$

Table 1. Receptor Mediated Chemotactic Activity (Chemotactic
          Index) and Receptor Independent Random Migration
          (Leading Front) of PMN Incubated for 3 Hours with
          Increasing Amounts of Sodium m-Arsenite

Arsenite Concentration	Chemotactic Index	Leading Front
Control	83.5 ± 9.5	31.3 ± 9.8
1×10^{-8} mol/l	77.8 ± 10.6	32.8 ± 8.9
5×10^{-8} mol/l	65.3 ± 9.7*	Not Done
1×10^{-7} mol/l	36.5 ± 11.9*	29.7 ± 8.0
1×10^{-6} mol/l	17.6 ± 14.6*	31.3 ± 10.2
1×10^{-5} mol/l	21.2 ± 10.7*	33.5 ± 7.9
1×10^{-4} mol/l	20.0 ± 12.5*	28.3 ± 8.8

* $P \leq 0.01$ Student's t versus control.

$I^=$ and I^+ are the fluorescence intensities detected with the excitation polarizer in horizontal position and the analyzer in
horizontal and vertical position respectively. The polarization of
fluorescence is directly related to the rotational relaxation time
of the fluorophore; thus, its magnitude provides a quantitative
index of the freedom of rotation of the fluorescent probe. A
change in the local packing of the molecules surrounding the probe
results in an altered polarization fluorescence value.

Circular Dichroism Measurements

The circular dichroic spectra were recorded at room temperature on a Jasco J-500A spectropolarimeter (Jasco, Tokyo, Japan)
under Nitrogen flux. Scans were performed in duplicate, using a
quartz cell of 0.2 cm path length. All the measurements were made
at a protein concentration of 40 µg/ml. The molar ellipticities
were calculated from spectra following Urry[12].

RESULTS AND DISCUSSION

The effect of the increasing arsenic concentrations on
chemotaxis is shown in Table 1. Chemotaxis was already inhibited
at the concentration of 5×10^{-8} mol/l, however increasing arsenic
concentration to 10^{-7} mol/l and 10^{-6} mol/l, the chemotactic index
decrease was respectively of 56 and 79%. No further decrease of
the chemotactic activity was observed at arsenic concentrations
higher than 10^{-6} mol/l and this remark suggests a dose dependent
toxic effect of arsenic.

Experiments performed in order to verify whether arsenic also
affects the random migration have not shown any effect induced by
arsenic (Table 1).

Since chemotaxis, i.e. a receptor mediated locomotion, is
inhibited by arsenic, whereas random migration, i.e. a receptor
indipendent locomotion, is not, our results might support the
hypothesis that arsenic modifies the plasma membrane structure.

Table 2. Fluorescence Polarization of Diphenylhexatriene (DPH) in
          Erythrocyte Membranes in Presence and in Absence of
          Sodium m-Arsenite

		Polarization
Control		0.250 ± 0.010
Arsenite	3.5 µmol/l	0.235 ± 0.003
"	6.9 µmol/l	0.239 ± 0.006
"	10.4 µmol/l	0.239 ± 0.001
"	13.9 µmol/l	0.254 ± 0.001
"	2.0 mmol/l	0.249 ± 0.013

In order to study the membrane physical properties in a
second series of experiments, we evaluated the action of the arsenic on the main components of cell membranes: the lipid bulk and
the proteins. We used the membrane of red blood cells which is a
well known experimental model for this type of study[13], very easy
to obtain and to manipulate.

With fluorescence polarization of DPH, we can acquire information about the physical state of membrane lipids which play an
important role on membrane functions and can be modified by
xenobiotics, such as anaesthetics[14]. Table 2 shows no modification
of DPH fluorescence polarization values after arsenic incubation
of ghosts, therefore arsenic does not modify membrane fluidity at
concentrations around 10^{-6} mol/l. Previous experiments had also
shown no effect of arsenic on the membrane viscosity in artificial
lipidic membranes[15].

Finally we have measured the influence of arsenic on the
structure of proteins of the membrane using the Circular Dichroism
technique. This technique can rapidly approximate the percentages
of the conformations present in solutions of proteins measuring
the tertiary structure of proteins alpha helix or beta conformation. Table 3 shows that arsenic at concentrations around
10^{-6} mol/l decreases the values of ellipticities, this decrease
being related to a decrease of the content of the alpha helical
content in the protein.

We suppose that the observed conformational change of the
membrane proteins can affect the biological processes mediated by
membrane proteic activities. The chemotactic response, which is
mediated by a membrane receptor, can be one of these activities.
Other Authors showed that at low concentrations trivalent arsenic
has effects on several cellular functions such as limphocyte

Table 3. Decrease of Molar Ellipticity (%) Induced by Sodium
          m-Arsenite on Erythrocyte Membranes

		224 nm	210 nm
Arsenite	4.5 µmol/l	- 12.5 %	- 9 %
"	11.0 µmol/l	- 11.0 %	- 8 %

blastogenesis[3], macrophage cytotoxic activity[5,16], fibroblasts pro-
liferation and viability[4]. These activities could also be affected
through an action of arsenic on the proteic structure of the
membrane.

CONCLUSION

The locomotion of PMN is affected by trivalent arsenic at
concentration higher than 5×10^{-8} mol/l, however our assays show
that the xenobiotic mainly acts on the receptor mediated cellular
function such as chemotactic activity, whereas it does not affect
the receptor indipendent locomotion, i.e. random migration. The
results obtained with erythrocyte ghosts membranes, a well acknowledged model for studying plasma membranes, show that arsenic
modifies the conformation of membrane proteins without affecting
the physical state of the lipid bulk. These effects were detected
at low concentrations of arsenite, i.e. 10^{-6} mol/l. Such
concentrations are in the same order of magnitude of those found
in biological fluids (e.g. urine). Even if our results are not
sufficient to exhaustively explain the in vivo biological effects
of arsenic, they are nonetheless indicative of a possible
mechanism of cellu-lar damage induced by exposure to low arsenic
concentrations.

Acknowledgments

This work was supported by Ente Regione Marche, Ancona,
Italy.

REFERENCES

1. M. Vahter, E. Marafante, A. Lindgren and L. Dencker, Tissue
   distribution and subcellular binding of arsenic in marmoset
   monkeys after injection of 74 As-Arsenite, Arch. Toxicol.
   51:65 (1982).
2. I. Nordenson, A. Sweins and L. Beckman, Chromosome aberra-
   tions in cultured human lymphocytes exposed to trivalent and
   pentavalent arsenic, Scand. J. Work Environ. Health 7:277
   (1981).
3. M. McCabe, D. Maguire and M. Nowak, The effects of Arsenic
   compounds on human and bovine lymphocyte mitogenesis in vitro,
   Environ. Res. 31:323 (1983).
4. A.B. Fischer, J.P. Buchet and R.R. Lauwerys, Arsenic uptake,
   cytotoxicity and detoxification studied in mammalian cells in
   culture, Arch. Toxicol. 57:168 (1985).
5. C. Aranyi, D.E. Gardner and J. Lewtas Huisingh, Evaluation of
   potential inhalation hazard of particulate silicious compounds
   by in vitro rabbit alveolar macrophage tests-Application to
   industrial particulates containing hazardous impurities, in:
   "Health Effects of Synthetic Silica Particulates", D.D.
   Dunnom, ed., American Society for Testing and Materials,
   Philadelphia (1981).
6. M. Governa, M. Valentino, I. Visonà and M. Rocco, Impairment
   of human polymorphonuclear leukocyte chemotaxis by
   2,5-hexanedione, Cell Biol. Toxicol. 2:33 (1986).
7. E.G. Maderazo and C.L. Woronich, A modified micropore filter
   assay of human granulocyte leukotaxis, in "Leukocyte

Chemotaxis",J.I. Gallin and P.G. Quie ,eds., New York (1978).

8. H.R. Hill, N.A. Hogan and T.G. Mitchell, Evaluation of a cytocentrifuge method for measuring neutrophil granulocyte chemotaxis, J. Lab. Clin. Med. 86:703 (1975).

9. S.H. Zigmond and J.G. Hirsch, Leukocyte locomotion and chemotaxis - News in methods for evaluation and demonstration of a cell-derived chemotactic factor, J. Exp. Med. 137:387 (1973).

10. T.L. Steck and J.A. Kant, Preparation of impermeable ghosts and inside-out vesicles from human erythrocyte membranes, Methods Enzymol. 31:172 (1974).

11. W.J. vanBlitterswijk, R.P. vanHoeven and B.W. vanDer Meer, Lipid structural order parameters (reprocal of fluidity) in biomembranes derived from steady-state fluorescence polarization measurements, Biochim. Biophys. Acta 644:323 (1981).

12. D.W. Urry, L. Masotti and J.R. Krivacic, Circular dichroism of biological membranes. I. Mitochondria and red blood cell ghosts, Biochim. Biophys. Acta 241:600 (1971).

13. D. Schachter, R.E. Abbott, R. Cogan and M. Flamm, Lipid fluidity of the individual hemileaflets of human erythrocyte membranes, Ann. N.Y. Acad. Sci. 414:19 (1983).

14. L. Mazzanti, G. Curatola, G. Zolese, E. Bertoli and G. Lenaz, Lipid protein interactions in mitochondria. VIII Effect of general anaesthestics on the mobility of spin labels in lipid vesicles and mitochondrial membranes, J. Bioenerg. Biomembr. 11:17 (1979).

15. G. Zolese, G. Curatola and E. Bertoli, Effects of sodium arsenite on liposomes and erythrocyte membranes, Soc. Ital. Biol. Sper., 62:1177 (1986).

16. C. Aranyi, N.J. Bradof and W. O'Shea, Effects of arsenic trioxide inhalation exposure on pulmonary antibacterial defenses in mice, J. Toxicol. Environ. Health 15:163 (1985).

IN VITRO STUDIES ON THE ACETYLATION OF SULPHAMETHAZINE BY

HUMAN WHOLE BLOOD FROM HEALTHY AND DIABETIC SUBJECTS

R.M. Lindsay, J.D. Baty, and N.R. Waugh*

Department of Biochemical Medicine
Ninewells Hospital and Medical School
* Department of Community Medicine
Tayside Health Board
Dundee, U.K.

INTRODUCTION

N-acetylation is the principal metabolic pathway for many compounds containing either arylamine or hydrazino groups. The reaction is catalysed by an N-acetyltransferase (NAT) enzyme and requires acetyl coenzyme A (acetyl-CoA) as the acetyl group donor. Drugs such as isoniazid and the sulphonamide sulphamethazine (SMZ) are acetylated by man in vivo (Fig. 1). A genetic polymorphism allows classification of individuals as either "rapid" or "slow" acetylators of these compounds. This has important therapeutic and toxic implications. Most in vitro studies on drug acetylation have involved liver NAT preparations although extrahepatic sites of NAT activity have also been reported[1].

Previous studies on the in vitro capacity of human blood to acetylate antibacterial sulphonamide drugs in vitro have been hindered by the lack of a sensitive and specific analytical method for measurement of the acetylated compound. The most common method is the Bratton-Marshall colorimetric procedure[2] for arylamides. This involves the diazotisation of free N4-amino groups with nitrous acid and coupling of this diazonium salt with a chromagen such as N-naphthylethylenediamine to form a complex which absorbs at 540 nm. After acidic hydrolysis at 100°C of the amide bond in the acetylated compound, the above procedure is repeated to allow determination of the total sulphonamide present. The amount of the acetylated sulphonamide may be calculated by subtracting the sulphonamide level before acidic hydrolysis from that after hydrolysis. This method is therefore not specific for the acetylated compound since it is measured indirectly. The presence of other arylamines and their acetylated derivatives may also interfere with this method. Using a modification of this colorimetric assay, Mandelbaum-Shavit and Blondheim[3] reported low, and in some cases undetectable, levels of NAT activity in human blood lysates towards SMZ.

We have developed a reverse phase high performance liquid chromatographic (HPLC) assay which separates SMZ and acetylsul-

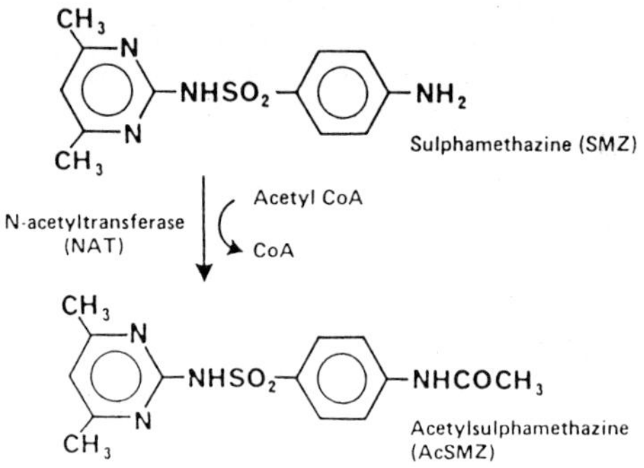

Fig. 1. Acetylation of sulphamethazine.

phamethazine (AcSMZ) and therefore enables study of the acetylation of SMZ by human whole blood in vitro[4]. We have previously used this assay to estimate the kinetic parameters V_{max} and K_m of the enzyme and to study the variation of the enzyme activity amongst healthy individuals of known acetylator phenotype.

No previous in vitro investigations on the acetylation reaction in diabetics and non-diabetics have been reported. Several groups have investigated the effect of added glucose on the in vitro acetylation of drugs by various NAT preparations from healthy subjects. Increasing the glucose concentration of human blood[5,6] and rat blood[6] has been shown to increase the in vitro acetylation of p-aminobenzoic acid. However, adding glucose to either rat or rabbit liver parenchymal cells did not alter the in vitro acetylation of either p-aminobenzoic acid, SMZ or sulphanilamide[7]. We report here the use of our HPLC method to study the enzyme activity present in blood samples from healthy and diabetic subjects and the effect of added glucose on the in vitro acetylation capacity.

MATERIALS AND METHODS

Materials

Sulphamethazine and acetic anhydride were obtained from Aldrich (Gillingham, U.K.). Sulphapyridine (SPD) was purchased from Sigma (Poole, U.K.). Potassium dihydrogen phosphate, disodium hydrogen phosphate and D-glucose were obtained from BDH (Poole, U.K.). HPLC-grade acetonitrile was obtained from Rathburn (Peebleshire, U.K.). Ethyl acetate (Pronalysis grade) was purchased from May and Baker (Dagenham, U.K.). Acetylsulphamethazine was synthesised from sulphamethazine by heating with acetic anhydride and was recrystallised from aqueous ethanol (m.p. 254-255°C) Its structure was confirmed by mass spectrometry and its purity was 100% by HPLC analysis.

Instrumentation

A Gilson 302 reciprocating pump (Gilson Medical Electronics, Villiers-le-Bel, France) was used to pump a mobile phase of acetonitrile:33.3 mmol/l phosphate buffer (pH 7.4) (15:85) at a flow-rate of 1 ml/min. Samples were injected using an Altex Model

500 autosampler (Beckman-RIIC, High Wycombe, U.K.) fitted with an Altex 210 valve (20 μl loop). The separation was performed on an Altex Ultrasphere ODS column (250 x 4.6 mm I.D., 5 μm particle size). Detection was achieved using a Pye Unicam LC3 UV detector (Pye Unicam, Cambridge, U.K.) set at 240 nm and the detector signal was recorded by a Hewlett-Packard 3388A integrator (Hewlett-Packard, Altrincham, U.K.). A Spherisorb ODS2 column (250 x 4.6 mm I.D., 5 μm particle size) was used with a 10:90 mixture of the same mobile phases for analyses involving SPD as internal standard. The detection limit for free and acetylated SMZ on these systems (at a signal-to-noise ratio of 2) was approximately 50 ng/ml.

<u>Subjects Studied</u>

Inter-individual variation of blood NAT activity in vitro was studied using samples from 13 healthy and 40 Type 2 (non insulin dependant or maturity onset) diabetic subjects. The latter were patients attending a monthly clinic and consisted of 22 patients treated by diet alone, 8 patients on insulin treatment and 10 treated by oral hypoglycaemic drugs. The age ranges of the healthy and diabetic subjects were 20-58 years and 39-66 years respectively.

<u>Incubation Protocol</u>

A volume of 0.2 ml of heparinised whole blood and SMZ (72 nmoles) in 0.2 ml 33.3 mmol/l phosphate buffer (pH 7.4) were mixed in 5 ml glass blood bottles. In a further set of experiments, 0.2 ml of a solution containing SMZ (72 nmoles) and glucose (2,4 or 20 μmoles) was added to 0.2 ml heparinised whole blood.

The samples were then incubated at 37°C for periods up to 24 hours, then extracted with 3 ml of ethyl acetate. After vortex mixing (20 seconds) and an equilibration time of 90 min, 2 ml of the organic layer was decanted and evaporated to dryness under nitrogen at room temperature. The residue was dissolved in the HPLC eluent prior to injection. For analyses using an internal standard, a solution of SPD (100 nmol) in 3 ml of ethyl acetate was used to extract the samples after incubation. To calibrate the assay, a range of AcSMZ and SMZ standards covering the range 0-21.7% acetylation (containing 0-15.62 nmol AcSMZ respectively) were added to whole blood and extracted immediately. The ratio of the AcSMZ/SMZ peak heights was then plotted against the known amount of AcSMZ present. For standards extracted using SPD as internal standard the ratio of the amounts of AcSMZ/SPD in the standards was plotted against the ratio of the AcSMZ/SPD peak heights.

A Beckman glucose analyser 2 was used to measure the plasma glucose levels. Red cell counts were performed using an Ortho-Diagnostics Systems coulter counter (model ELT-800/WS) by the Department of Haematology at Ninewells Hospital and Medical School, Dundee.

Statistical analysis was performed by a Student's t-test (paired t-test where appropriate) using the separate variances of the population samples being compared.

RESULTS

HPLC Analysis of AcSMZ and SMZ

There were no interfering peaks at any of the retention times of interest in any of the samples studied. Chromatograms of blank samples (i.e. no SMZ present) which had glucose added prior to incubation were not significantly different from those without added glucose.

The order of elution of SMZ and AcSMZ was dependent on the analytical column used (Fig. 2 and 3). When the Ultrasphere column was used, SPD eluted between AcSMZ and SMZ and therefore it was not possible to use this compound as an internal standard with this system. Attempts to improve the resolution of the three compounds (by varying the % acetonitrile in the mobile phase) were unsuccessful. The ability of the Spherisorb column to completely separate these analytes allowed SPD to be used as internal stardand in this case. Since the correlation between analyses performed with and without an internal standard was good (r>0.99), samples were subsequently analysed without adding SPD.

The calibration line of the amount of AcSMZ present against

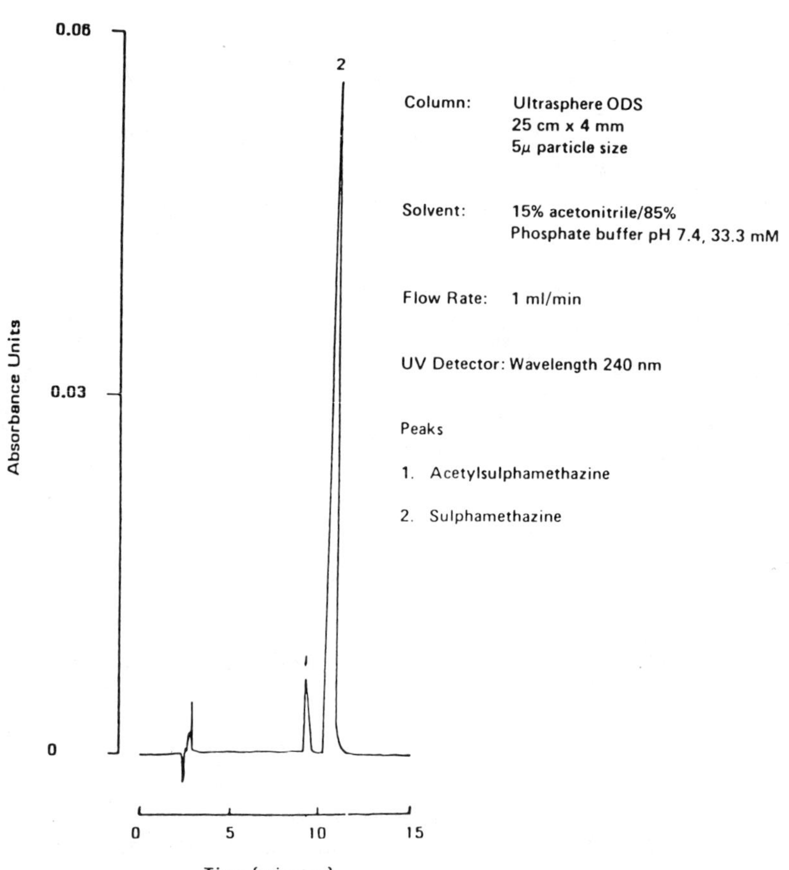

Fig. 2. HPLC separation of sulphamethazine and acetylsulpha-methazine (human whole blood extract).

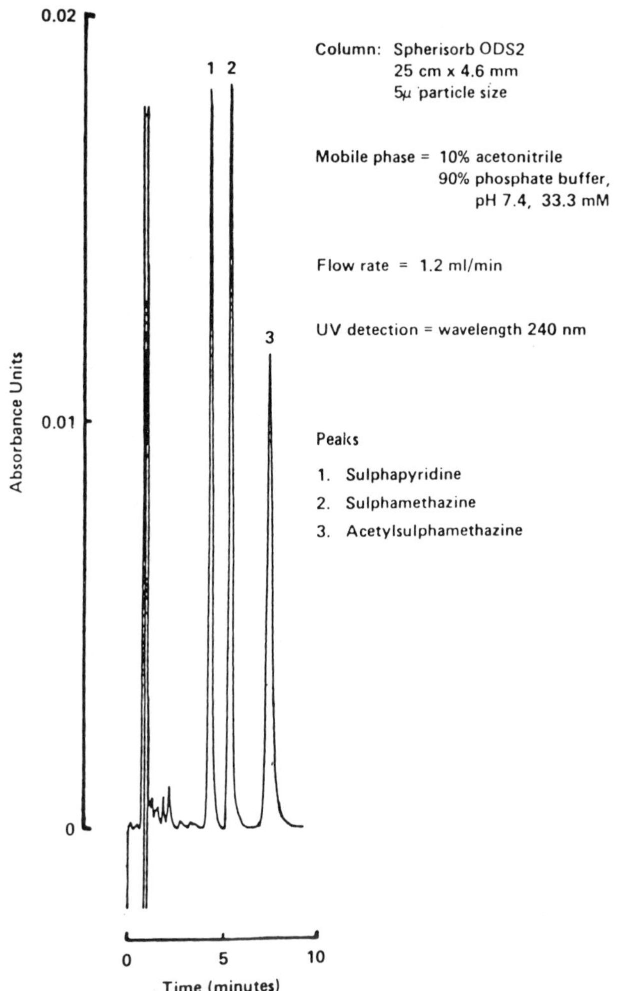

Fig. 3. HPLC separation of sulphamethazine, acetylsulpha-
methazine and sulphapyridine (human whole blood extract).

the ratio of the peak heights of the parent compound and its
acetylated metabolite was linear (r>0.99) over the range studied.
Only a small variation (<5%) was found in the slopes of calibration lines performed during a 12 month period. The precision of
the assay (n=5) at 0.62 and 15.62 nmoles AcSMZ was 4.0% and 2.3%
respectively.

<u>In Vitro Blood N-acetyltransferase Activity of Healthy and
Diabetic Subjects</u>

Table 1 shows the amounts of AcSMZ produced by blood samples
from healthy and diabetic volunteers, together with their plasma
glucose concentrations. The amount of AcSMZ produced in 24 hours
using blood from diabetic subjects was significantly higher
(p<0.001, two tailed Student's t-test) than that from healthy
volunteers. The correlation between the amount of AcSMZ produced
and the plasma glucose concentration of all the subjects studied
was 0.77 (n=53).

Table 1. In Vitro Acetylation of Sulphamethazine by Blood
          Obtained from Healthy and Diabetic Subjects (non age
          matched).

	Mean (range) plasma glucose concentration (mmol/l[+])	Mean (range) AcSMZ produced (nmol/10^9 rbc[++])
Healthy volunteers (n = 13)	5.1 (3.6 - 6.8)	3.05 (1.96 - 4.03)
Diabetics outpatients (n = 40)	8.7 (5.8 - 19.0)	4.18 (1.83 - 8.04)

Initial SMZ concentration = 180 µmol/l (equivalent to 72 nmol per
sample). Incubation time = 24 hours.
The value in the two groups are significantly different using the
Student's t-test. [+]p<0.0005 (one tailed test), [++]p<0.001 (two
tailed test).

The Effect of Added Glucose on Blood Acetylation in Vitro

        The increase in the amounts of AcSMZ produced following the
addition of glucose to blood from both a) healthy and b) diabetic
individuals is shown in Table 2a. Increasing the incubation
glucose concentration by 5 mmol/l approximtely doubled the amount
of AcSMZ produced by each sample population. Increasing the
incubation glucose concentration by a further 5 mmol/l caused only
a slight further increase in the amount of AcSMZ produced by each
group although the difference was statistically significant
(p<0.05, paired t-test). In each of these cases, the amount of SMZ
acetylated in vitro by blood samples from the diabetics was
significantly higher (p<0.01, two tailed Student's t-test) than
that by the healthy volunteers. Blood from diabetic subjects, in
which the glucose concentration had been increased in vitro by
5 mmol/l, acetylated more SMZ (p<0.02, two tailed Student's
t-test) in vitro than samples from healthy volunteers, in which
the in vitro glucose level was increased by 10 mmol/l (Table 2b)
despite having a significantly lower mean incubation glucose
concentration (p<0.0005, one tailed Student's t-test).

The Effect of Added Glucose on the Initial Rate of Acetylation

        Using blood samples donated by healthy (n=6) and diabetic
(n=6) subjects, it was shown that in vitro elevation of their
initial glucose concentration by 5 and 50 mmol/l did not increase
the initial rate of AcSMZ production (Tables 3a, 3b) but extended
the linearity of the reaction. Increasing the original glucose
concentration by just 5 mmol/l extended the reaction linearity to
24 hours. Fig. 4 illustrates the prolonged period of reaction
linearity observed following glucose addition to a blood sample
from a healthy volunteer.

Table 2a.   Effect of Added Glucose on the in Vitro Acetylation
            of Sulphamethazine by Blood obtained from Healthy and
            Diabetic Subjects (non age matched)

	Mean (range) initial plasma glucose concentration (mmol/l[+])	Increase in incubation glucose concentration (mmol/l)	Mean (range) AcSMZ produced (nmol/10^9 rbc[++])
Healthy volunteers (n = 12)	5.2 (3.4 - 6.8)	0	3.01 (1.96 - 3.98)
		5	6.21 (4.29 - 8.29)
		10	6.54 (4.59 - 8.60)
Diabetic outpatients (n = 15)	9.0 (6.8 - 14.2)	0	3.96 (2.79 - 5.27)
		5	7.94 (2.79 - 5.27)
		10	8.27 (6.51 -13.67)

Initial SMZ concentration = 180 μmol/l (equivalent to 72 nmol per
sample).
Incubation time = 24 hours.
The values in the two groups are significantly different using
Student's t-test. [+]p<0.0005 (one tailed test), [++]p<0.01 (two
tailed test).

Table 2b.   Effect of Added Glucose on the in Vitro Acetylation
            of Sulphamethazine by Blood obtained from Healthy and
            Diabetic Subjects (non age matched)

	Mean (range) incubation glucose concentration (mmol/l[+])	Mean (range) AcSMZ produced (nmol/10^9 rbc[++])
Healthy volunteers (n = 12)	12.6 (11.7 - 13.4)	6.54 (4.59 - 9.60)
Diabetic outpatients (n = 15)	9.5 (8.0 - 12.1)	7.94 (6.49 - 11.82)

Initial SMZ concentration = 180 μmol/l (equivalent to 72 nmol per
sample).
Incubation time = 24 hours.
The values in the two groups are significantly different using the
Student's t-test. [+]p<0.0005 (one tailed test), [++]p<0.02 (two
tailed test).

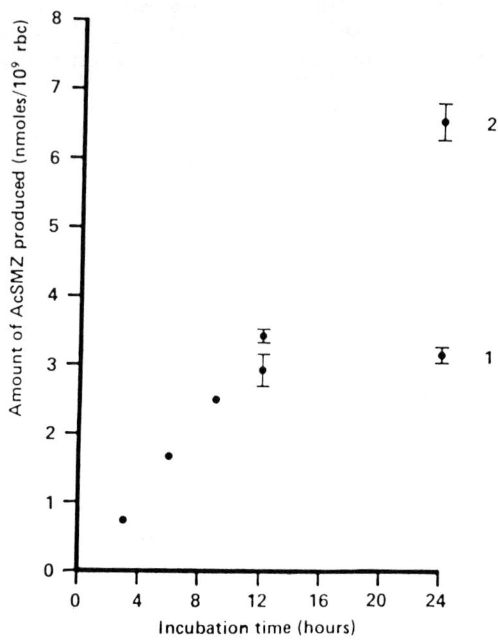

Fig. 4. Time course of acetylsulphamethazine production by human
whole blood incubated with sulphamethazine
(initial concentration = 180 µmol/l)
1 = no added glucose
2 = initial incubation glucose concentration increased by
    5 mmol/l.
The amount of AcSMZ produced at points shown without
error bars (95% confidence limits) were not significantly
different (p > 0.1, Students t-test).

DISCUSSION

The HPLC assay described allows both specific detection and
reliable quantitation of AcSMZ produced in vitro following
incubation of human whole blood with SMZ. This is in contrast to a
previous study using a modification of the Bratton-Marshall
colorimetric assay which failed to detect activity in all the
samples studied[3]. A further advantage of the chromatographic
assay which we have previously exploited is that the superior
sensitivity and specificity permits investigation of the in vitro
acetylation reaction at substrate concentrations down to
18 µmol/l[4].

The in vitro NAT activity of blood from healthy and diabetic
subjects has never been previously reported. We have found
significantly higher NAT activity towards SMZ in blood donated by
diabetic subjects than that present in samples from healthy
volunteers. The activities reported have been corrected for the
number of red blood cells since previous experiments demonstrated
that plasma does not acetylate SMZ in vitro and therefore
variation of the red cell count could result in differences in the
observed NAT activity[8].

Table 3a. Effect of Àdded Glucose on the Initial Rate of
           Acetylsulphamethazine Production in Vitro by Human
           Whole Blood from Healthy Volunteers

Subject	Initial plasma glucose (mmol/l)	Increase in incubation glucose concentration (mmol/l)	Rate of AcSMZ production (nmol/h/10^9 rbc)	Incubation time (h)
ML	5.5	0	0.29 (0.27 - 0.31)	0,3,6,9
		5	0.29 (0.28 - 0.30)	
		50	0.30 (0.29 - 0.31)	
JB	5.4	0	0.18 (0.16 - 0.19)	0,3,6,9
		5	0.17 (0.16 - 0.18)	
		50	0.18 (0.17 - 0.19)	
JW	5.9	0	0.24 (0.22 - 0.26)	0,3,7
		5	0.23 (0.21 - 0.25)	
		50	0.25 (0.24 - 0.26)	
MLe	6.0	0	0.31 (0.29 - 0.33)	0,4,8
		5	0.30 (0.28 - 0.32)	
		50	0.31 (0.28 - 0.33)	
RM	4.7	0	0.20 (0.19 - 0.22)	0,3,6,9
		5	0.21 (0.19 - 0.23)	
		50	0.21 (0.20 - 0.22)	
BC	4.3	0	0.32 (0.29 - 0.35)	0,4,5,7
		5	0.32 (0.30 - 0.34)	
		50	0.34 (0.33 - 0.35)	

Initial SMZ concentration = 180 μmol/l (72 nmol per sample).
Rate values are the mean and 95% Confidence Limits of at least
four replicates performed at the indicated times.

The correlation between the amount of AcSMZ produced in vitro
and the plasma glucose concentration of all the subjects studied
is highly significant (r=0.77, n=53, p<0.001). This result con-
trasts the in vivo studies of Burrows[9] who reported that values
for the % acetylation of SMZ by 6 non-selected diabetics measured
on 2 separate occasions were numerically similar although the glucose levels of the subjects each time were quite different.

We have previously reported minor intra-individual variation
of in vitro blood NAT activity in a single healthy volunteer[8]. The
available data suggest only slight variation (coeffecient of variation <4%) in the fasting plasma glucose concentration of this
subject during this period. It may be that the greater variability of in vitro blood NAT activity in the diabetic subjects
(Table 1) is a reflection of their more variable fasting glucose
values.

Table 3b.   Effect of Added Glucose on the Initial Rate of
              Acetylsulphamethazine Production in Vitro by Human
              Whole Blood from Diabetic Volunteers

Subject	Initial plasma glucose (mmol/l)	Increase in incubation glucose concentration (mmol/l)	Rate of AcSMZ production (nmol/hr/10^9 rbc)	Incubation time (h)
D1	7.5	0	0.24 (0.21 - 0.26)	0,4.25,8.5
		5	0.24 (0.21 - 0.26)	
		50	0.24 (0.23 - 0.25)	
D2	7.4	0	0.34 (0.32 - 0.35)	0,3,5,9
		5	0.33 (0.32 - 0.35)	
		50	0.33 (0.31 - 0.35)	
D3	6.3	0	0.16 (0.14 - 0.18)	0,3.5,9.25
		5	0.17 (0.15 - 0.19)	
		50	0.17 (0.15 - 0.19)	
D4	7.4	0	0.43 (0.39 - 0.46)	0,3.5,9
		5	0.42 (0.39 - 0.45)	
		50	0.44 (0.40 - 0.48)	
D5	5.8	0	0.22 (0.19 - 0.24)	0,3.25,8
		5	0.22 (0.19 - 0.25)	
		50	0.22 (0.19 - 0.24)	
D6	11.5	0	0.18 (0.16 - 0.20)	0,3,8
		5	0.19 (0.17 - 0.21)	
		50	0.19 (0.17 - 0.20)	

Initial SMZ concentration = 180 µmol/l (72 nmol per sample).
Rate values are the mean and 95% Confidence Limits of duplicate
performed at the indicated times.

Increasing the endogenous glucose concentration by either 5
or 10 mmol/l significantly raised the amount of SMZ acetylated in
vitro by each of the two groups. The effect of added glucose
appears to be saturable since the increases in AcSMZ produced by
increasing the original glucose level by 10 mmol/l are only
slightly greater than that obtained following an elevation of only
5 mmol/l (Table 2a). Before and after each of these treatments,
the amounts of AcSMZ produced in vitro by samples from the
diabetics are both significantly higher than that obtained from
the non diabetic group (p<0.01, two tailed Student's t-test).
Similarly, the incubation glucose concentrations of the diabetics' samples are significantly higher (p<0.005, one tailed
Student's t-test) in each case. However comparison of the amounts
of AcSMZ produced by blood samples from diabetics, in which the in
vitro glucose concentration was increased by 5 mmol/l, with that
by samples from healthy volunteers, in which the in vitro glucose
level was raised by 10 mmol/l indicates that the higher blood
glucose values of the diabetics is not the sole explanation for
their increased NAT activity in vitro (Table 2b). In these cases,

the samples from the diabetics acetylate more SMZ in vitro yet
have a significantly lower total glucose concentration. This
suggests that although the in vitro NAT activity shows some
dependance on the initial blood glucose level, other factors must
also operate to affect the final amount of AcSMZ produced.

At an initial SMZ concentration of 180 µmol/l, the rate of
AcSMZ production is constant up to at least 7 hours but not for as
long as 24 hours[4]. Our results demonstrate that the addition of
glucose has no effect on the initial acetylation rate (Tables 3a,
3b) but extended the reaction linearity, in some cases up to 24
hours (Fig. 4).

The explanation for the enhanced production of AcSMZ in the
presence of added glucose in unclear. Blondheim[5] proposed that
glucose may act as both an energy source and as a source of acetate. The addition of acetate to human whole blood had only a
slight, if any, effect on the in vitro acetylation of p-aminobenzoic acid, a result which was later obtained using rat blood[6].
However, the in vitro acetylation capacity of washed human blood
cells was significantly increased by adding acetate and supple-
menting samples with both acetate and glucose raised the in vitro
acetylation of p-aminobenzoic acid further[5].

In conclusion, the hypothesis that elevated levels of
glucose increase the extent of drug acetylation in vitro is supported by our results. They suggest that blood samples from diabetics may contain elevated levels of acetyl-CoA or a precursor of
this compound. Further experiments to investigate the metabolism
of glucose by human blood in vitro need to be conducted in order
to more fully interpret the results reported here.

REFERENCES

1.   W.W. Weber, H.E. Radtke and R.H. Tannen, Extrahepatic
     N-acetyltransferases and N-deacetylases, in: "Extrahepatic
     Metabolism of Drugs and Other Foreign Compounds", T.E. Gram,
     ed, MTP Press Limited, Lancaster (1980).
2.   A.C. Bratton and E.K. Marshall jr., A new coupling component
     for sulfanilamide determination, J. Biol. Chem. 128:537 (1939)
3.   F.Mandelbaum-Shavit and S.H. Blondheim, Acetylation of
     p-aminobenzoic acid by human blood, Biochem. Pharmacol. 30:65
     (1981).
4.   R.M.Lindsay and J.D. Baty, In vitro studies on the acetylat-
     ion of human whole blood, Br. J. Pharmacol. 85:378P (1985).
5.   S.H. Blondheim, In vitro acetylation of drugs by human blood
     cells, Archiv. Biochem. Biophys. 55:365 (1955).
6.   R.Subrahmanyam, R. Quentin Blackwell and L.S. Fosdick, Factors
     that influence the acetylating activity of blood effect of
     addition of certain metabolites and physiologically active
     substances, J. Periodontol. 32:139 (1961).
7.   E.M.Suolinna, Metabolism of sulphanilamide and related drugs
     in isolated rat and rabbit liver cells, Drug Metab. Disp.
     8:205 (1980).
8.   J.D.Baty, R.M. Lindsay and S. Sharp, Use of high performance
     liquid chromatography in the measurement of in vitro
     acetylation in man, J. Chromatogr. 353:329 (1986).
9.   A.W.Burrows, T.D.R. Hockaday, J.I. Mann and J.G. Taylor,
     Diabetic dimorphism according to acetylator status, Br. Med.
     J. 1:208 (1978).

CONTROL OF GLUCOSE AND LIPID METABOLISM BY GEL-FORMING FIBER IN

DIABETES: MEDIUM TERM EFFECTS OF GUAR-GUM

A. Montani and M.E. Vimercati

Clinical Chemistry Laboratory
Civic Hospital
Casalpusterlengo, Milan, Italy

INTRODUCTION

Diabetes mellitus is a major health problem in the western world because it affects about 5% of the population, its prevalence is rapidly increasing and it is associated with severe complications.

Dietary fiber seems to have marked effects on blood glucose concentration in both normal and diabetic subjects. Actually, arterial disease appear uncommon in those countries where dietary fiber intake is high. In fact various diets rich in certain types of dietary fiber, or containing purified dietary fiber preparations, have been shown to lower the serum cholesterol levels in normal individuals[1].

Dietary fiber has been defined as the residue of plant cells resistant to digestion by the alimentary enzymes[2]. Cellulose, hemicellulose, pectins and lignins present in the cell walls are the main components. The interior storage polysaccharides, such as the water soluble galactomannans present within leguminous seeds must be added[3].

Post-prandial hyperglycaemia and glycosuria are reduced when meals are enriched with guar-gum; the reduction induced by gel-forming fiber seems to depend on delayed gastroenteric absorption[4] and not on impaired absorption of carbohydrates[5]. An improvement in the control of diabetes for a long period has already been observed in long term studies[6,7].

In order to elucidate medium-term effects of gel-forming fiber in diabetes, we have observed some in-patients supplying their diets with 15 grams of guar-gum per day over a period of 15 days.

MATERIALS AND METHODS

Seven diabetic men ranging 52-74 years were studied. Six were insulin-independent (Type II diabetes) and one had a Type I

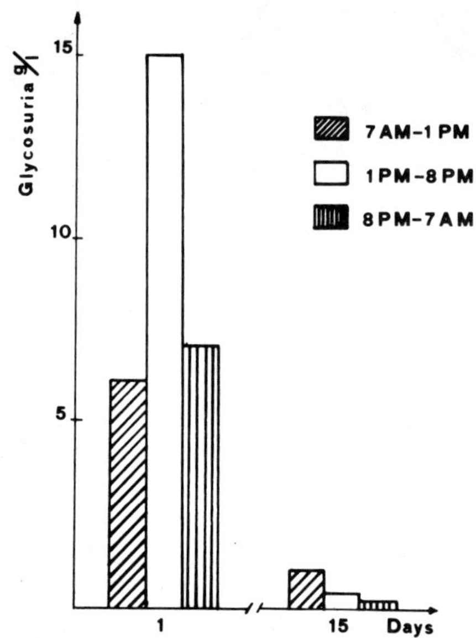

Fig. 1. Behaviour of glycosuria during guar gum treatment.

insulin-dependent diabetes. They were fed for a period of 15 days on a diet containing 15 grams of guar-gum (Guarem, Remeda Pharmaceuticals, Kuopio, Finland) without varying their dietoterapic and pharmacological treatment throughout the study.

Informed consent was obtained from each subject before the beginning of the present research.

The values of serum glycaemia, glycosuria, copper, hemoglobin A_1, sodium, potassium, chloride, uric acid, protrombin time, calcium, phosphorus, magnesium, iron, chlolesterol, HDL-cholesterol and triglycerides have been examined at the beginning, after seven days and at the end of the dietary treatment.

During this period, each subject was given 15 grams of the granules per day with the three main meals; the guar-gum was supplied at the beginning of breakfast, lunch and dinner with sachets of 5 grams mixed with 100-200 ml of water and rapidly ingested.

All assays were performed with enzymatic and photometric methods by using commercial kits; sodium and potassium were assayed by flame photometry and chloride by coulometric silver generation method.

RESULTS

Not any subjects stopped guar treatment because of side-effects; hypoglycaemia has never been shown and it was not necessary to change pharmacological treatment during the period of our research. The great decrease of glycosuria, which practically disappeared after 15 days of the treatment described, is the most significant datum (Fig. 1).

Table 1. Mean Values of Serum Glucose (mg/100 ml) and Glycosilated
         Hemoglobin (% of HBA1)

| | Mean Value ± S.D. | | $\Delta\%$ | p |
	(a)	(b)		
Glycaemia	192.57 ± 71.41	167.29 ± 87.34	-13	<0.4
HBA1c	11.27 ± 2.85	10.36 ± 2.12	- 8	<0.05

(a) before and (b) after treatment with guar-gum; n = 7.

The determination of fractionated glycosuria (7 a.m.-1 p.m.;
1 p.m.-8 p.m.; 8 p.m.-7 a.m.) allows a more precise remark than
glycaemia, which could suffer from various interferences.
Glycaemia shows a slight decrease, and glycosilated hemoglobin
too, according to the decrease of glycosuria (Table 1). The
glycosilation of hemoglobin is a fact insulin-dependent which occurs in red blood cells permeable to glucose, so that the levels
of glycosilated hemoglobin depend upon glucose concentration of
each subject.

The decrease of total cholesterol, triglycerides and the
slight increase of HDL-cholesterol in serum seem to be very
interesting (Table 2).

According to these data, guar-gum treatment could be very
important in the regulation of a main risk factor, in subjects
with frequent and early vascular diseases[8]. Guar has practically
no effect with regard to calcium, phosphorus, uric acid,
protrombin time (i.e. vitamin K) and plasmatic electrolytes. Last
but not least, it is significant that the values of serum iron and
copper show no decrease at all (Table 3); this fact could be very
in-teresting for dietary supplementation with guar-gum.

DISCUSSION

In the present study, dietary supplementation with guar-gum
in diabetics, either insulin-dependent or insulin-independent, is
efficient in glycometabolic regulation, with a further positive

Table 2. Mean Values of Serum Triglycerides, Cholesterol and
          HDL-Cholesterol (mg/100 ml)

| | Mean Value ± S.D. | | $\Delta\%$ | p |
	(a)	(b)		
Triglyc.	173.43 ± 34.66	150.86 ± 30.44	-13	<0.2
Cholest.	222.71 ± 55.40	196.86 ± 75.18	-12	<0.4
HDL-Chol.	27.00 ± 7.75	30.29 ± 5.88	+12	<0.2

(a) before and (b) after the treatment with guar-gum; n = 7.

Table 3. Mean Values of Serum Copper and Iron (μg/100 ml)

	Mean Value $\pm$ S.D.		Δ%	p
	(a)	(b)		
Copper	112.86 $\pm$ 32.32	124.57 $\pm$ 59.23	+10	>0.9
Iron	87.71 $\pm$ 11.95	83.29 $\pm$ 18.25	- 5	<0.7

(a) before and (b) after the treatment with guar-gum; n = 7.

effect on cholesterol and triglyceride levels.

The use of gel-forming fiber could be very useful in the dietotherapy of diabetes. Even if the effects of fibers on glucose regulation are not yet fully clear, we wish their constant utilization in dietoterapy as main components.

Diet can play a particular role in preventing diabetes and it could avoid some complications in diabetic subjects. From the histopathologic point of view, many of these complications are constituted of micro and macro vascular diseases. Cholesterol has an atherogenic effect[9], except HDL-cholesterol, which seems to play a protective role. Also serum glucose levels seem related to microvascular diseases[10].

We can conclude that the dietary supplementation with guar-gum improves glycometabolic control in diabetes with combined favourable effects on glucose and lipid metabolism. This kind of dietary enrichment seems to be very useful in the therapy of the Type II diabetics in the presence of coexistent hypercholeste-rolaemia and hypertriglyceridaemia.

Acknowledgements

The authors wish to thank Dr. A. Viglezio of A. Gazzoni & C. (Bologna, Italy) for supplying the Guarem granules.

REFERENCES

1. H.C. Trowell, Ischaemic heart disease, atheroma and fibrinolysis, in: "Refined carbohydrate food and disease", D.P. Burkitt and H.C. Trowell, eds., Academic Press, London (1975).
2. H.C. Trowell, Dietary fibre, ischaemic heart disease and diabetes mellitus, Proc. Nutr. Soc. 32:151 (1973).
3. H.C. Trowell, D.A.T. Southgate, T.M.S. Wolever, M.A. Gassull and D.J.A. Jenkins, Dietary fibre re-defined, Lancet i:964 (1976).
4. S. Holt, R.C. Heading, D.C. Carter, L.F. Prescott and P. Tothill, Effect of gel fibre on gastric emptying and absorption of glucose and paracetamol, Lancet i:636 (1979).
5. D.J.A. Jenkins, T.M.S. Wolever, A.R. Leeds, M.A. Gassull, P. Haisman, J. Dilawari, D.U. Goff, G.L. Metz and K.G.M.M. Alberti, Dietary fibres, fibre analogues and glucose

tolerance: importance of viscosity, Br. Med. J. 1:1392 (1978).

6. J.W. Anderson and K. Ward, Long-term effects of high
   carbohydrate, high fiber diets on glucose and lipid
   metabolism: a preliminary report on patients with diabetes,
   Diabetes Care 1:77 (1978).

7. D.J.A. Jenkins, T.M.S. Wolever, R. Nineham, R. Taylor,
   G.L. Metz, S. Bacon and T.D.R. Hockaday, Guar cribspread in
   the diabetic diet, Br. Med. J. 2:1744 (1978).

8. A. Aro, M. Uusitupa, E. Voutilainen, K. Hersio, T. Korhonen
   and O. Siitonen, Improved diabetic control and
   hypercholesterolaemic effect induced by long-term dietary
   supplementation with guar-gum type 2 (insulin-independent)
   diabetes, Diabetologia 21:29 (1981).

9. W.B. Kannell, W.P. Castelli and T. Gordon, Cholesterol in the
   prediction of atherosclerotic disease. New prospectives based
   on the Framingham study, Ann. Int. Med. 90:85 (1979).

10. G.F. Lenti and M. Trovati, Attualità in tema di diabete, Fed.
    Med. 35:660 (1982).

ADDRESSES OF SENIOR AUTHORS

Ascalone V.   Laboratorio di Farmacocinetica
                       L.I.R.C.A. - Synthèlabo
                       via Rivoltana 35
                       20090 Limito (MI), Italy

Borriello R.   Istituto di Medicina Legale
                       I Facoltà di Medicina e Chirurgia
                       via L. Armanni 5
                       80138 Napoli, Italy

Campanella L.   Dipartimento di Chimica
                       Università di Roma "La Sapienza"
                       piazzale A. Moro 5
                       00185 Roma, Italy

Centini F.   Istituto di Medicina Legale
                       Università di Siena
                       via delle Scotte
                       53100 Siena, Italy

Chiarotti M.   Istituto di Medicina Legale
                       Università Cattolica del Sacro Cuore
                       Largo Francesco Vito 1
                       00168 Roma, Italy

Delva P.   Istituto di Clinica Medica
                       Università di Verona
                       Policlinico Borgo Roma
                       37134 Verona, Italy

Dorizzi R.   Laboratorio Analisi
                       Ospedale Civile
                       37045 Legnago (VR), Italy

Ensing K.   Laboratorium voor Analytische Chemie
                       en Toxicologie
                       Rijksuniversiteit
                       Antonius Deusinglaan 2
                       NL-9713 AW Groningen, The Netherlands

Fenoil R.

Laboratorio Analisi
Ospedale Martini Nuovo ULSS 1-23
via Tofane 71
10141 Torino, Italy

Ferrara S.D.

Istituto di Medicina Legale
Università di Padova
via Falloppio 50
35100 Padova, Italy

Froldi R.

Istituto di Medicina Legale
Università di Macerata
via Don Minzoni 9
62100 Macerata, Italy

Giachetti C.

Istituto "Antoine Marxer" R.B.M.
Casella Postale 226
10015 Ivrea, Italy

Governa M.

Clinica Medicina del Lavoro
Università di Ancona
Nuovo Ospedale Regionale
60020 Torrette (AN), Italy

Grossi G.

Laboratorio Centralizzato
Policlinico "S. Orsola"
via Massarenti 9
40138 Bologna, Italy

Jain N.C.

Director of Toxicology
Rancho Hospital
7601 East Imperial Highway
Downey, CA 90242, USA

Lhoest G.,

Mass Spectrometry Section
School of Pharmacy-U.C.L.
7246 av. E. Mounier
1200 Brussels, Belgium

Lindsay R.M.

Department of Biochemical Medicine
University of Dundee
Ninewells Hospital and Medical School
Dundee DD1 9SY, U.K.

Lisboa B.P.

Universitats-Frauenklinik und Poliklinik
Universitat Hamburg
Martinistrasse 52
D-2000 Hamburg 20, FRG

Macek K.   Institute of Physiology
                            Czechoslovak Academy of Sciences
                            Budéjovickà 1083
                            Praha 4-KBC, Czechoslovakia

Mari F.   Istituto di Medicina Legale
                            Policlinico di Careggi
                            50134 Firenze, Italy

Massé R.   Institut National de la Recherche
                            Scientifique, Université du Québec
                            Pavillon Gamelin
                            7401 rue Hochelaga
                            Montréal, Québec, Canada H1N 3M5

McDowall R.D.   Department of Drug Analysis
                            Smith Kline and French Research Ltd.
                            The Frythe
                            Welwyn, Herts, AL6 9AR, U.K.

Minuz P.   Istituto di Clinica Medica
                            Università di Verona
                            Policlinico Borgo Roma
                            37134 Verona, Italy

Montagna M.   Istituto di Medicina Legale
                            Università di Pavia
                            via Forlanini 12
                            27100 Pavia, Italy

Montani A.   Laboratorio Analisi
                            Ospedale Civico, ULSS 54
                            piazza Cappuccini 2
                            20071 Casalpusterlengo (MI), Italy

Musumarra G.   Dipartimento di Scienze Chimiche
                            Università di Catania
                            viale A. Doria 8
                            95125 Catania, Italy

Piemonte G.   Laboratori Universitari Ricerche Mediche
                            Università di Verona
                            Policlinico Borgo Roma
                            37134 Verona, Italy

Prosek M.   Analytical Laboratory-Research Department
                            LEK Pharmaceutical and Chemical Works
                            PO Box 81
                            61107 Ljubljana, Yugoslavia

Rizzo V.

Centro Cefalee, Fondazione "C. Mondino"
Università di Pavia
via Palestro 3
27100 Pavia, Italy

Ròzanski L.

Instytut Ochrony Roslin
Miczurina 20
60-318 Poznan, Poland

Sanguinetti F.

Centro "Alberto Sanguinetti"
via T. Gulli 34
48100 Ravenna, Italy

Scotto di Tella A.

Istituto di Medicina Legale
I Facoltà di Medicina e Chirurgia
Università di Napoli
via L. Armanni 5
80138 Napoli, Italy

Tagliaro F.

Istituto di Medicina Legale
Università di Verona
Policlinico di Borgo Roma
37134 Verona, Italy